STREETLIFE

Urban Retail Perspectives and Prospects

Edited by Conrad Kickert and Emily Talen

Our street-level economy is undergoing dramatic change. Retailers are reeling from the rise of e-commerce, rising rents, and increasing storefront vacancies, along with a cultural shift from material to experiential consumerism. Today, the COVID-19 pandemic is contributing to economic upheaval as commercial corridors and the small businesses they house face sweeping closures, bankruptcy, and job losses.

Streetlife brings together scholars who have been trying to make sense of the changing retail landscape at street level and what it means for urbanism's future. *Streetlife* pays special attention to the varied responses and policies that have emerged to address the competing realities of small business loss and neighbourhood needs. With case studies from the United States, as well as contributions covering Canada and Europe, this book demystifies the logic behind street-level urban retail and calls for better plans, designs, policies, and innovations to bolster sales.

Streetlife shows that now, more than ever before, we need to understand what makes our storefronts tick, what awaits them, and what we can do as planners, designers, developers, entrepreneurs, and policymakers to maintain retail as integral to urban lifestyle.

CONRAD KICKERT is an assistant professor in the School of Architecture and Planning at the University of Buffalo.

EMILY TALEN is a professor in the Social Sciences Division at the University of Chicago.

Streetlife

*Urban Retail Perspectives
and Prospects*

EDITED BY CONRAD KICKERT
AND EMILY TALEN

UNIVERSITY OF TORONTO PRESS
Toronto Buffalo London

ISBN 978-1-4875-0713-8 (cloth)   ISBN 978-1-4875-3564-3 (EPUB)
ISBN 978-1-4875-2481-4 (paper)   ISBN 978-1-4875-3563-6 (PDF)

Library and Archives Canada Cataloguing in Publication
Title: Streetlife : urban retail perspectives and prospects /
   edited by Conrad Kickert and Emily Talen.
Names: Kickert, Conrad, editor. | Talen, Emily, 1958–, editor.
Description: Includes bibliographical references and index.
Identifiers: Canadiana (print) 20220417067 | Canadiana (ebook)
   20220417075 | ISBN 9781487507138 (cloth) | ISBN 9781487524814 (paper) |
   ISBN 9781487535643 (EPUB) | ISBN 9781487535636 (PDF)
Subjects: LCSH: Stores, Retail – United States – Case studies. | LCSH: Small
   business – United States – Case studies. | LCSH: Street life – United States –
   Case studies. | LCGFT: Case studies.
Classification: LCC HF5429.3 .S77 2023 | DDC 381/.10973 – dc23

We wish to acknowledge the land on which the University of Toronto Press
operates. This land is the traditional territory of the Wendat, the Anishnaabeg,
the Haudenosaunee, the Métis, and the Mississaugas of the Credit First Nation.

This book has been supported by KTH Royal Institute of Technology,
Stockholm, and its Centre for the Future of Places.

University of Toronto Press acknowledges the financial support of the
Government of Canada, the Canada Council for the Arts, and the Ontario
Arts Council, an agency of the Government of Ontario, for its publishing
activities.

Canada Council   Conseil des Arts
for the Arts   du Canada

ONTARIO ARTS COUNCIL
CONSEIL DES ARTS DE L'ONTARIO
an Ontario government agency
un organisme du gouvernement de l'Ontario

Funded by the   Financé par le   Canada
Government   gouvernement
of Canada   du Canada

Contents

Figures and Tables

Figures

Tables

Business Profiles

STREETLIFE

Introduction: The Urban Retail Predicament

CONRAD KICKERT AND EMILY TALEN

Our street-level economy is undergoing one of the most profound transformations in history. Even before the COVID-19 pandemic, retailers on both sides of the Atlantic were already reeling from the rise of e-commerce, ever-loosening regulations protecting them from suburbanization, and a cultural shift from material to experiential consumerism. Where retail behemoths are already shaken up, small retailers – the lifeblood of our urban main streets – are toppling like dominoes. Today, the global pandemic is accelerating store and mall closures, bankruptcies, and job losses, resulting in an economic upheaval that may well match the fallout from deindustrialization that occurred in the latter half of the twentieth century (Corkery, 2017; Peterson, 2017; Thompson, 2017).[1] More than ever before, we need to understand what makes our storefronts tick, what awaits them, and what we can do as planners, designers, developers, entrepreneurs, and policymakers to maintain these key elements of urban life.

The purpose of this book is not to reiterate grim retail statistics, but to provide some reflection on the purpose, value, and meaning of our long-valued urban storefronts that are often taken for granted. Furthermore, the book will demystify the logic of urban retail streets, substantiating better plans, designs, policies, and innovations to bolster these social, cultural, and economic hubs. *New Yorker* staff writer Adam Gopnik coldly but rightfully proclaimed that what sustained Jane Jacobs' Greenwich Village "street ballet" was not a "mysterious equilibrium of types," or "magic folk dancing," but simply market forces (Gopnick, 2016).

In Jacobs' time, Hudson Street's sidewalk life was sustained by compassionate, independently owned retailers, bars, and restaurants. While we describe these small businesses under various terms on either side of the Atlantic – from the more universal "independent" or "local" to the distinctly American "mom-and-pop" colloquialism – we universally understand their key role in the liveability and sociability of our cities. So how do we respond to the fact that these independent storefront businesses are becoming increasingly rare in the United States, Canada, and Europe? Jacobs' corner grocer of 1961 has made way for the chain drugstore, the fast-casual restaurant, or for nothing at all. From her Hudson Street apartment window, Jacobs would now oversee dozens of empty street-level windows waiting for new life as New York's storefront vacancy has doubled over the past decade alone (Stringer, 2019). Whether in Greenwich Village or Greenwich, London, urban storefronts struggle to find relevance in today's perfect storm of economic volatility, retail consolidation, the meteoric rise of e-commerce – and now, a global pandemic. This transatlantic existential crisis of storefronts raises questions about urbanism in general. If cities are their streets, and streets are their uses, what do the profound changes in urban retail mean for the future of our cities?

In search of some answers, 19 authors from a variety of backgrounds discuss the opportunities and challenges facing storefronts on either end of the Atlantic. Focusing mainly on the United States but including Canada and Europe, we explore the dynamics of urban retail as it relates to economics, sociability, culture, and place. Our goal is to deepen understanding of the dynamics that drive street-level commerce, and in so doing, bolster our appreciation of the key role of retail in fostering urbanism. We are not ready to relegate retail-activated streetlife to the historical dustbin, despite current trends or calamities.

The contemporary challenges facing urban retail have unleashed a deep sense of loss – at times even mourning. This is because urban storefronts signify far more than points of purchase. They serve as centres of social and cultural life and economic opportunity, and are the foundation of thriving urban neighbourhoods. Shifting far more rapidly than urban structure and form itself, storefronts are excellent if ephemeral barometers of the vitality of their surroundings, predicting and perpetuating growth or decline. This is why, more than ever, urban storefronts deserve our understanding, appreciation, and hope.

There are some large questions at stake in this urban retail sea change. Do the struggles of retail and its storefronts predicate the decline of urbanity, or is the loss just a natural step in urban evolutionary change? Will we regret letting our streetscapes descend into homogenization or vacancy, or will we come to rely on their predictability and see new potential? Is there something to blame for this – perhaps a failure of policy – and would a more proactive approach be able to turn the tide? In New York, these questions have become the topic of a lively public, professional, and academic debate, yielding critical visual narratives like *Store Front: The Disappearing Face of New York* (Murray & Murray, 2008), crowdsourced vacancy surveys (vacantNewYork.com), and heated blogs (Jeremiah's Vanishing

New York). In the United Kingdom, this debate has prompted national inquiries on the future of the high street, complete with recommendations of life after retail (Burt et al., 2010; Portas, 2011). Fingers quickly point to greedy landlords, ruthless chains, overbearing governments, a disinterested public, or Silicon Valley aggression, but there is little consensus on main drivers of retail decline, let alone the prospects for our cities at eye level. Perhaps we are overwhelmed by the demise of hundreds of American main streets (Talen & Jeong, 2019), and we even see long-protected European high streets gasping for air (Carmona, 2015). If even Manhattan and London struggle to maintain urban commerce, is there hope for Middle America and small-town Europe?

In times of upheaval, it is tempting to reminisce. The iconic image of the shopping street as a place of economic, social, and cultural exchange has fuelled nostalgic recreations in theme parks, movies, and models, often theatrically harking back to a past that never fully existed. Stripped of its rough edges, Main Street America is immortalized in Disneyland and EuroDisney, and main streets and high streets on both sides of the Atlantic are reinvented as lifestyle centres. High street Britain lives on in the nation's model villages and train sets (Gibbs, 2012; Coleman, 2006; Sorkin, 1992; Francaviglia, 1996; Baudrillard, 1994). A desired retail past percolates into developer-driven shopping "experiences," not unlike Victor Gruen's recreation of the Viennese inner city in the post-war American shopping mall (Gruen & Smith, 1960; Hardwick, 2004). Some commercial hyperrealities needn't even venture that far: Los Angeles plays itself at Universal City Walk and Berlin recreates itself at the Potsdamer Platz (Hannigan, 2010; Roost, 1998). A lingering unease reminds us that while we may walk through an imagined commercial past, our ever-changing consumption patterns no longer fit their facades. Walmart and Tesco may disguise themselves as the village they pillaged (LeCavalier, 2016), but the world has irreversibly changed behind their front doors.

While we do not see the demise of traditional storefront retail as a fait accompli, there is plenty of grey area about what urban streets in a post-pandemic world will be like. There are a number of questions to be considered – a pursuit to which this book is dedicated. One is the degree to which the current urban retail predicament should be seen as part of a centuries-long sequence of relentless supersedence in retail. Is it, in fact, the latest chapter in a mass-market victory, which has consistently relied on organizational efficiency, technological innovation, and economic and cultural manipulation (Savitt, 1989; Stobart, 2010; Brown, 1987)? The contemporary Western standard of living is undeniably the outcome of the popular "democratization of luxury," which has brought continuously improving goods and services via cunning advertising

Figure 0.1. The result of centuries of retail consolidation, decades of virtualization, and years of isolation: vacant storefronts dot our cities (*above*). Storefronts can hold more than just retailers; they can be reused as social, productive, expressive, and residential spaces as well. This "attention centre" in The Hague repurposes a storefront to invite passers-by in for a chat (*left*). (Credit: Images by C. Kickert.)

combined with production and distribution efficiencies that result in an ever-shrinking number of larger and larger retailers (Satterthwaite, 2001; Strasser, 2004). The quest for increasing retail efficiency has fuelled a continuous cycle of creative destruction, as retail "innovators" optimized their way out of small, independent shops over a century ago, and out of many cities decades ago. Now, they seem to be optimizing their way out of physical space altogether. The cutthroat succession of retail formats and spaces is both systematic and cyclical – for every urban storefront that is closing today, dozens have gone before it (Evans, 2011; Kickert, 2016).

A Brief History of Urban Retail's Decline

We need to understand how urban retail has evolved to fully grasp the uphill battle we face in our quest to sustain it. In the continuum of retail evolution, the active street-life-supporting storefronts we often seek represent but a snapshot product of the Industrial Era, superseding centuries of bazaars and weekly markets, travelling salespeople, and direct sales by artisans. In the nineteenth century – the era of industrial urbanization – food supply once rooted in temporary markets and itinerant retail evolved into a fine-grained pattern of neighbourhood corner stores. Also at this time, central urban streets came to host newly mass-produced durable items like fashion, luxury, and home goods, fuelling the desires of a wealthy urban "leisure class" strolling along newly built sidewalks and newly invented plate glass storefronts (Veblen, 1899; Davis, 1966; Lesger, 2013).

Within mere decades, new models of merchandizing and distribution would already challenge this street-level hegemony. Discovering the benefits of store clustering, central management, and marketing, individual luxury stores began to cluster into arcades aimed at the urban bourgeoisie. The iron-and-glass Parisian arcade, interrogated by Walter Benjamin in *The Arcades Project*, was the forerunner of the department store, a similarly Parisian invention that spread to both sides of the Atlantic by the mid-nineteenth century (Benjamin et al., 1999 [1940]; Geist, 1983). Large enough to control their own suppliers and manufacturers, progressive enough to supersede bargaining by fixed prices, but most importantly shrewd enough to drastically lower prices on mass-market goods, department stores unleashed a veritable "retail revolution," multiplying from their roots as *magasins des nouveautés* to dominate urban markets (Davis, 1966).

While it is easy to be awed by the splendour of the nineteenth-century department store, they put several destructive processes in motion. First, their business model did not only hinge on economies of scale and pricing; it also relied on drastically reducing the need for skilled clerks, separating the hitherto close relationship between consumers and retailers. Furthermore, the increased efficiency of department stores fuelled the first wave of independent store closures. Emile Zola's vivid 1883 description of the fictional Au Bonheur des Dames department store did not just muse about the temptations of modern consumerism, but also exposed department stores' relentless replacement of independent competitors, described as a "triumph over the annoying obstinacy of the infinitely small" (Zola, 1883/1886, 342). Despite considerable opposition from manufacturers and independent retailers, department stores ultimately prevailed as urban palaces of consumerism on both sides of the Atlantic, propelled by novel escalators, elevators, and public transportation systems (Howard, 2015, 31–36; Coleman. 2006, 36–38; Davis, 1966, 290–295).

Gradually, however, the downtown department store fell victim to retailing's next step toward efficiency – the chain store. Instead of centralizing their operations, several Victorian-era British retailers sought economies of scale by multiplying as chain stores, targeting low prices and high sales volumes while synchronizing their outward appearance to provide customers with a sense of dependability and efficiency. Slowly, these chains chipped away at the hegemony of department stores, and also further deteriorated the position of remaining independent retailers. British chains like Lipton and W. H. Smith and American chains like A&P and F. W. Woolworth began to dominate an increasing number of markets within the first decades of the twentieth century. Their rise accelerated with self-service grocery stores, an American invention that drastically reduced operating expenses. The associated savings were eagerly embraced by Americans suffering from rampant inflation during and after the First World War. Government efforts to intervene in the resultant retail establishment erosion were mixed. Nations like the United States initially chose the side of manufacturers and independent retailers by fixing prices, unsuccessfully capping store sizes, and limiting large-scale purchasing (Levinson, 2012). Dutch legislation, on the other hand, chose to weed out smaller stores to encourage retailers' viability and consumers' buying power. In either case, smaller retailers were losing the battle for the urban street. Working against them, too, were the impulses of a newly formed city planning profession intent on using zoning to prohibit neighbourhood stores adjacent to or below housing. Urban street retail fell victim to a simplistic understanding of urban life (Jacobs, 1961, 558–585; Davis, 1966, 280–284; Coleman, 2006, p. 39; Nijs & Knoester, 2007).

Another blow against urban street retail was the rise of mail-based consumerism, which goes further back than many realize. In the United

States especially, improvements in mail and parcel shipping propelled mail order companies like Montgomery Ward and Sears Roebuck to market dominance – a remarkable parallel with the contemporary rise of e-commerce. These companies were also able to leverage public investments in postal infrastructure to perfect a superior distribution system. Augmented with increasingly refined advertisement campaigns, mail order giants left small-town main streets gasping for air by the turn of the twentieth century. As suburbanization took hold, companies like Sears understood that automobile-based consumption in physical stores had an increasingly strong appeal, and they skipped over downtown department stores to build America's first large network of auto-centric suburban stores. These operated alongside mail order warehouses – predating the increasingly popular contemporary "omnichannel" mixture of in-store and online shopping options by nearly a century (Raff & Temin, 1999).

Sears' success and the ongoing rise of automobility and suburbanization set the tone for retail decentralization of retail across America and Canada. Hybrids between walkable and drivable retail like J. C. Nichol's Country Club Plaza in Kansas City ultimately gave way to fully auto-centric American shopping malls in the 1950s, which soon found their way across the Atlantic (Satterthwaite, 2001; Hardwick, 2004). Small retailers hardly stood a chance in this transformation. America, Canada, and Europe's post-war rebirth hinged on the rise of mass-market consumerism enabled by significant efficiency improvements in manufacturing, advertising, and retailing. Encouraged by public policies and subsidies, national retailers focused on serving an increasingly suburban middle class (Cohen, 1996; Gosseye & Avermaete, 2017).

For well over a century, success in retailing hinged on economies of scale that urban independent retailers simply could not achieve. Benefiting from a boom during the Great Depression, the high-volume, low-price model of the American self-service store prompted an increasing number of smaller grocers to expand into full-fledged supermarkets where one-stop shopping could fill ever-larger car trunks and refrigerators. By mid-century, this model had become an (often-suburban) staple on both sides of the Atlantic. In the decades that followed, consolidation ensured that only a handful of retailers dominated most national and even global grocery markets (Miellet & Voorn, 2001; Longstreth, 2000; Wrigley, 1993; Rutte & Koning, 1998). In addition, an increasing number of discount retailers realized that the low-cost model could also work for non-food items, and began to open stores on cheap land at the urban fringe. Their supremacy was sealed in the United States with the eradication of the Fair Trade Law in 1975 (Boyd, 1997). This fuelled the subsequent rapid growth of "big-box" or "hypermarket" retailers like Walmart, Kmart, and

Target, which soon spread into Canada as well (Howard, 2015; Jones & Hernandez, 2000; LeCavalier, 2016). Just over three decades after the shopping mall had replaced the urban core, these category-killing discounters clustered in "power centres" threatened their mid-century predecessor by the 1990s (Gibbs, 2012). Over time, discounters virtually eliminated the remaining American urban department store and significantly challenged European counterparts (Howard, 2015; Miellet & Voorn, 2001). While European retail policies at first curbed the rise of out-of-town retailers, legal loopholes and deregulation due to political shifts allowed a growing number of hypermarkets and shopping centres to encircle cities (Davies, 1995; Evers et al., 2005).

With the rise of e-commerce, mass-market retailers' quest for efficiency has now come full circle. Starting with easy-to-ship merchandise like books and music, online-first retailers like Amazon now dominate the market for almost all merchandise categories. Ironically, the e-commerce trajectory can be predicted, as e-tailers follow the same cycle of creative destruction that spawned their physical mass-market predecessors, and may ultimately render them obsolete. After all, the same relentless drive for efficiency of brick-and-mortar chain retailers have taken away mass-market retailers' unique benefits over online retailers, such as skilled staff, store ambiance, and a strong service image (Verhoef et al., 2015; Satterthwaite, 2001). The current dynamic is eerily reminiscent of previous retail evolutions. In the United States alone, department stores have closed hundreds of outlets over 2018 alone, and have lost more than half a million jobs since 2001 (US Bureau of Labor Statistics, 2018). Chains are rapidly following in their wake of decline. The trends are similar if not more concerning across the Atlantic, where stronger regulatory protection for retailers can be circumvented by e-retailers, which also benefit from high-quality postal services that allow for next-day and even same-day delivery. As a result, for example, British online sales accounted for more than twice the market share of the United States before the onset of COVID-19 (eMarketer, 2019). As a result, the deterioration of European urban retail has alarmed the general public, professionals, academics, and regulators alike, amplifying the American narrative of a "retail apocalypse" to an outright "death of the High Street" (Centre for Retail Research, 2018; Locatus, 2018; Carmona, 2015).

Independent retailers that still line our urban streets risk falling behind in the age of e-commerce if they cannot adopt online sales fast enough, for example, deterred by high start-up costs and managerial resistance (Parker & Castleman, 2009; GfK, 2019). On the other hand, some retail start-ups are skipping the opening of physical stores altogether, instead offering their merchandise via marketplaces like Amazon, Etsy, and eBay (Wang et al., 2008; Bezos, 2017). Many physical retailers understand that in order to survive, they must adopt online sales strategies, and are slowly transforming their

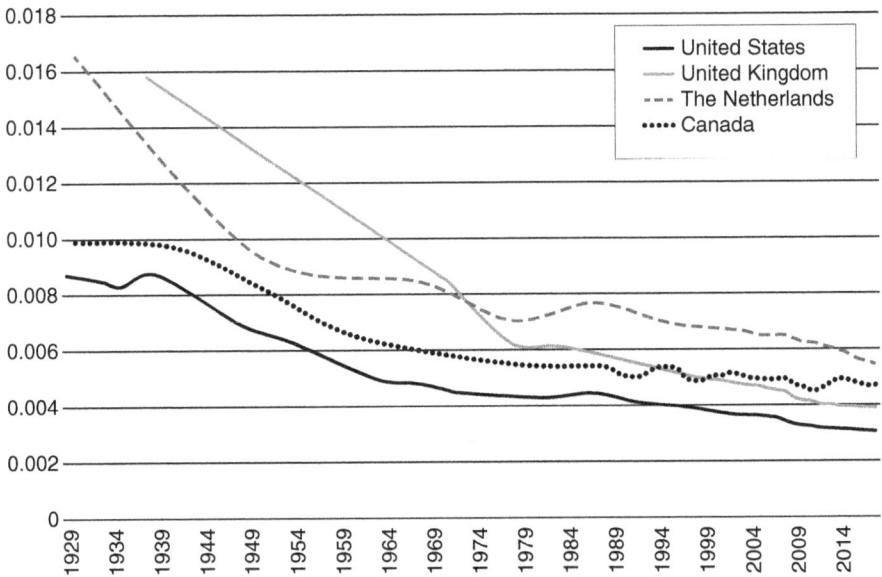

Figure 0.2. Retail establishments per capita in the United States, the United Kingdom, The Netherlands, and Canada, 1929 and 2018.

stores into e-commerce distribution centres. In the battle for mass-market dominance, there seems little time for the nostalgic experience of physical retail, as ominously reflected in a statement by a Walmart senior manager: "The misconception is that we're in the retail business, [but really] we're in the distribution business" (Dunne et al., 2016, p. 55).

We must not forget that the move toward e-commerce is only the latest strike in a centuries-long battle for mass-market dominance. Our current urban retail predicament reflects centuries of attritional retail evolution, in which a small number of mass-market innovators have consistently gained market share at the cost of their competitors. The outcome of this pattern is especially visible in nations with few regulatory barriers. The United States may be known for its overbuilt retail floor space, but the *number* of shops per capita there is almost half that of the highly regulated Netherlands – a far more salient metric for fine-grained, walkable retail (Figure 0.2). Figure 0.2 shows that the decline of retail establishments per capita over the past 90 years has been a nearly constant decline, and actually far less severe in recent times than in the mid-twentieth century when retail "innovations" were especially dramatic.[2] Shrinking household sizes in most urban neighbourhoods only exacerbate the national trends depicted in this graph. The households of urban planner Clarence Perry's idealized neighbourhood unit of 1929 would now support only 20 per cent of the original number of envisioned stores. In other words, four-fifths of the storefronts in even our most stable urban neighbourhoods have lost their retail function, looking for new lives ever since. Then again, in cities like Chicago, the number of retail establishments per capita has actually gone up, as Emily Talen's chapter in this book demonstrates. Clearly, the urban retail picture is more complex.

Urban Retail Redux

While the decline of urban street retail seems inevitable, it is also true that street-level commerce is ingrained in the very physical, social, and cultural life of our cities. Given this central role, it is not surprising that notions of "urban sustainability" tend to include fine-grained networks of destinations fronting commercial streets. The market, too, bears out the visceral appeal of a thriving retail street. Urban dwellers pay a premium to be near walkable urban retail, fuelling developer interest in maintaining or creating retail as an amenity (Leinberger & Alfonzo, 2012; Pivo & Fisher, 2011; McLinden, 2018; Beske & Dixon, 2018; Novak, 2019). Surveys show that Britons' main reasons for selecting an urban neighbourhood is proximity to restaurants, leisure, and cultural facilities, more than proximity to the workplace or even the aesthetics of a neighbourhood (Thomas et al., 2015).[3] Furthermore, urban storefronts present employment and entrepreneurship opportunities, especially for immigrants and ethnic minorities (Fairlie & Robb, 2010; Zukin et al., 2016).

While urban retailers struggle to find their footing within a new post-pandemic social and economic reality, the pandemic is also spawning new ideas about the role of retail. There are even optimistic views that urban retail streets will be able to build off their significant "strengths and advantages" (Alter, 2020). One view holds that, since more and more people will continue to work from home, neighbourhood retail streets will become more prominent as they service home workers throughout the day. A closer integration between work space and the public realm – "pop-up offices, meeting pods, and technology centers linked to town squares" – will motivate more investment in neighbourhood public spaces such as retail streets (Woods, 2020). Especially businesses that had already been integrating online delivery and storefront operations can benefit from neighbourhood-based co-working, as they can service more customers throughout the day. And there is renewed emphasis on localizing and avoiding larger gatherings (including large workplaces like office towers), which is likely to be good for urban retailers. Shopping and working on a retail street might even start to feel like "an act of civic engagement" (Wilson, 2018).

Urban retail also has the potential to yield far more than economic value. For example, urban retailers have been shown to support citizens' health by anchoring walkable neighbourhoods and providing access to healthy food options (Beaulac et al., 2009; Walker et al., 2010; Forsyth, 2015; Rundle et al., 2008). Perhaps even more significantly, retailers support healthy urban social life. Walkable shops function as neighbourhood hubs that build a sense of community and social capital, access that is especially important to disadvantaged, disabled, or elderly consumers (Stroope et al., 2014; Goetz & Rupasingha, 2006). The urban ground floor – bars, restaurants, and coffee shops especially – are key spaces for the social life of cities beyond the home and the workplace, coined by sociologist Ray Oldenburg as "third places" to meet friends, encounter strangers, and negotiate cultural and social roles (Silver & Clark, 2016; Oldenburg, 1999). Increasingly, these foodservice establishments have become innovation hubs as informal places to exchange knowledge and collaborate (Katz & Wagner, 2014). The popularity of foodservice establishments has grown to outnumber any other category of ground floor commerce in many cities in America, Canada, and Europe, and food away from home now takes up the majority of American nutritional spending (McLaughlin & Dicken, 2018; Office for National Statistics, 2019; United States Census Bureau, 2019) – even if health crises can disrupt these trends. Surveys continue to document that neighbourhood grocery stores, restaurants, gyms, and coffee shops are key amenities that lead to a greater quality of urban life, positively influencing social connections and trust (Cox & Streeter, 2019).

Despite the contemporary discourse on the "apocalyptic" rise of e-commerce, and the fraught history of retail as we've laid out here, urban storefront commerce is far from dead. E-commerce market shares may be accelerating, but physical retail has unique advantages over online venues that ensure their continued relevance – a topic we will discuss throughout the book. Beyond facilitating sales, storefronts serve as social and cultural communicators; they enable face-to-face interactions and knowledge exchange; and they offer house services that can happen only in real life. Storefronts uniquely feed social interaction and cultural negotiation through sidewalk and aisle encounters, coffee table conversations, and plate glass glances. They provide streets with local relevance, a sense of community and safety, and visual excitement for passers-by (Gehl, Johansen Kaefer, & Reigstad, 2006; Kickert, 2016). Our consumer necessities may increasingly hail from online stores, yet we continue to quench our social and cultural thirst on main and high streets, feeding a symbiosis between economic opportunity, sociability, and psychological stimulation.

Solidification of the retail experience will require diligence in warding off the common liabilities associated with any urban upgrade: gentrification leading to displacement, and the growth of urban inequality. Opportunities for greater inclusion may come from the transformation of storefronts as passive places of consumption to places of production: coffee shops and lunchrooms serving as knowledge-sharing hubs in the innovation economy, and food clusters directly connecting consumers with food producers, including ethnic entrepreneurs (Katz & Wagner, 2014; Florida, 2002; Brown, 2019; Kershen, 2017). An increasing number of small-batch and artisanal producers are starting storefront businesses, often mixing manufacturing and sales in hubs or even veritable districts in cities like New York, Detroit, London, and Barcelona (Wolf-Powers et al., 2017; Anderson, 2012). Blending manufacturing and retailing, agnostic about selling its products virtually or in physical stores, this Maker Movement demonstrates on the continued relevance of storefronts. The reign of retailers on our main and high streets may well give way to a more diverse storefront future that blends transaction and experience, production and consumption, and local community and global culture – ironically reminiscent of the urban commercial experience of centuries past.

Reading Guide

This book introduces the challenges and opportunities facing urban storefronts with three chapters on retail trends and transformations. Most chapters focus on the United States; some focus on Canada and Europe. Luc Anselin and Irene Farah analyse the demographic characteristics of the US retail sector over a 25-year period, from 1990 to 2015. They show that the retail sector has become more corporate, even though small-scale, independent retailers retain an important market share and show great robustness. On the other hand, they demonstrate that there is great volatility in the retail sector, with extreme changes in birth and death rates of the number of establishments from year to year. Fitting with the long-term trend of retail decline explained previously in this chapter, all retailers experienced a net decline in establishments, but small retailers are declining the most rapidly, and retail decline has become more pronounced since 2010. While small retailers on average have shorter life spans than their larger peers, it seems that once establishments survive the first three or four years, they are more likely to last far longer.

Next, Kevin Credit, Luc Anselin, and Irene Farah help us understand the different definitions and dynamics between retailers catering to our daily needs, our staple durable goods, and our increasing desire for experiences. As might be predicted, especially the experience economy has grown the most in recent years, while the other retail categories have declined. Still, cities differ: "Sunbelt" areas have seen higher rates of experience economy growth, while "Rustbelt" areas have seen lower rates of decline in all categories. They then relate these retail fluctuations to occupational structure, finding that cities with more "creative class" occupations have lower decreases in basic needs and general merchandise businesses and increases in experience economy businesses.

Rachel Meltzer investigates the relationship between neighbourhood demographic change and neighbourhood commercial life. What are the strains on retailers in areas undergoing rapid population shifts, especially neighbourhoods experiencing significant external investment and gentrification? The most common narrative is that, for incumbent residents, such changes are threatening; services become less affordable and less relevant to their daily needs. Furthermore, legacy business owners struggle to keep up with changes in their neighbourhood clientele. Yet Meltzer argues that there is another side to the gentrification story: the increasing number and quality of services that arrive in a gentrifying neighbourhood are welcomed. As cities try to respond to neighbourhood commercial shifts in rapidly gentrifying areas, there is a need to balance "diversity and localness of consumption and production" with pro-investment strategies that improve life in urban neighbourhoods.

The Case of E-Commerce

As mentioned before, a significant part of the challenges for urban retailers hails from the rise of e-commerce. Three chapters describe some of these challenges, focusing on three different markets on both sides of the Atlantic. Liz Mack takes a deep look at the history and current implications of e-commerce, the latest in a long line of retail transformations. While e-commerce experienced a slow start in the mid-1990s, the world saw a nearly 25 per cent increase in online purchases in just one recent year (2017 to 2018), setting up a "David and Goliath battle of small independent entities against online-only juggernauts like Amazon." How can independent retail survive? Mack argues that small retailers need to leverage their competitive advantages (like the physical shopping experience and customer service), but also embrace the growth of e-commerce. They need to jump out ahead on this hybrid format. The pace of retail innovation is extremely fast, and new technologies are seeking ways to emulate whatever advantages independent store owners might currently possess. Small can mean nimble; in the digital age, small retailers need to think strategically.

Christopher Daniel and Tony Hernandez study the rise of e-commerce and its implications for the Canadian urban context, specifically focusing on the Toronto urban region. Using a Delphi interview method, they survey retail experts on the expected impact of the rise of e-commerce on the needs for retail space in the Toronto area, discovering that rapid urban growth will continue to fuel more, yet different retail construction. They discuss the changing roles of stores in the consumer journey to purchase in the era of e-commerce, shifting from a place of transaction to an experiential setting. As lifestyles and consumer behaviours and preferences change, retail will need to be nimble and responsive.

Colin Jones looks at the influence of online retailing on the real estate portfolio of British retailers, and the resultant retail hierarchy. Building from the regulatory, social, and economic context of retailing in the United Kingdom, Jones argues that the current decline in British retail establishments could be seen as the tail end of a shakeout that began in the late 2000s. He argues that there is a symbiotic relationship between physical stores and online sales, but also that e-commerce affects the physical hierarchy of retail locations and clusters – where smaller towns suffer especially.

The Survival of Independent Retail

In the onslaught of online retail, it would be easy to see that independent retailers are the easiest victims. Yet the following three chapters paint a very different picture. As Vikas Mehta shows, independent stores are not dead, and in fact many of them thrive. While starting out his quest on *how* small retailers survive, Mehta increasingly discovered the answer lies in *why* they survive. Rather than mere spaces of transaction, retailers fulfil crucial roles in urban neighbourhoods, as they contribute to social connection and quality of life. Stores that last play an advocacy role, supporting events and procuring neighbourhood improvements, providing a platform for messaging and social exchange, for collective life and neighbourhood "centredness." In short, there is ample proof that small retailers are community builders.

Similar to Vikas Mehta's findings, Emily Talen's survey of independent stores in Chicago shows that most small retailers were optimistic about the future of retail. Few felt particularly threatened by e-commerce. Instead, they highlighted their positive role in the life of the neighbourhood – despite challenges such as regulations and gentrification. Conversely, governments have significant potential to support small businesses through a variety of policies.

Countering popular narratives, Jeffrey Parker does not see chain stores and independent stores as necessarily in opposition. He asks whether it would be possible for chain stores and independent businesses to coordinate efforts to maintain neighbourhood "hipness." Indeed, using a case study of Chicago's Wicker Park neighbourhood, he is able to show that chain stores play a major role in the maintenance of neighbourhood hipness. Chain stores can still threaten neighbourhood hipness, but it is also possible for them to provide strategic resources for "reputational hipness." Ultimately, these resources help sustain the independent retailer and provide a kind of "prophylactic" against a more complete takeover of chain stores.

Retail, Place, and Place-Making

Much can be gained by analysing retail in context. The next set of chapters investigate the place qualities of varying aspects of retail. While retailers are often studied as economic drivers that produce turnover, the consumer side is an equally valuable perspective. As such, Hyesun Jeong and Terry Clark explore clusters of retail activities that combine to create retail "scenes," which define place quality. They show that there are limits to treating retail as an individual transaction measured by sales volume – and gains to be made by looking at retail in its spatial surroundings. In other words, the context of a cafe matters.

Rosa Danenberg investigates the morphology of commercial corridors, and its contribution to the viability, diversity, and resilience of its retailers. The configuration and form of retail streets intersect with social, economic, and cultural dimensions to create commercial corridors with very different character. This character, in turn, has a major impact on long-term viability and resilience in the face of change. Furthermore, commercial streets and their retail units can support a diversity of retailers, when they comprise independently owned plots and mixed ground floor functions. Conversely, large blocks with larger ground floor units tend to deactivate street frontages.

Matthew Carmona analyses the complexity of the mixed street corridor in London. He shows how retail is but one dimension of an intricate urban structure, consisting of physical fabric, social activity and economic exchange, communication and movement, and real estate. He shows how we need to understand the "needs, conflicts and synergies" that operate within and between these four dimensions. In other words, revitalization of a main street cannot just focus on retail alone. It needs to consider all dimensions. Most often, however, the dimensions are treated in isolation, as no single agency is responsible to oversee the complex whole.

Toward Solutions

We conclude with three chapters that focus on solutions to the retail predicament. Michael Mehaffy and Tigran Haas present strategies for successful main street retail development in new urban districts, using a case study development near Portland, Oregon. They demonstrate the value of the right location and design of retail streets, but also the importance of business recruitment, development, support, and management, all of which constitute the "curation" of commercial corridors. In other words, successful main streets require careful planning, based on a clear understanding of main street dynamics.

Conrad Kickert's chapter presents the role of retail research in substantiating a better future for urban retail. While the experience economy and the maker economy are bringing new kinds of retail investment to cities, risk-averse retail organizations are hesitant about returning to cities. Kickert ascribes this to an imbalance in retail research. In its quest to mitigate development risk, marketing-based retail research has become less spatial, less contextual, and less urban. However, the retail dynamics specific to spatial and urban (as opposed to suburban or virtual) locations are knowable. Kickert shows this by illustrating the internal spatial logic of urban retail from the supply- and demand-side perspectives, using American and European examples. This logic supports urban retail because it substantiates the viability of new urban retail locations beyond aggregated consumer profiles, taking spatial configuration and interaction into account. Knowing this spatial logic helps planners, retailers, and developers assess the risk of urban retail success or failure, which may in turn encourage urban retail reinvestment.

From her perspective in American retail real estate consultancy, Heather Arnold surveys the future of retail real estate. The rapid changes in retailing can be, as Arnold argues, "confounding" to everyone involved: from architects, developers, and planners to lenders and business owners. Arnold gives us an on-the-ground, streetwise assessment of the impacts of current market trends, demonstrating their profound effects on retail rents, lease terms, store sizes, and site selection.

NOTES

1 The US Department of Commerce reported a decline of 6.2 per cent in retail trade sales between February and March 2020 (US Department of Commerce, 2020). Retail sales in clothing and accessories are now down 50.7 per cent between 2019 and 2020, according to Bartik et al. (2020).

2 The statistical review shown in Figure 0.2 compares the number of physical retail establishments that sell non-automotive goods to consumers, which excludes non-store retailers, car sales and repairs, fuel sales, bars, and restaurants. For the United States, the United States Census Bureau's County Business Patterns and Census of Retail Trade has been used. For the United Kingdom, the Census of Distribution, Retail Sales Inquiry, Annual Abstract of Statistics, and Business Monitor of 1980 and 1984 of the Office for National Statistics has been used, along with Eurostat Enterprise Statistics for Trade (NACE Rev. 1.1G and Rev. 2). Furthermore, the 1938 retail count is estimated in Jefferys et al. (1950). For The Netherlands, statistics are used from the Kamer van Koophandel, Centraal Bureau voor de Statistiek and the Hoofdbedrijfschap Detailhandel, along with Locatus data, and data from Borchert (1998). For Canada, statistics are used from the decennial Census of 1931, 1941, and 1951, Wholesaling and Retailing in Canada (1994–1998), Retailing in Canada, Retail Store Survey, Business Patterns (1988–2014), and Business Counts (2014 onwards). Differences in measuring methodologies and definitions of retail between the sources and collection times of consulted statistical data have likely introduced a margin of error in this review, as frequent gaps in values and data comparability have required manual normalization.

3 After a YouGov poll of 1,725 residents of four British urban regions.

REFERENCES

Alter, L. (2020, 20 April). The coronavirus and the future of main street, post-pandemic. TreeHugger. https://www.treehugger.com/urban-design/coronavirus-and-future-main-street.html

Anderson, C. (2012). *Makers: The new industrial revolution.* Crown business.

Bartik, A. W., Bertrand, M., Cullen, Z. B., Glaeser, E. L., Luca, M., & Stanton, C. T. (2020). *How are small businesses adjusting to COVID-19? Early evidence from a survey* (Working Paper No. 26989). National Bureau of Economic Research. https://doi.org/10.3386/w26989

Baudrillard, J. (1994). *Simulacra and simulation: The body, in theory.* University of Michigan Press.

Beaulac, J., Kristjansson, E., & Cummins, S. (2009). Peer reviewed: A systematic review of food deserts, 1966–2007. *Preventing Chronic Disease, 6*(3), 1–10.

Benjamin, W., Eiland, H., McLaughlin, K., & Tiedemann, R. (1999). *The arcades project.* The Belknap Press of Harvard University Press.

Beske, J., & Dixon, D. (2018). *Suburban remix: Creating the next generation of urban places.* Island Press.

Bezos, J. P. (2017). Shareholder letter. Seattle, WA.

Borchert, J. G. (1998). Spatial dynamics of retail structure and the venerable retail hierarchy. *GeoJournal, 45*(4), 327–336.

Boyd, D. W. (1997). From "mom and pop" to Wal-Mart: The impact of the Consumer Goods Pricing Act of 1975 on the retail sector in the United States. *Journal of Economic Issues, 31*(1), 223–232. https://doi.org/10.1080/00213624.1997.11505899

Brown, G. H. (2019). *Food halls of North America*. Cushman & Wakefield.

Brown, S. (1987). Institutional change in retailing: A review and synthesis. *European Journal of Marketing, 21*(6), 5–36. https://doi.org/10.1108/eum0000000004701

Burt, S., Sparks, L., & Teller, C. (2010). Retailing in the United Kingdom: A synopsis. In *European Retail Research*, 173–194. Springer. https://doi.org/10.1007/978-3-8349-8938-3_8

Carmona, M. (2015). London's local high streets: The problems, potential and complexities of mixed street corridors. *Progress in Planning, 100*, 1–84. https://doi.org/10.1016/j.progress.2014.03.001

Centre for Retail Research. (2018). Who's gone bust in retailing 2010–18? http://www.retailresearch.org/whosegonebust.php

Cohen, L. (1996). From town center to shopping center: The reconfiguration of community marketplaces in postwar America. *The American Historical Review, 101*(4), 1050–1081. https://doi.org/10.2307/2169634

Coleman, P. (2006). *Shopping environments: Evolution, planning and design.* Architectural Press.

Corkery, M. (2017, 15 April). Is American retail at a historic tipping point? *The New York Times.* https://www.nytimes.com/2017/04/15/business/retail-industry.html

Cox, D., & Streeter, R. (2019). *The importance of place: Neighborhood amenities as a source of social connection and trust.* American Enterprise Institute.

Davies, R. L. (1995). *Retail planning policies in Western Europe.* Routledge.

Davis, D. (1966). *A history of shopping.* University of Toronto Press.

Dunne, P., Lusch, R., & Carver, J. (2016). *Retailing.* Cengage Learning.

eMarketer. (2019). *Retail ecommerce sales penetration, by country.*

Evans, J. R. (2011). Retailing in perspective: The past is a prologue to the future. *The International Review of Retail, Distribution and Consumer Research, 21*(1), 1–31. https://doi.org/10.1080/09593969.2011.537817

Evers, D., Van Hoorn, A., Van Oort, F., & Noorman, N. (2005). *Winkelen in megaland.* Rijksplanbureau (Ed.). NAi Uitgevers.

Fairlie, R. W., & Robb, A. M. (2010). *Race and entrepreneurial success: Black-, Asian-, and white-owned businesses in the United States.* MIT Press.

Florida, R. L. (2002). *The rise of the creative class, and how it's transforming work, leisure, community and everyday life.* Basic Civitas Books.

Forsyth, A. (2015). What is a walkable place? The walkability debate in urban design. *Urban Design International, 20*(4), 274–292. https://doi.org/10.1057/udi.2015.22

Francaviglia, R. V. (1996). *Main street revisited: Time, space, and image building in small-town America*. University of Iowa Press.

Gehl, J., Johansen Kaefer, L., & Reigstad, S. (2006). Close encounters with buildings. *Urban Design International, 11*, 29–47. https://doi.org/10.1057/palgrave.udi.9000162

Geist, J. F. (1983). *Arcades: The history of a building type*. MIT Press.

GfK. (2019). 2018 small business owners' survey. https://nrf.com/insights/small-business/small-business-data-center

Gibbs, R. J. (2012). *Principles of urban retail planning and development*. John Wiley & Sons.

Goetz, S. J., & Rupasingha, A. (2006). Wal-Mart and social capital. *American Journal of Agricultural Economics, 88*(5), 1304–1310. https://doi.org/10.1111/j.1467-8276.2006.00949.x

Gopnik, A. (2016, 26 September). Jane Jacobs's street smarts. *The New Yorker.*

Gosseye, J., & Avermaete, T. (2017). *Shopping towns Europe: Commercial collectivity and the architecture of the shopping centre, 1945–1975*. Bloomsbury.

Gruen, V., & Smith, L. (1960). *Shopping towns USA: The planning of shopping centers*. Reinhold Publishing Corporation.

Hannigan, J. (2010). *Fantasy city: Pleasure and profit in the postmodern metropolis*. Routledge.

Hardwick, M. J. (2004). *Mall maker: Victor Gruen, architect of an American dream*. University of Pennsylvania Press.

Howard, V. (2015). *From main street to mall: The rise and fall of the American department store*. University of Pennsylvania Press.

Jacobs, J. (1961). *The death and life of great American cities*. Random House.

Jefferys, J. B., MacColl, M., & Levett, G. L. (1950). *The distribution of consumer goods: A factual study of methods and costs in the United Kingdom in 1938*. Cambridge University Press.

Jones, K., & Hernandez, T. (2000). Dynamics of the Canadian retail environment. In T. E. Bunting & P. Filion (Eds.), *Canadian cities in transition* (pp. 404–422). Oxford University Press.

Katz, B., & Wagner, J. (2014). The rise of innovation districts: A new geography of innovation in America. Brookings Institution.

Kershen, A. J. (2017). *Food in the migrant experience*. Routledge.

Kickert, C. C. (2016). Active centers – interactive edges: The rise and fall of ground floor frontages. *Urban Design International, 21*(1), 55–77. https://doi.org/10.1057/udi.2015.27

Leinberger, C. B., and Alfonzo, M. (2012). *Walk this way: The economic promise of walkable places in Metropolitan Washington, DC* (Metropolitan Policy Program, Ed.). Brookings Institution.

Lesger, C. (2013). *Het winkellandschap van Amsterdam. Stedelijke structuur en winkelbedrijf in de vroegmoderne en moderne tijd, 1550–2000*. Uitgeverij Verloren.

Levinson, M. (2012). The great A&P and the struggle for small business in America. Hill and Wang.

Locatus. (2018). Retail counts. Woerden, Netherlands.

Longstreth, R. W. (2000). *The drive-in, the supermarket, and the transformation of commercial space in Los Angeles, 1914–1941.* MIT Press.

McLaughlin, P. W., & Dicken, C. (2018). Evolution of the food-away-from-home industry: Recent and emerging trends. In M. J. Saksena, A. M. Okrent, & K. S. Hamrick (Eds.), *America's eating habits: Food away from home.* United States Department of Agriculture.

McLinden, S. (2018). Retail is the vital hub for successful mixed-use development, experts say. International Council of Shopping Centers.

Miellet, R., & Voorn, M. (2001). *Winkelen in weelde: warenhuizen in West-Europa 1860–2000.* Walburg Pers.

Moss, J. (2022). Jeremiah's vanishing New York [blog]. vanishingnewyork. blogspot.com

Murray, J., & Murray, K. L. (2008). *Store front: The disappearing face of New York.* Gingko Press.

Nijs, T. de, & Knoester, J. (2007). *Op zoek naar de verdwenen middenstand.* Verloren.

Novak, M. (2019). Assessing the long-term performance of the urban cores in four New Urbanist communities. *Journal of Urban Design, 24*(3), 368–384. https://doi.org/10.1080/13574809.2019.1568186

Office for National Statistics. (2019). Annual business survey: 2017 revised results.

Oldenburg, R. (1999). *The great good place: Cafes, coffee shops, bookstores, bars, hair salons, and other hangouts at the heart of a community.* Da Capo Press.

Parker, C. M., & Castleman, T. (2009). Small firm e-business adoption: A critical analysis of theory. *Journal of Enterprise Information Management, 22*(1/2), 167–182. https://doi.org/10.1108/17410390910932812

Peterson, H. (2017, 21 March). The retail apocalypse has officially descended on America. Business Insider. http://www.businessinsider.com/the-retail -apocalypse-has-officially-descended-on-america-2017-3

Pivo, G., & Fisher, J. D. (2011). The walkability premium in commercial real estate investments. *Real Estate Economics, 39*(2), 185–219. https:// doi.org/10.1111/j.1540-6229.2010.00296.x

Portas, M. (2011). *The Portas review: An independent review into the future of our high streets.* Department for Business, Innovation and Skills.

Raff, D., & Temin, P. (1999). Sears, Roebuck in the twentieth century: Competition, complementarities, and the problem of wasting assets. In N. R. Lamoreaux, D. M. G. Raff, & P. Temin (Eds.), *Learning by doing in markets, firms, and countries* (pp. 219–252). University Of Chicago Press.

Roost, F. (1998). Recreating the city as entertainment center: The media industry's role in transforming Potsdamer Platz and Times Square. *Journal of Urban Technology, 5*(3), 1–21. https://doi.org/10.1080/10630739883804

Rundle, A., Neckerman, K. M., Freeman, L., Lovasi, G. S., Purciel, M., Quinn, J., Richards, C., Sircar, N., & Weiss, C. (2008). Neighborhood food environment and walkability predict obesity in New York City. *Environmental Health Perspectives, 117*(3), 442–447. https://doi.org/10.1289/ehp.11590

Rutte, G. T. M., & Koning, J. (1998). *Zelfbediening in Nederland: geschiedenis van de supermarkttoekomst.* De Prom.

Satterthwaite, A. (2001). *Going shopping: Consumer choices and community consequences.* Yale University Press.

Savitt, R. (1989). Looking back to see ahead: Writing the history of American retailing. *Journal of Retailing, 65*(3), 326–355.

Silver, D. A., & Clark, T. N. (2016). *Scenescapes: How qualities of place shape social life.* University of Chicago Press.

Sorkin, M. (1992). *Variations on a theme park: The new American city and the end of public space.* Hill and Wang.

Stobart, J. (2010). A history of shopping: The missing link between retail and consumer revolutions. *Journal of Historical Research in Marketing, 2*(3), 342–349. https://doi.org/10.1108/17557501011067860

Strasser, S. (2004). *Satisfaction guaranteed: The making of the American mass market.* Smithsonian Books

Stringer, S. M. (2019). *Retail vacancy in New York City: Trends and causes, 2007–2017.* Bureau of Budget.

Stroope, S., Franzen, A. B., Tolbert, C. M., & Mencken, F. C. (2014). College graduates, local retailers, and community belonging in the United States. *Sociological Spectrum, 34*(2), 143–162. https://doi.org/10.1080/02732173.2014.878612

Talen, E., & Jeong, H. (2019). Does the classic American main street still exist? An exploratory look. *Journal of Urban Design, 24*(1), 78–98, https://doi.org/10.1080/13574809.2018.1436962

Thomas, E., Serwicka, I., & Swinney, P. (2015). *Urban demographics: Why people live where they do.* Centre for Cities.

Thompson, D. (2017, 9 August). Restaurants are the new factories. *The Atlantic.* https://www.theatlantic.com/business/archive/2017/08/restaurant-jobs-boom/536244/

United States Census Bureau. (2019). Advance monthly sales for retail and food Services.

US Bureau of Labor Statistics. (2018). All employees: retail trade: department stores.

US Department of Commerce. (2020, March). Advance monthly sales for retail and food services, March 2020. Accessed November 2020.

Veblen, T. (1899). *The theory of the leisure class.* Macmillan.

Verhoef, P. C., Kannan, P. K., & Inman, J. J. (2015). From multi-channel retailing to omni-channel retailing: Introduction to the special issue on multi-channel retailing. *Journal of Retailing, 91*(2), 174–181. https://doi.org/10.1016/j.jretai.2015.02.005

Walker, R. E., Keane, C. R., & Burke, J. G. (2010). Disparities and access to healthy food in the United States: A review of food deserts literature. *Health & Place, 16*(5), 876–884. https://doi.org/10.1016/j.healthplace.2010.04.013

Wang, S., Zheng, S., Xu, L., Li, D., & Meng, H. (2008). A literature review of electronic marketplace research: Themes, theories and an integrative framework. *Information Systems Frontiers, 10*(5), 555–571. https://doi.org/10.1007/s10796-008-9115-2

Wilson, J. L. (2018). Shopping locally: An exploration of motivations and meanings in the context of a revitalized downtown. PhD dissertation, University of North Carolina, Greensboro, NC.

Wolf-Powers, L., Doussard, M., Schrock, G., Heying, C., Eisenburger, M., & Marotta, S. (2017). The maker movement and urban economic development. *Journal of the American Planning Association, 83*(4), 365–376. https://doi.org/10.1080/01944363.2017.1360787

Woods, S. (2020, 8 April). How urban places can adapt after the coronavirus. *Public Squar.* https://www.cnu.org/publicsquare/2020/04/08/how-urban-places-can-adapt-after-coronavirus

Wrigley, N. (1993). Retail concentration and the internationalization of British grocery retailing. In R. D. F. Bromley & C. J. Thomas (Eds.), *Retail change: Contemporary issues* (pp. 41–68). UCL Press.

Zola, E. (1886). *Au bonheur des dames* (The ladies' paradise). Vizetelly & Co. (Original work published 1883)

Zukin, S., Kasinitz, P., & Chen, X. (2016). *Global cities, local streets: Everyday diversity from New York to Shanghai.* Routledge.

PART ONE

Retail Trends and Transformations

1 The Life and Death of Retail: Insights from Firm Demography

LUC ANSELIN AND IRENE FARAH

Introduction

Firm demography employs the methods of human demography to analyse events in the lifespan of establishments and institutions, such as their birth, death, aging, and movement patterns. The field has its roots in organizational sociology, especially in the work of Stinchcombe (1965), who suggested a demography of organizations and studied the birth and death processes of a range of different types of establishments (see also Carroll & Hannan, 2000, for a classic reference). This collection of methods is variously referred to as business demography, industrial demography, and even firmography (van Wissen, 2002).

This chapter applies firmography to the retail sector. What is the life and death pattern of retailers, and how does it vary by category, like size, and ownership structure? Looking at a 25-year period – 1990 to 2015 – we present the demographic characteristics of the US retail sector using detailed data on retail establishments. Our analysis uses a dataset known as the National Establishment Time Series (NETS), a proprietary database developed by Walls and Associates.

To our knowledge, this is the first such study for US retail. While a formal analysis of birth and death processes, age structure, and movement patterns of establishments has seen considerable application in sociology, business economics, economic geography, and regional science (e.g., van Dijk & Pellenbarg, 2002; Coad, 2009; Brown et al., 2013; Bishop & Shilcof, 2017), there are no recent empirical research results using firm demographics that pertain to the US retail sector. Most applications of firm demography focus solely on the birth and death processes, with much less attention paid to the survival probabilities during the life of a firm (e.g., Coad, 2018). Also, the empirical context typically pertains to manufacturing firms, and very little work has been done that deals with the retail sector. Exceptions are the study of births and deaths in the UK retail sector in the 1980s by Johnson and Parker (1994), and of selected retail establishments in Belgium in 1996–2000 by Dejardin (2004).

One caveat to note: as argued in van Wissen (2002), the retail "life and death" analogy with human populations is not perfect. For example, whereas for humans, birth and death are unambiguous, this is not the case for commercial establishments, where the exact start date is seldom (e.g., first time in a database, date of registration, date of incorporation, etc.). Similarly, death is not necessarily unavoidable (some firms have existed for hundreds of years and still exist), and the end of a firm in one form can coincide with the start of another (e.g., through merger or acquisitions).

Despite this caution, we believe the analysis of retail life and death using these methods offers significant insight about important trends and what they might portend for the future of urban retail.

Data and Definitions

NETS

Our data come from the National Establishment Time Series (NETS) database, which contains information on more than 60 million individual establishments over the 25-year period 1990–2015. The data are derived from the Dun and Bradstreet files and curated by Walls and Associates to create a consistent time series of establishments in existence in January of each year. For each establishment, several pieces of information are recorded, such as its name, address, latitude and longitude, code for the Standard Industrial Classification (SIC) and North American Industry Classification System (NAICS), type ("standalone," "branch," or "headquarters"), number of employees, sales, year the business started, and the last year the business is in the database. For the purposes of our analysis, we follow the definition of the retail sector used by Walls (2018) as establishments with eight-digit SIC codes ranging from 52 to 59.[1] For the 25-year period, the database contains 8,880,203 separate establishments in this category.

NETS data have been used in previous examinations of the retail sector, for example, by Schuetz et al. (2012) and Sutton (2014). Some limitations of NETS have been reported, especially the accuracy of the employment and sales information, business characterization, and locational precision.[2] It should also be noted that NETS does not record events that happen within a 1-year time span, although since this pertains to establishments that fail almost immediately, we can safely ignore it.

Table 1.1. Retail employment in the United States, descriptive statistics, selected years

Year	N	Min	Max	Mode	Mean	Q1	Median	Q3
1990	2,076,300	1	1,000	3	9.14	2	3	7
1995	2,310,510	1	1,000	3	9.41	2	3	7
2000	2,558,189	1	1,000	1	9.62	2	3	8
2005	3,009,822	1	1,000	2	8.83	2	3	7
2010	3,099,412	1	1,000	2	9.01	2	3	7
2015	2,985,950	1	1,000	2	9.65	2	3	8

In our current analysis, we do not track the movement of establishments. More precisely, we record the location when the firm is first present in the database and ignore later potential relocations. Since our analysis pertains to the United States as a whole (and most retail firms move within the same city, if they move at all), this is not a serious limitation. After some data cleaning, our final data set consists of 8,858,502 establishment records.[3]

Categories of Retail Establishments

We are particularly interested in how the life and death of retail varies by categories of establishments, looking especially at small, independent retailers, relative to larger establishments and chains. Since there is no consistent definition of small independent establishments (the Census, Small Business Administration, and local agencies all use different size thresholds), we base our classification largely on the actual distribution of employment and the number of establishments in the observations in our sample.

We considered the characteristics of the distribution of employment in all US retail establishments in 6 years (1990, 1995, 2000, 2005, 2010, and 2015), shown in Table 1.1. The number of establishments increased slowly from about 2 million in 1990 to a peak of about 3 million in 2010 (a 49 per cent growth relative to 1990), after which it decreased slightly.

The median employment is remarkably stable at three employees, throughout all reported time periods. The mode decreases slightly, from three employees in the first two periods, with a low of one employee in 2000 to a value of two employees in the remaining years. The mean is somewhat misleading, since it is heavily affected by the observations at the tail of the distribution. In each instance, the mean number of employees is larger than the value for the 75-percentile (Q3). It is reported here only for the sake of completeness.

In order to define our categories of retail, we use a set of criteria based on the number of employees, the number of establishments with the same company name, and whether an establishment is "standalone." This classification allows a clear distinction between the smaller firms and the chain stores. Our four categories are the following:

- Small: at most three employees in any year, unique company name, and classified as "standalone"
- Mid-size: can exceed three employees in some years, company name appears at most twice
- Regional chain: company name appears 3 times or more, but fewer than 40 times
- Corporate chain: company name appears 40 times or more

This classification yields the largest share for the small category (3,629,325 establishments, or 41.0 per cent), followed by the mid-size category (2,697,530 establishments, or 30.5 per cent). There are 1,582,664 establishments in the regional chain category (17.9 per cent) and 948,983 (10.7 per cent) in the corporate chain category. Note that these shares are for the complete 25-year period that contains all establishments in the United States that were at one time part of the data, and they may vary considerably from year to year. We turn to this next.

Evolution of Retail Categories

The shares of the four categories in the total number of retail establishments in the United States is shown in Table 1.2, for the same years as in Table 1.1 (note that the total number of retail establishments is listed in the second column of Table 1.1). The selected years may mask some of the fluctuations in the intermediate period, but they capture the overall trend.

Of the four categories, the mid-size category takes the largest share in each year, although it declines consistently throughout the years, from about 42 per cent in 1990 to 35 per cent in 2015. Almost 72 per cent of the establishments in 1990 were in the small or mid-size categories, most closely associated with the notion of "independent" businesses. There was a slight decline in 2000, but afterward their share remained more or less stable.

The main change between 1990 and 2015 is the increase in the share of the corporate category, going from 9.5 per cent in 1990 to almost double at 16 per cent in 2015. Regional chains peaked in 1995, thereafter decreasing to roughly the same share as the corporate chains by 2015 – significant because in 1990 their share was almost double that of the corporate chains. Note that the classification of establishments by category is fixed over time (i.e., it is based on their size and type over their lifetime

Table 1.2. Share in number of retail establishments, by retail category, selected years

Year	Small	Mid-size	Regional	Corporate
1990	29.11	42.44	18.98	9.48
1995	28.10	40.15	20.25	11.51
2000	29.48	37.35	19.85	13.33
2005	33.86	34.70	18.52	12.92
2010	34.47	34.32	17.15	14.06
2015	33.02	35.20	15.72	16.06

Table 1.3. Share in retail employment, by retail category, selected years

Year	Total	Small	Mid-size	Regional	Corporate
1990	18,982,812	6.91	45.57	20.77	26.75
1995	21,738,555	6.02	40.70	20.92	32.36
2000	24,604,119	5.54	38.78	20.52	35.16
2005	26,584,819	6.86	38.00	18.90	36.23
2010	27,912,030	6.79	36.29	16.90	40.02
2015	28,816,184	6.32	35.60	15.85	42.23

in the database), so that a shift in share should not be interpreted as a change in the size of an establishment (e.g., moving to more than three employees), but rather as a change in the mix of establishments in existence in January of a given year.

The share in retail employment for each category is listed in Table 1.3, with total employment given in the second column. The latter increased by over 50 per cent during the 25 years, almost double the increase of overall employment in the United States over the same period.

As is to be expected, the share of small retail in employment is much smaller than its share in the number of establishments, whereas the opposite is true for the corporate category. The latter rises from 27 per cent in 1990 to 42 per cent in 2015, a sizeable increase. Here again, the mid-size category has the largest share in 1990 but loses that lead to the corporate chains by 2010. The combined employment share of the "independents" (small and mid-size) decreases throughout – from 52 per cent to 42 per cent. Although the independents maintain an important share of both establishments and employment in retail, the corporatization of this sector is clearly evident in evolution over the past 25 years.

In the remainder of this chapter, we focus on the number of establishments.

Retail Births and Deaths

The annual birth and death rates by retail category are reported in Table 1.4. These rates are computed as the number of births/deaths for each category during the year, divided by the number of establishments existing in the corresponding category at the beginning of the year. Since the NETS data are snapshots for January in each year, the last year for which we can compute these rates is 2014 (the births/deaths between January 2014 and January 2015 relative to the number of establishments in January 2014).

The pattern in birth rates shows large changes from year to year, with both very large and very small values. The small retail category shows the greatest difference between minimum and maximum rates, with a range of 26.16, followed by the corporate chains with a range of 19.88, and the regional chains with a range of 16.89. The mid-size category has a narrower range in birth rates of 10.73, largely due to a much smaller maximum. Interestingly, both small and mid-size categories have their highest birth rates in 2010, right after the recession of 2009, whereas the regional chains have their second-highest and the corporate chains their third -highest rates in that year. Throughout, the small category tends to have the highest birth rates (in all years, except three), sometimes considerably higher than the other categories.

The time series for small, mid-size, and regional chains show a fairly strong correlation (0.88 for small-mid-size, 0.87 for small-regional, and 0.86 for mid-size-regional), but the corporate chains are less correlated with the others (0.59 with small, 0.61 with mid-size, and 0.81 with regional chains).

In order to extract the broad patterns between the four categories, Figure 1.1 depicts a locally estimated scatterplot smoothing (LOESS) fit (Cleveland, 1979) to the annual time series of points with the birth rate values. This graph emphasizes major directions of change, by smoothing out the extreme variations from year to year. All except the regional chains show a similar pattern of increased rates in the early to late 1990s, followed by a continuous slow decrease. The latter is temporarily halted by the uptick in 2010 that followed the 2009 recession, but after that, the downturn is more pronounced. The regional chains buck the trend somewhat, with a generally increasing rate until the same downturn is encountered as in the other sectors. The simplified view of the LOESS curve brings out the high rates for the small category and the high degree of volatility (much steeper up and down slopes) for the corporate category.

The death rates similarly show a high range between minimum and maximum values for each category. Again, the largest range is for the small category (22.67), but in this instance the corporate chains have the

Table 1.4. Annual retail birth and death rates, by category

Year	Small		Mid-size		Regional		Corporate	
	Births	Deaths	Births	Deaths	Births	Deaths	Births	Deaths
1990	10.33	6.91	5.56	5.16	6.70	5.90	2.79	3.66
1991	7.71	12.91	5.52	7.56	6.98	9.69	4.86	6.46
1992	14.25	8.88	9.30	7.01	17.43	8.73	22.05	6.14
1993	9.84	9.00	5.91	5.93	8.46	7.78	7.62	5.45
1994	21.53	18.43	11.02	6.36	19.98	9.44	20.56	3.60
1995	12.19	11.32	6.57	6.44	9.40	8.66	9.85	5.99
1996	23.91	12.72	10.02	7.05	17.04	9.17	16.79	5.06
1997	16.32	9.97	7.18	6.09	11.08	8.20	9.59	4.16
1998	12.27	12.09	6.02	6.52	8.10	9.54	7.03	5.32
1999	9.96	12.77	5.53	6.22	7.63	9.11	8.64	5.59
2000	17.46	10.04	7.32	5.55	11.22	7.62	11.25	5.27
2001	27.04	10.83	11.26	5.77	14.48	7.94	8.99	5.86
2002	14.37	9.46	6.22	5.18	7.80	7.41	6.75	6.86
2003	9.92	8.43	4.82	4.93	6.01	6.77	5.22	4.25
2004	14.46	12.79	6.94	6.05	8.36	8.54	8.69	5.18
2005	15.46	9.81	9.78	5.39	9.27	7.25	5.19	4.07
2006	9.12	6.45	9.85	3.83	7.24	4.73	4.47	2.74
2007	15.10	4.43	7.07	5.15	10.05	4.62	5.30	2.91
2008	11.74	4.03	6.53	5.55	7.30	4.77	5.88	4.96
2009	7.79	26.70	5.38	15.96	5.51	21.17	16.68	11.28
2010	32.15	8.34	13.97	6.01	18.65	8.01	19.04	4.85
2011	10.09	12.18	6.17	6.41	6.70	8.56	7.48	3.96
2012	7.01	12.03	4.64	5.60	4.05	8.37	3.46	4.99
2013	6.19	18.57	4.29	8.57	3.35	12.43	2.17	7.72
2014	5.99	14.50	3.24	6.47	3.09	9.57	5.91	5.86

smallest range (8.54). Not surprisingly, all categories reach their maximum death rate in 2009, the year of the recession, but more interestingly, the minimum was in one of the preceding years (2008 for small, 2007 for regional chains, and 2006 for mid-size and corporate chains), resulting in major year-to-year changes around 2010. The correlation between the time series for the different categories is stronger than for the birth rates, again with a slightly smaller correlation between the first three and the corporate chains (correlation of 0.71 small-corporate, 0.85 mid-size-corporate, and 0.86 regional-corporate).[4]

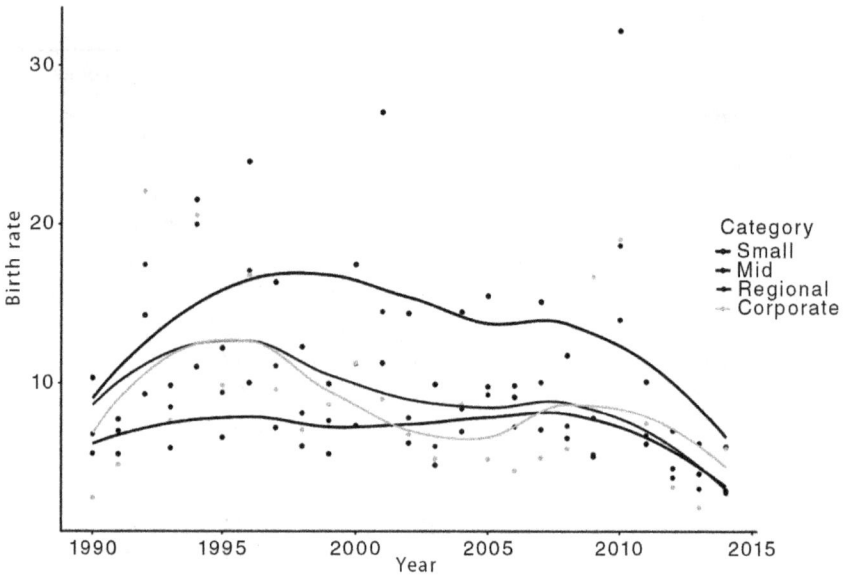

Figure 1.1. Retail establishment birth rates, by category. (Category lines in 1990, from top to bottom: "Small," "Regional," "Corporate," and "Mid.")

The LOESS curves shown in Figure 1.2 bring out the overall tendencies. As they were for the birth rates, the death rates for the small category are the highest. They show a clear pattern of an increase in the 1990s, followed by a decrease until around 2005 that turns into a clear (and steep) upward trend afterwards. Both the mid-size and regional chains show a similar pattern, but less pronounced. The smoothed curve for the corporate chains is almost flat, although it also has an upward slope starting around 2005. The tendency toward higher death rates in the recent years is clear in all four curves.

The net effect of birth and death rates is depicted in Figure 1.3, showing both the values (as points) and the smoothed curve. During most of the period, the net effect is positive, except for four years before 2010 where all sectors were in the negative (1991 and 2009 arguably coinciding with economic recessions). The positive spike in 2010 is followed by almost exclusively negative net rates afterwards. If the overall shape of the local fit is any indication, the situation after 2010 is particularly ominous for the small category and may be less so for the corporate category, although the slope of the curve for the latter is steeply downward in recent years.

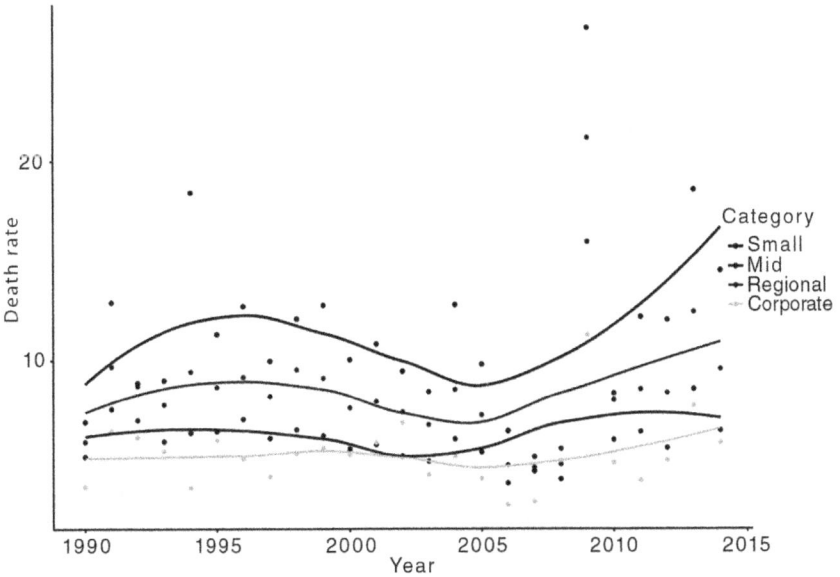

Figure 1.2. Retail establishment death rates, by category. (Category lines in 1990, from top to bottom: "Small," "Regional," "Mid," and "Corporate.")

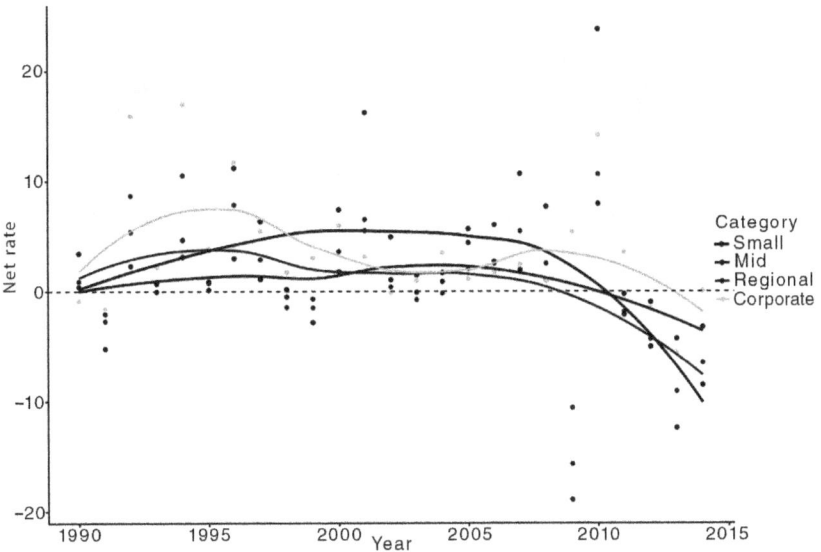

Figure 1.3. Retail establishment net change rates, by category. (Category lines in 1990, from top to bottom: "Corporate," "Regional," "Small," and "Mid.")

Overall, the birth and death rates suggest considerable churning in the retail sector, especially for the small category, which may gain almost a third in new establishments during any given year, as well as lose up to a quarter of existing establishments in another year. The general trend seems to be for the birth rates to decline since the early 2000s, followed by an increase in the death rates a few years later. The net change, after many years of growth, seems to have turned into a consistent pattern of decline in recent years.

Age Structure

As mentioned earlier, the NETS data also include the year each establishment started. This allows us to compute the age structure of the retail sector, by category, for each year we have data. The age of an establishment in any given year is then derived as the difference between the year in question and the year the business started. For businesses that started in 1990 and later, we have complete coverage (since we can verify the year started). However, for businesses that started in 1989 and earlier, we have many missing data, resulting in left-censoring. We only consider those establishments for which we have the actual start date.[5] The loss of information as a result of this censoring is the most extreme in 1990 and gradually decreases as time goes on (and businesses die off and are replaced by new ones).[6]

The age structure that results from the complex interplay between birth and death rates yields the age distributions reported in Table 1.5 for the "independents" (small and mid-size) and in Table 1.6 for the "chains" (regional and corporate). The values in the table show the percentage of the retail establishment population for that year that is in a given age category. We reported the same six years as before, which are representative of the overall evolution, although there remains considerable year-to-year variation as a result of the volatility in birth rates (and, to a lesser extent, death rates). The variability in the birth rates affects the lower age categories immediately, and its effect slowly lessens as the firms in question die out.

The characteristics of the age distribution are summarized in Table 1.7. The dominant characteristic of these distributions is that they are highly skewed, with a very long tail to the right (the difference between median and Q3 is much larger than the difference between Q1 and median). The inter-quartile range (a measure of spread) is fairly stable over the years (except for the corporate category in 1995 and 2000), around 20 years for the mid-size category, and in the low to mid-teens for the others.

Table 1.5. Age structure for selected years, independents

Age	Small						Mid-size					
	1990	1995	2000	2005	2010	2015	1990	1995	2000	2005	2010	2015
1	8.8	21.9	10.4	14.3	9.6	6.6	6.9	11.1	5.7	7.0	6.1	3.4
2	6.8	9.1	10.3	9.0	10.1	7.0	5.1	5.7	5.6	4.7	6.5	4.5
3	6.9	11.6	10.3	10.9	11.2	7.0	5.3	8.2	5.8	5.5	6.5	4.6
4	6.6	6.6	11.0	10.7	5.5	7.8	5.2	4.7	6.8	7.0	3.9	5.4
5	6.5	5.2	6.4	6.7	8.0	15.5	5.1	4.0	4.7	4.7	5.8	10.2
6	5.8	4.6	7.0	4.2	6.9	4.6	4.6	4.9	5.8	3.5	4.5	4.4
7	4.8	3.0	4.1	4.0	3.9	5.1	4.2	3.2	3.6	3.5	2.7	4.7
8	4.1	3.0	4.5	4.2	4.0	4.9	3.6	3.3	4.7	3.7	2.8	4.4
9	3.7	2.7	2.9	4.7	5.9	2.8	3.3	3.1	3.0	4.3	4.6	2.7
10	3.9	2.7	2.7	3.1	3.5	4.1	3.4	3.0	2.7	3.2	3.0	3.9
11	3.1	2.3	3.0	3.3	2.4	3.3	3.0	2.8	3.8	3.9	2.3	3.0
12	3.0	1.9	1.9	2.1	2.3	2.3	3.0	2.5	2.4	2.5	2.3	1.8
13	2.7	1.8	2.0	2.2	2.4	2.2	2.9	2.3	2.5	3.2	2.4	1.9
14	2.7	1.6	1.8	1.6	2.8	3.3	2.9	2.1	2.3	2.2	2.9	3.1
15	2.5	1.9	1.8	1.5	1.9	2.1	2.5	2.2	2.3	2.0	2.2	2.1
16	1.9	1.4	1.5	1.7	2.0	1.5	2.1	2.0	2.1	2.9	2.7	1.5
17	1.8	1.4	1.3	1.1	1.3	1.5	2.1	2.0	1.9	1.8	1.8	1.6
18	2.0	1.3	1.2	1.1	1.4	1.5	2.3	1.9	1.8	1.9	2.2	1.7
19	1.5	1.3	1.1	1.0	1.1	1.7	1.9	2.0	1.6	1.7	1.6	2.0
20	1.8	1.2	1.3	1.0	1.0	1.2	2.0	1.7	1.7	1.7	1.5	1.6
21	1.3	0.9	1.0	0.9	1.3	1.3	1.6	1.5	1.6	1.6	2.3	1.9
22	1.1	0.9	1.0	0.7	0.8	0.9	1.4	1.4	1.6	1.4	1.4	1.3
23	0.9	1.0	0.8	0.7	0.8	1.0	1.2	1.6	1.5	1.3	1.5	1.6
24	0.8	0.7	0.9	0.6	0.7	0.8	1.1	1.3	1.5	1.2	1.4	1.2
25	1.0	0.9	0.8	0.7	0.7	0.7	1.2	1.4	1.3	1.3	1.3	1.1
26	0.8	0.6	0.6	0.6	0.6	1.0	1.0	1.1	1.1	1.2	1.2	1.8
27	0.7	0.6	0.6	0.6	0.5	0.6	1.0	1.0	1.1	1.2	1.1	1.0
28	0.7	0.4	0.7	0.5	0.5	0.6	1.0	0.8	1.2	1.1	1.0	1.1
29	0.6	0.4	0.5	0.5	0.5	0.5	0.9	0.7	1.0	1.2	1.0	1.0
[30, 40]	5.2	3.3	3.2	3.0	3.5	3.7	7.4	6.5	6.7	7.5	8.5	8.1
[40, 50]	3.3	2.2	1.6	1.2	1.3	1.7	5.2	5.0	4.1	4.0	4.2	4.8
[50, 60]	1.3	0.8	1.0	0.8	0.6	0.7	2.4	2.0	2.9	3.0	2.7	2.6
[60, 70]	0.7	0.4	0.4	0.3	0.4	0.4	1.5	1.4	1.4	1.2	1.9	2.0
[70, 80]	0.3	0.2	0.2	0.2	0.2	0.1	0.7	0.8	0.8	0.8	0.9	0.8
[80, 90]	0.2	0.1	0.1	0.1	0.1	0.1	0.4	0.4	0.4	0.5	0.6	0.6
[90, 400]	0.2	0.2	0.1	0.1	0.1	0.1	0.6	0.6	0.6	0.6	0.7	0.8

Table 1.6. Age structure for selected years, chains

Age	Regional chains						Corporate chains					
	1990	1995	2000	2005	2010	2015	1990	1995	2000	2005	2010	2015
1	8.3	21.3	8.4	8.9	6.8	3.4	9.8	33.1	11.7	10.4	18.3	6.6
2	6.3	8.4	7.8	5.9	6.8	3.7	7.0	11.7	8.7	6.0	5.9	2.4
3	6.6	15.0	8.7	6.9	8.3	4.0	7.6	28.8	10.5	7.1	4.7	3.5
4	6.3	5.7	10.6	8.1	4.0	5.4	6.9	5.7	15.8	7.4	3.6	6.6
5	6.1	4.0	6.1	6.5	5.4	11.2	6.9	2.6	8.9	9.5	4.0	12.3
6	5.6	4.2	9.2	4.6	5.0	4.2	6.4	2.1	14.3	7.0	6.7	12.6
7	4.8	2.7	4.4	4.2	3.1	4.3	5.4	1.2	5.2	5.1	3.8	4.1
8	4.0	2.8	7.3	4.8	3.3	4.9	4.3	1.3	12.7	6.2	4.4	3.0
9	3.8	2.6	3.0	6.0	5.2	2.6	3.9	1.1	2.4	9.2	4.7	2.4
10	3.8	2.5	2.3	3.7	4.2	3.5	4.1	1.1	1.1	5.3	6.0	2.7
11	3.2	2.2	2.9	5.7	3.1	3.2	3.2	1.0	1.0	8.5	4.6	4.7
12	3.2	2.0	1.8	2.8	2.8	2.2	3.2	0.8	0.6	3.0	3.0	2.7
13	2.9	1.8	1.9	4.7	3.2	2.3	2.9	0.7	0.6	7.5	4.0	3.1
14	2.9	1.7	1.8	2.0	4.1	3.5	2.8	0.7	0.5	1.4	6.1	3.1
15	2.6	1.7	1.7	1.6	2.6	3.1	2.6	0.8	0.5	0.7	3.4	4.2
16	2.1	1.5	1.5	2.1	4.0	2.3	2.0	0.6	0.5	0.6	5.4	3.2
17	1.9	1.5	1.4	1.3	2.0	2.1	1.8	0.6	0.4	0.4	1.8	2.0
18	2.1	1.4	1.2	1.4	3.3	2.3	1.8	0.5	0.4	0.4	4.8	2.7
19	1.6	1.4	1.1	1.3	1.6	3.0	1.5	0.5	0.3	0.3	0.9	4.3
20	1.7	1.2	1.2	1.2	1.3	2.0	1.5	0.5	0.4	0.3	0.4	2.3
21	1.3	1.0	1.0	1.1	1.8	3.2	1.2	0.4	0.3	0.3	0.4	3.8
22	1.2	0.9	1.1	1.0	1.1	1.7	0.9	0.3	0.3	0.3	0.2	1.3
23	1.0	1.0	1.0	0.9	1.2	2.7	0.8	0.4	0.3	0.2	0.3	3.3
24	0.8	0.8	1.0	0.8	1.0	1.3	0.8	0.3	0.3	0.2	0.2	0.6
25	0.9	0.8	0.9	0.9	1.0	1.0	0.8	0.3	0.2	0.2	0.2	0.3
26	0.8	0.7	0.7	0.8	0.9	1.5	0.6	0.2	0.2	0.2	0.2	0.3
27	0.8	0.6	0.7	0.8	0.8	0.9	0.6	0.2	0.2	0.2	0.2	0.2
28	0.7	0.5	0.7	0.7	0.7	1.0	0.5	0.2	0.2	0.2	0.1	0.2
29	0.6	0.4	0.6	0.7	0.7	0.9	0.4	0.2	0.1	0.2	0.1	0.1
[30, 40]	5.2	3.4	3.5	4.3	5.5	6.3	3.6	1.0	0.8	0.9	1.0	1.0
[40, 50]	3.2	2.3	1.9	1.9	2.3	3.2	2.0	0.6	0.3	0.3	0.4	0.4
[50, 60]	1.5	0.9	1.2	1.2	1.2	1.4	0.8	0.2	0.2	0.2	0.1	0.2
[60, 70]	0.9	0.6	0.5	0.5	0.8	1.0	0.5	0.1	0.1	0.1	0.1	0.1
[70, 80]	0.4	0.3	0.3	0.3	0.4	0.4	0.2	0.1	0.1	0.0	0.0	0.0
[80, 90]	0.2	0.1	0.2	0.2	0.2	0.3	0.2	0.0	0.0	0.0	0.0	0.0
[90, 400]	0.4	0.2	0.2	0.2	0.3	0.3	0.2	0.1	0.0	0.0	0.0	0.0

Table 1.7. Summary of age distribution, by category

Year	Quartile*	Small	Mid-size	Regional	Corporate
1990	Q1	4	5	4	4
	Q2	8	12	9	8
	Q3	17	23	18	15
1995	Q1	2	4	2	1
	Q2	5	10	4	3
	Q3	13	22	13	4
2000	Q1	3	5	4	3
	Q2	6	11	6	5
	Q3	13	23	14	8
2005	Q1	3	5	4	4
	Q2	5	11	9	7
	Q3	11	23	15	11
2010	Q1	3	5	4	3
	Q2	6	12	10	8
	Q3	13	25	18	14
2015	Q1	4	5	5	5
	Q2	7	11	11	8
	Q3	14	25	21	16

* The value of the first quartile (25%), the median (50%), and the third quartile (75%).

The mid-size category has the highest median age throughout (consistently 11–12 years), although by 2015 it is matched by the regional chains (at 11 years). The other categories show an interesting pattern of almost equal median ages up to 2000 (although going down and up), but after 2000 the small category starts to have a slightly smaller median age (7 years, compared to 8 for the corporate chains). Somewhat surprisingly, there is no pronounced effect of the 2009 recession on the median age, suggesting that it affected all categories similarly.[7] By 2015, the third quartile ranges from 14 and 16 years for small and corporate chains, compared to 21 for the regional chains, and 25 for mid-size. Of the four categories, the last seems to have the greatest longevity.

Cohort Analysis

An alternative view of the age distribution in the different categories of the retail sector is to follow a single cohort of births over time. Unlike the previous analysis, this is not affected by left-censoring. This so-called life table

Table 1.8. Evolution of 1990 retail establishment cohort by category

Year	Age	Small		Mid-size		Regional		Corporate	
		Pop.	Deaths	Pop.	Deaths	Pop.	Deaths	Pop.	Deaths
1991	0	62,456	6	48,959	25	26,711	29	5,494	5
1992	1	62,450	901	48,934	790	26,682	578	5,489	145
1993	2	61,549	3,439	48,144	2,893	26,104	2,252	5,344	433
1994	3	58,110	25,900	45,251	10,370	23,852	8,049	4,911	1,257
1995	4	32,210	7,818	34,881	4,562	15,803	3,020	3,654	569
1996	5	24,392	1,764	30,319	1,988	12,783	730	3,085	142
1997	6	22,628	835	28,331	1,084	12,053	441	2,943	70
1998	7	21,793	797	27,247	1,107	11,612	438	2,873	109
1999	8	20,996	733	26,140	937	11,174	384	2,764	85
2000	9	20,263	706	25,203	907	10,790	378	2,679	84
2001	10	19,557	1,273	24,296	1,208	10,412	547	2,595	135
2002	11	18,284	936	23,088	951	9,865	447	2,460	140
2003	12	17,348	900	22,137	905	9,418	441	2,320	82
2004	13	16,448	1,093	21,232	906	8,977	467	2,238	94
2005	14	15,355	895	20,326	901	8,510	411	2,144	70
2006	15	14,460	656	19,425	543	8,099	270	2,074	53
2007	16	13,804	294	18,882	420	7,829	178	2,021	53
2008	17	13,510	279	18,462	407	7,651	157	1,968	63
2009	18	13,231	2,303	18,055	2,435	7,494	1,083	1,905	239
2010	19	10,928	592	15,620	805	6,411	345	1,666	89
2011	20	10,336	764	14,815	831	6,066	386	1,577	83
2012	21	9,572	517	13,984	671	5,680	278	1,494	63
2013	22	9,055	1,121	13,313	1,085	5,402	519	1,431	89
2014	23	7,934	704	12,228	659	4,883	306	1,342	74
2015	24	7,230	–	11,569	–	4,577	–	1,268	–

approach (see, e.g., Carroll & Hannan, 2000, Ch. 6) is analogous to the models developed for human demography, although with the caveats mentioned in the introduction. An important underlying assumption is that the given cohort is representative of a "typical" cohort, which cannot be assessed with our data. Nevertheless, it remains interesting to consider the patterns of the evolution of the population for the different categories of retail.

The principle is simple: we follow the cohort of establishments that came about in 1990–1991 throughout the time period in the database. This is illustrated in Table 1.8. The "births" during 1990–1991 constitute the establishments that form the population in January 1991 and were not in the database in 1990. Consequently, 1991 corresponds to year 0 in Table 1.8.[8] Some of these establishments do not make it until January 1992, which we label as "deaths" during the first period (year 0). For our data, as shown in the first row of the table, there were 62,456 new small establishments in January 1991, of which 6 did not make it until January 1992. Similarly, the figures for the other categories are 48,959 mid-size establishments with 25 deaths, 26,711 regional establishments with 29 deaths, and 5,494 corporate establishments with 5 deaths. The population at the next time point is obtained directly by subtracting the number of deaths, or:

$$\text{Pop}\ (t+1) = \text{Pop}(t) - \text{Deaths}(t, t+1).$$

The sequence continues until the population in January 2015 is reached, the latest point for which we have data (year 24).

For each year, we can compute a number of statistics from this information, the most relevant of which are the annual death rate and the survival function. The death rate is the ratio of deaths in the period t to $t+1$ over the population at time t. In the interest of conserving space, we have not listed the detailed rates here, but they can be easily computed from the information in Table 1.8.

All four categories show a similar pattern over time, although the rates differ considerably. All have the highest death rate in years 3 and 4, followed by a gradual decline until a peak associated with the recession of 2008–2009 (year 18). The latter is an anomaly that we do not expect a "typical" cohort to experience. However, the peak in mortality after just a few years (in our case 3–4) is well documented and referred to in the literature as the "liability of newness" (Stinchcombe 1965).

The highest death rates are experienced in year 3 for all four categories of retail, but the actual rates differ considerably: 45 per cent for small, 23 per cent for mid-size, 34 per cent for regional, and 26 per cent for corporate chains. They drop considerably in year 4, although they are still much higher than other years (all in single digits): 24 per cent for small, 13 per cent for mid-size, 19 per cent for regional, and 16 per cent for corporate chains. The spike in mortality in year 18 is in the low to mid teens, smaller than the rate in year 4 in all instances.

The mortality information can be employed to compute a survival function. The latter is the complement of the cumulative distribution of deaths up to time t, or $F(t)$. The complement gives the fraction of the initial population that survives up to year t:

$$G(t) = 1 - F(t).$$

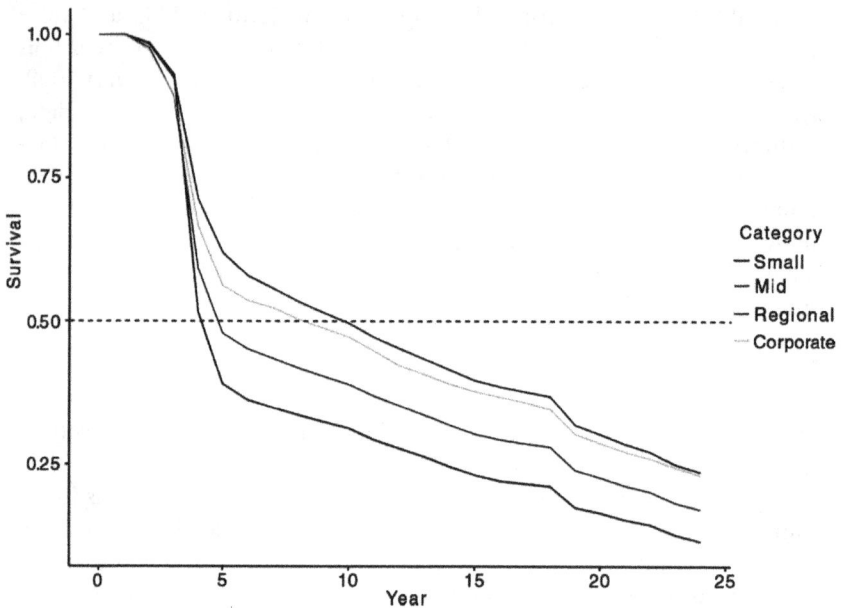

Figure 1.4. Survival function by retail category. (Category lines in year 5, from top to bottom: "Mid," "Corporate," "Regional," and "Small.")

An important summary characteristic associated with the survival function is the half-life of the population, that is, how many years it takes until only 50 per cent of the initial population survives (or 50 per cent of the initial population has died off).

Figure 1.4 illustrates the survival function for the cohort "born" in 1990–1991, for each of the retail categories. All four functions exhibit the same general shape, fairly flat in the beginning, with a sharp drop-off in years 3 and 4, due to the peaks in death rates for those years. The drop-off is the steepest for the small category, but the least steep for the mid-size group. Interestingly, the two typical components of "independents" thus show quite different patterns. In part, this may be due to the mid-size entities having acquired a critical mass, which allows them to weather the liability of newness more than the truly small establishments. After the big drop, all four curves show a gradual, almost linear decline, until the recession in year 18, which briefly interrupts this trend.

The half-life differs considerably among categories. The lowest value is again obtained for the small establishments, which are reduced to less than half their initial size by year 4. Next are the regional chains, with a half-life of 5 years, followed by corporate chains with 8, and finally mid-size with 10. The ranking of the half-life is similar to what we found for the median age in the previous section, but the values are somewhat smaller. In part this is because the life table is the pure evolution of a single cohort, whereas the cross-sectional age distribution mixes the birth and death experiences of different cohorts.

Conclusions

The initial description of the demographic characteristics of the US retail sector offered here reveals some interesting patterns over the last 25 years. Clearly, the sector has become more corporate, even though "independents" (which we classified as small and mid-size) retain an important share of establishments as well as employment.

The most striking finding is the volatility in the sector, with extreme changes in birth and death rates of the number of establishments from year to year. This affects the small category the most (with birth rates of up to 32 per cent as well as death rates up to 26 per cent), but the other categories show similar year-to-year variability. Over the 25-year period of our study, the prospects evidenced in the most recent years are not good, with all categories showing net declines, a trend that seems to start in the early 2000s and becomes more pronounced after 2010. In addition, while there was a significant rebound in 2010 following the 2009 recession, the rebound was not sustained.

The changes in birth and death rates translate into a complex age structure for establishments that varies slowly from year to year, mostly in the early age categories. Nevertheless, there are some interesting patterns. For all categories, the age distribution is highly skewed, with a very long right-hand tail around a median age in the single digits for all but the mid-size category (in the lower teens), and, since 2010, also the regional chains. The most vulnerable are the small independent establishments, whose median age varies between 5 and 8 years, contrasting with 11–12 years for the mid-size category. The long tail in the distribution suggests that once an establishment reaches a critical age, it can last a long time afterwards. This is supported by the life table analysis of the cohort of establishments "born" in 1990. All retail categories show very high mortality rates in years 3 and 4 of their existence (the highest for the small category) after which they slowly drop off. Again, the mid-size category turns out to be the most robust, with an estimated half-life of 10 years (roughly matching the median age estimated from the age distribution), contrasting with only 4 years for the small category.

Overall, the retail sector shows considerable churning over the period observed, with sometimes significant differences between the four categories we distinguished. Even though the sector has become more corporate, the mid-size establishments (the slightly larger-sized "independents") show great robustness and seem to be able to survive and thrive in this changing environment.

This chapter was limited to a description of the patterns using the methodology of firm demography. Further work needs to be done to understand what drives these patterns, nationally and regionally. Furthermore, we have considered only broad categories of retail, and it remains to be seen whether these patterns apply uniformly to the many subcategories that can be distinguished (e.g., food services, eating and drinking establishments, bookstores).

NOTES

1 The included categories are 52 (building materials, hardware, garden supply, and mobile home dealers), 53 (general merchandise stores), 54 (food stores), 55 (automotive dealers and gasoline service stations), 56 (apparel and accessory stores), 57 (home furniture, furnishings, and equipment stores), 58 (eating and drinking places), and 59 (miscellaneous retail).

2 For a detailed discussion of the pros and cons of NETS relative to other data sources, see, for example, Neumark et al. (2005), Neumark et al. (2011), Kaufman et al. (2015), Barnatchez et al. (2017), and Echeverri-Carroll and Feldman (2019).

3 We eliminated all establishments from our sample with more than 1,000 employees at any point in the time series, since very high numbers in one retail establishment were likely due to coding error or misclassified headquarters of large corporations. This removed 1,159 establishments from the data. In addition, we removed 20,501 establishments with an unspecified company name ("DLISTED") and 41 establishments with a year start in 2016, which reduced the overall size of the data set to 8,858,502 unique establishments over the 25-year period.

4 The correlations among the other categories are, respectively, 0.83 for small-mid-size, 0.93 for small-regional chains, and 0.96 for mid-size-regional chains.

5 We also carried out an analysis using "imputed" start dates, that is, assuming that the age distribution of the missing data is the same as of those observed. As is to be expected, the main results of interest, such as the median age, are not affected. If the age structure of the missing data were somehow very different from the observed pattern, the results would obviously be affected, but we have no way to ascertain this. However, even in the worst case, given

the mortality among firms, the effect of the missing observations would become negligible by the later years in the data set.

6 In 1990 the censoring results in a loss of 669,121 establishments, or 32.3 per cent out of the original 2,076,300, but by 2015, the share has decreased to only 2.4 per cent (72,108 out of the original 2,985,950). The effect of the censoring also differs by retail category, since it is affected in part by the differential birth and death rates between the categories. It is the most severe for the corporate category, where in 1990, 82 per cent of the establishments in this category were censored, with 10 per cent censoring remaining in 2015. The smallest degree of censoring is for the mid-sized establishments, with only 16 per cent affected in 1990 and less than 1 per cent by 2010. The effect on the small businesses similarly drops very quickly, going from 34 per cent in 1990 to 1 per cent and less starting in 2000, reflecting the shorter lifespan of such businesses. Regional chain censoring similarly starts at round 40 per cent in 1990 and drops to 4 per cent by 2010.

7 The largest share of 1-year-old establishments is found in 1995, and in all instances this value is quite extreme compared to the other years. This effect is felt for all categories, with the lowest median age occurring in that year, as small as half the value for other years (e.g., 3 versus 7–8 in most other years).

8 Since we don't know the number of establishments in January 1989, the only reliable starting point is January 1991, which allows a direct comparison with January 1990.

REFERENCES

Barnatchez, K., Crane, L., & Decker R. (2017). An assessment of the National Establishment Time Series (NETS) database. Finance and Economics Discussion Series 2017-110. Board of Governors of the Federal Reserve System.

Bishop, P., & Shilcof, D. (2017). The spatial dynamics of new firm births during an economic crisis: The case of Great Britain, 2004–2012. *Entrepreneurship and Regional Development, 29*, 215–237. https://doi.org/10.1080/08985626.2016.1257073

Brown, J. P., Lambert, D. M., & Florax, G. M. (2013). The birth, death, and persistence of firms: Creative destruction and the spatial distribution of US manufacturing establishments, 2000–2006. *Economic Geography, 89*, 203–226. https://doi.org/10.1111/ecge.12014

Carroll, G. R., & Hannan, M. T. (2000). *The demography of corporations and industries.* Princeton University Press.

Cleveland, W. S. (1979). Robust locally weighted regression and smoothing scatterplots. *Journal of the American Statistical Association, 74*, 829–836. https://doi.org/10.1080/01621459.1979.10481038

Coad, A. (2009). *The growth of firms: A survey of theories and empirical evidence.* Edward Elgar.

Coad, A. (2018). Firm age: A survey. *Journal of Evolutionary Economics, 28,* 13–43. https://doi.org/10.1007/s00191-016-0486-0

Dejardin, M. (2004). Sectoral and cross-sectoral effects of retailing firm demographies. *Annals of Regional Science, 38,* 311–334. https://doi.org/10.1007/s00168-004-0197-6

Echeverri-Carroll, E. L., & Feldman, M. P. (2019). Chasing entrepreneurial firms. *Industry and Innovation, 26,* 479–507. https://doi.org/10.1080/136627 16.2018.1475220

Johnson, P., & Parker, S. (1994). The interrelationship between births and deaths. *Small Business Economics, 6,* 283–290. https://doi.org/10.1007/bf01108395

Kaufman, T. K., Sheehan, D. M., Rundle, A., Neckerman, K M., Bader, M. D., Jack, D., & Lovasi, G. S. (2015). Measuring health-relevant businesses over 21 years: Refining the National Establishment Time-Series (NETS), a dynamic longitudinal data set. *BMC Research Notes, 8,* 507. https://doi.org/10.1186 /s13104-015-1482-4

Neumark, D., Wall, B., & Zhang, J. (2011). Do small businesses create more jobs? New evidence for the United States from the National Establishment Time Series. *The Review of Economics and Statistics, 93,* 16–29. https:// doi.org/10.1162/rest_a_00060

Neumark, D., Zhang, J., & Wall, B. (2005). Employment dynamics and business relocation: New evidence from the National Establishment Time Series. *Research in Labor Economics, 26,* 39–83. https://doi.org/10.1016/s0147 -9121(06)26002-3

Schuetz, J., Kolko, J., & Meltzer, R. (2012). Are poor neighborhoods "retail deserts"? *Regional Science and Urban Economics, 42,* 269–285. https://doi.org /10.1016/j.regsciurbeco.2011.09.005

Stinchcombe, A. L. (1965). Social structure and organizations. In J. March (Ed.), *Handbook of Organizations* (pp. 142–193). Rand McNally.

Sutton, S. A. (2014). Are BIDs good for business? The impact of BIDs on neighborhood retailers in New York City. *Journal of Planning Education and Research, 34,* 309–324. https://doi.org/10.1177/0739456x14539015

van Dijk, J., & Pellenbarg, P. H. (2002). Spatial perspectives on firm demography. *Papers in Regional Science, 79,* 107–110. https://doi.org/10.1111/j.1435-5597.2000 .tb00763.x

van Wissen, L. J. (2002). Demography of the firm: A useful metaphor? *European Journal of Population, 18,* 263–279. https://doi.org/10.1023/a:1019750727018

Walls, D. W. (2018). *2014 National Establishment Time Series (NETS) Retail Database.* Walls and Associates.

2 The Ups and Downs of Retail, 2000–2015

KEVIN CREDIT, IRENE FARAH, AND LUC ANSELIN

In the years between the Great Recession and the COVID-19 pandemic, large numbers of retail store closures prompted concern over a growing retail "apocalypse." However, academic research on the spatial and temporal extent of such an apocalypse – and its relationship to regional economic vitality – is lacking. This chapter uses an extensive national dataset on retail business establishments – National Establishment Time Series, or NETS – to explore temporal and spatial trends in the total number of retail stores by Core-Based Statistical Area (CBSA) from 2000 to 2015. We separate retail into three categories: "experience economy," "basic needs," and "general merchandise," and correlate them with CBSA-level occupational structure and population. The results indicate that experience economy businesses have increased since 2011, while basic needs and general merchandise have decreased, lending support to the idea that a downturn in physical (or product-focused) retailing has indeed occurred in recent years. However, there is significant heterogeneity in the pattern by CBSA. Larger regions with higher proportions of workers in creative class occupations tend to have larger recent increases in experience economy establishments and lower decreases in physical retailers, while smaller regions with higher proportions of workers in working class occupations have seen a significant decline in independent retailers and moderately poor performance in physical retail more generally. Interestingly, the most-resilient regions for physical retail in the post-Recession period are smaller, institutionally anchored regions that often have strong traditional main street design. Although the direction of causality is not explored here, it appears that occupational-economic structure and physical main street design may both play significant roles in the growth of experience economy businesses and resilience to the recent retail downturn, including the ability to support independently owned retailers.

Introduction

The previous chapter has focused primarily on the birth, death, and age structure of retail in the United States from 1990 to 2015, using the NETS business data. While the analyses in Chapter 1 centre on differences across size of establishments for the United States as a whole, this chapter studies differences among metropolitan areas and retail subcategories from 2000 to 2015, making use of the same database but at a more fine-grained scale. Although both chapters emphasize the decline of retail in recent years, Chapter 1 underlines the volatility of the sector and the increase of corporatization in both the number of establishments and number of employees, while this chapter highlights the spatial heterogeneity within the sector and its association with local economic development.

Over the last 30 years, the retail landscape in the United States has changed significantly. The rise of big-box stores and e-commerce retailers like Amazon has threatened the more locally focused independent retailers that still make up a majority of retail establishments. Changes in the nature of retailing have significant implications for economic development and planning since small, independent retailers provide the kind of fine-grained land use fabric essential for vitality, social interaction, walking, and sense of place (Jacobs, 1961; Smith & Sparks, 2000). In addition, big-box and e-commerce firms generally export a significant share of retail revenue out of the regional economy while contributing to a more auto-oriented, unsustainable development pattern (Houston & Eness, 2009).

While there has been a surge of interest in the popular media in the so-called retail apocalypse in the years leading up to the COVID-19 pandemic and widespread economic shutdowns that it prompted, spurred in part by the closing of several major chain retailers (Peterson, 2019; Thompson, 2017; Montgomery, 2018), some financial analysts have responded by saying that the "reports of retail's death are premature" (Dutta, 2018). However, there is relatively little academic work that has clearly examined the extent of the post-Recession retail downturn on a national scale. Nor has there been much quantitative research on the most recent trends in retail location (Wrigley & Lambiri, 2015; Jones & Livingstone, 2018; Singleton et al., 2016; Wrigley & Dolega, 2011). Likely the Great Recession (2007–2009) accelerated retail change by forcing firms to increase their competitiveness in a generally poor macroeconomic climate. The Recession is also likely to have hastened the elimination of non-competitive or non-innovative businesses. Better understanding the temporal and spatial trends in retail vitality is particularly important for identifying characteristics that foster resilience for independent retailers, especially as we enter a new era of economic disturbance caused by the pandemic and its widespread effects.

To shed light on these issues, we use point-level establishment data from the National Establishment Time Series (NETS) for the entire country from 2000 to 2015. The data allow us to investigate temporal trends in retail activity among three different types of retailers: businesses in the "experience economy," and two types of physical (or product-focused) retail categories: "basic needs" (neighbourhood goods such as grocery and drug stores) and "general merchandise" (larger retailers, including department and big-box) stores. In addition, we use NETS data to analyse differences in retail vitality between chain and independent stores. These trends are assessed at the individual Core-Based Statistical Area (CBSA) level in order to identify which regions have had the largest (and smallest) downturns by retail category, and whether there is any spatial association between regions in regard to retail trends (i.e., are there national-scale concentrations of recent decline?). Finally, the chapter explores the relationship between retail trends and regional occupational structure (at the CBSA-scale) in order to ascertain whether regions with higher proportions of "creative class" workers have been more resilient to retail downturn.

The results of our analysis suggest that retail businesses in the basic needs and general merchandise categories have experienced a downturn in recent (pre-pandemic) years, while businesses in the experience economy category have seen growth; when viewed through the lens of ownership type, chain retailers have fared relatively better than independently owned businesses, though both have declined in recent years. The term "retail apocalypse" may be unnecessarily provocative, but this analysis does lend support to the idea that, on an aggregate scale, physical (or product-focused) retailers have declined in recent years, with the number of general merchandise stores in 2015[1] falling to 705,987 from a peak of 968,978 in 2009, closing in on the number of stores in 2000 (675,436), at the start of the time series. However, there is significant heterogeneity in these overall trends by CBSA, both *when* the most recent structural break in each trend has occurred, and the *extent of the decline* in independent, chain, basic needs, and general merchandise businesses (with some CBSAs demonstrating recent increases). In general, CBSAs with greater proportions of creative class occupations tend to have lower decreases in independently owned, basic needs, and general merchandise businesses (although the correlations are relatively weak). On the other hand, working class regions tend to see smaller recent increases in experience economy businesses, large declines in independently owned businesses, and smaller decreases (or even increases) in the number of chain stores. Larger CBSAs (in terms of population) tend to follow the national trend for business type (increasing experience economy and decreasing basic needs and general merchandise businesses), but also see smaller decreases or even increases in independently owned stores in recent years.

These findings, while exploratory, shed empirical light on the extent of the (pre-pandemic) recent retail downturn. First, structural changes are indeed happening to the retail industry, with the number of retailers selling physical goods declining in recent years, even as the number of experience economy businesses have increased. However, these structural changes are applied unevenly across the country – larger regions (like Atlanta, GA, and Boston, MA) and those with high proportions of creative class occupations (such as Austin, TX, and Kansas City, MO-KS) tend to have larger recent increases in experience economy and independently owned retail businesses, while smaller working class regions (like Elkhart, IN, and Rockford, IL) tend to have smaller increases in experience economy businesses, larger declines in independently owned businesses, and *smaller* declines in chain stores. Interestingly, there is also a third group of high performing small regions like Manhattan, KS, and Charleston, SC, that have demonstrated the highest level of physical retail and independently owned resilience in the post-Recession period.

This supports the idea that, in the post-Recession era, the experience of the small-town "main street" of independent retailers is bifurcated: in the working class regions of the country that have been increasingly left behind by the modern information economy, it is indeed disappearing; however, small regions with thriving economies (often anchored by large institutions) have demonstrated the most recent physical and independent retail resilience, indicating that perhaps the best hope for physical retail lies in institutionally anchored small towns and those with strong traditional main street urban design (like Charleston, SC). At the same time, regions with more creative economies have also shown strong overall retail resilience and an associated growth (or arrested decline) in experience economy and independently owned businesses. While it is not possible to know from this analysis whether these changes are primarily the result of these occupational-economic features of regions – or are contributing to them, or both – it does suggest that the future of independently owned "main street" retail is likely to lie in small, institutionally anchored and large, creative economy metros, and increasingly in businesses that provide in-person experiences that cannot be easily replicated online. To what extent the closure of many such experience-related businesses that has resulted from COVID-19, as well as widespread substitution of online for in-person interactions, influences this trend remains to be seen and will be a fascinating topic for future research.

Background

Retail "Apocalypse"?

The plight of independent retailers and main streets due to the rise of auto-oriented shopping opportunities, big-box stores, and the globalization of retail has been widely discussed since at least the 1980s (Gardner & Sheppard, 1989; Bromley & Thomas, 1993, 1995; Coe & Wrigley, 2007; Barata-Salgueiro & Erkip, 2014), although much of the literature covering the plight of the retail sector since the Great Recession has occurred in the popular media and "grey" literature (Peterson, 2019; Thompson, 2017; Montgomery, 2018; Dutta, 2018; Wrigley & Lambiri, 2015; Retail Economics & Squire Patton Boggs, 2017; KPMG, 2018). Interest in the recent "retail apocalypse" has expanded significantly since 2017, as several major retail chains have closed hundreds of stores or entered bankruptcy, including longstanding retail stalwarts like Sears, JCPenney, and Macy's. These closures have generated a flurry of articles and commentary on the future of "brick-and-mortar" retail in the age of Amazon and e-commerce competition (Peterson, 2019; Thompson, 2017; Montgomery, 2018; Dutta, 2018). As these (and many other) sources point out, from a national perspective, physical retailing appears to be in serious decline. According to Coresight Research (2019), as of Week 15 of 2019, year-to-date announced retail closures in the United States had already surpassed 2018's total (5,994 closures and 2,641 openings in the first 15 weeks of 2019, compared to 5,864 closures and 3,239 openings in all of 2018). In addition, as the square footage of retail closures has increased since 2016, data from the US Census indicates a steady increase in the percentage of US retail sales made online, increasing to over 10 per cent of all retail sales in the first quarter of 2019 (Montgomery, 2018; US Department of Commerce, 2019).

These trends appear to reflect changes in consumer preferences that have placed increasing value on "experiences" rather than commodities, goods, or services (Retail Economics & Squire Patton Boggs, 2017; KPMG, 2018; Pine & Gilmore, 1998). Indeed, industry research by the firms Retail Economics and Squire Patton Boggs (2017) has indicated that consumers are much more willing to shop at a given physical retail location if there are coffee shops or restaurants and bars in the same area, and that retail spending as a proportion of total consumer spending is projected to drop from around 26.5 per cent in 2004 to just above 20 per cent in 2028. These data underscore the idea that "retailers and shopping centres need to create opportunities for consumers to share experiences with friends and family, enabling the expression of social, commercial, political and cultural interests" (p. 12).

Others have argued that the retail "apocalypse" narrative greatly exaggerates the existing state of the physical retail sector. As the global consulting firm Klynveld Peat Marwick Goerdeler (KPMG) points out in their "Global retail trends 2018" report, "[P]hysical retail isn't actually dead, but boring retail is" (p. 8). From this perspective, the same US Census figures that show growing spending on e-commerce also indicate that 90 per cent of all retail is still done within physical stores (2019; KPMG, 2018). Relatedly, some have argued that the recent rash of retail closures has more to do with a structural adjustment due to overbuilt commercial space in the United States (Cowen, 2017; Thompson, 2017). By some estimates, in 2015, US shopping centres' gross leasable area (GLA) per capita was much higher than any similar country at 23.5 square feet, compared to Canada's 16.4, the United Kingdom's 4.6, and France's 3.8 (Cowen, 2017). And some analysts have argued that the stock prices for large retail chains already reflect the negative structural environment and thus won't drop further, pointing out that even e-commerce firms like Amazon have adopted a physical retail strategy with their chain of Amazon bookstores (Dutta, 2018). Another point made recently is that an "apocalypse" more accurately reflects a bifurcation in the retail industry, where boutique luxury retailers catering to high income shoppers and big-box and other discount retailers catering to price-conscious shoppers have squeezed out the retailing "middle class." In 2019, big-box retailers like Walmart, Target, and Lowe's announced unexpectedly high profits and associated share prices, while mid-range retailers like Abercrombie & Fitch and Gap continue to struggle (Peterson, 2019).

Several recent studies have evaluated the characteristics that make retail more or less resilient to downturn. Singleton et al. (2016) evaluated both supply (store/infrastructure characteristics) and demand (demographic and internet use characteristics) factors to create an overall "e-resilience" score for retail centres in the United Kingdom. In another study, retail resilience appeared to be based on the diversity of retail mix within a centre, including both "greater representation of small independent specialist stores" and the presence of corporate grocery stores, anchor stores, and a professional management structure (Wrigley & Dolega, 2011, p. 2357; Guimaraes, 2018).

In the remainder of this chapter, we add a new perspective on the retail landscape by taking a comprehensive, longitudinal look at retail trends and associated local economic impacts in the United States. We compare three main retail types: retail classified as being part of the "experience economy," versus two types of physical (product-focused) retail, "basic needs" and "general merchandise." What do the varying trends in these divergent retail types, together with their associated occupational profiles, mean for local economic development and resilience?

Occupational Structure and Retail Trends

The extent of the recent downturn in physical retailing is not comprehensively understood at the national or even regional scale. However, given that the elements that might theoretically influence recent retail trends – technological change and e-commerce adoption, demand for experience-oriented consumption, and traditional economic factors such as disposable income and education – exhibit regional heterogeneity, we would logically expect retail changes to exhibit similar heterogeneity. The connection between (and regional expression of) many of these factors might be explained by Richard Florida's theory of the "creative class" (2012), which posits that, as the macro-structure of the economy continues to shift to focus on more creative and knowledge-intensive forms of work, regions (and employers) that cater to the preferences of these creative workers will tend to perform better than others. In some ways this is similar to traditional notions of knowledge-based economic development, but one of the key ideas of Florida's work is that the preferences of today's creative workers are different from those of highly educated workers in previous generations:

> Creative people do not move for traditional reasons. The physical attractions that most cities focus on building – sports stadiums, freeways, urban malls, and tourism-and-entertainment districts that resemble theme parks – are irrelevant, insufficient, or actually unattractive to them. What creatives look for are high-quality amenities and experiences, an openness to diversity of all kinds, and above all else the opportunity to validate their identities as creative people. The communities that creatives are attracted to do not thrive for traditional economic reasons, such as access to natural resources or proximity to major transportation routes ... A big part of their success is that they are places where creative people want to live. (Florida, 2012, p. 186)

Florida's theory provides a useful framework for understanding why particular regions might exhibit higher or lower levels of retail decline: we would expect regions with high proportions of creative class workers to contain relatively higher levels of experience economy and independent establishments, and (perhaps) lower levels of general merchandise and chain establishments, although it is also possible that regions with high proportions of creative class workers are generally economically healthier than others and thus have higher levels of all types of retail businesses.

Data

This chapter primarily uses the National Establishment Time Series (NETS) database of individual establishments across the United States from Walls and Associates. This database provides annual snapshots of Dun and Bradstreet (D&B) data, providing the location of US retail establishments from 2000 to 2015. We classified businesses as chain (CHN) retailers if they were a branch location or if the name of the company was repeated more than twice. We defined independent (IND) retailers if they were classified as a "standalone" company, or if they were classified as a "headquarters" establishment but the company name was used at most twice.

In order to have a better understanding of retail trends by category, we categorize businesses as "experience economy" (or service economy) (EE), "basic needs" (BN), and "general merchandise" (GM) stores, according to the North American Industry Classification System (NAICS) (shown in Table 2.1). How did we come to this classification scheme? According to Pine and Gilmore's (1998) seminal article, the "experience economy" is industry-agnostic; rather, the idea is one of large-scale macroeconomic change such that the primary product of nearly *all* contemporary businesses is a positive, memorable experience, with the physical product or service being sold only a "prop" for the larger "performance" that is of primary importance (and value). Still, for the purposes of this retail analysis, it is useful to break out businesses that rely more heavily on in-person experiences than those that are primarily product-oriented. McKinsey & Company (Goldman et al., 2017) provide an oft-cited delineation of experience-type businesses, of which only "food service" fits within the larger category of retail businesses considered in this chapter,[2] so NAICS code 722 is used to classify "food services and drinking places" as "experience economy" businesses here. The distinction between "basic needs" and "general merchandise" businesses for retailers that sell physical goods primarily comes from the idea that (1) certain kinds of retail are necessary for the essentials of everyday life (and thus for providing a sustainable local retail environment), including grocery stores and pharmacies (Talen & Jeong, 2019), while others (2) are oriented primarily towards larger-scale general merchandise retailing, falling under the US Census Bureau designation of "General Merchandise Normally Sold

Table 2.1. NAICS industrial classifications for retail categories of interest

NAICS	Category name
Experience Economy	
722	Food Services and Drinking Places
Neighbourhood Goods ("Basic Needs") Retail	
445	Food and Beverage Stores
446	Health and Personal Care Stores
General Merchandise, Apparel, Furnishings, and Other (GAFO)	
442	Furniture and Home Furnishings Stores
443	Electronics and Appliance Stores
444	Building Material and Garden Equipment and Supplies Dealers
448	Clothing and Clothing Accessories Stores
451	Sporting Goods, Hobby, Musical Instrument, and Book Stores
452	General Merchandise Stores
4532	Office Supplies, Stationery, and Gift Stores

in Department Stores" or GAFO (US Census, 2020). Classifying retail businesses in this way[3] provides for a more fine-grained analysis of post-Recession trends – including those businesses in the so-called experience economy that might be expected to be less affected by recent structural changes to the retail sector (Retail Economics & Squire Patton Boggs, 2017; KPMG, 2018; Pine & Gilmore, 1998).

We are interested primarily in total counts of establishments because they provide a stable year-to-year trend that reflects the aggregate sum of business activity within a given region. While rates of retail births and deaths each year can be considered as in the previous chapter, the interpretation is not straightforward: births and deaths are often correlated, and business "churn" can often be an economic positive that reflects competition and innovation (Fotopoulos & Spence, 1998; Baptista & Karaöz, 2011). It is also important to note that establishment counts do not correlate well with employment totals. There may be a higher number of small (independent) stores, but they tend to have fewer employees, and there may be fewer (larger) chain stores, but they tend to have more employees.

These counts of existing establishments (by year) in each of the three retail categories of interest were then aggregated to Core-Based Statistical Area (CBSA) boundaries for 356 CBSAs. Longitudinal data on occupational structure from the Bureau of Labor Statistics (BLS) and population data from the Bureau of Economic Analysis (BEA) from 2015 were also joined

Table 2.2. OES occupational classifications for creative, working, and service class occupations

Occupation code	Occupation title	Class
11-0000	Management Occupations	Creative
13-0000	Business and Financial Operations Occupations	Creative
15-0000	Computer and Mathematical Occupations	Creative
17-0000	Architecture and Engineering Occupations	Creative
19-0000	Life, Physical, and Social Science Occupations	Creative
23-0000	Legal Occupations	Creative
25-0000	Education, Training, and Library Occupations	Creative
27-0000	Arts, Design, Entertainment, Sports, and Media Occupations	Creative
29-0000	Health-Care Practitioners and Technical Occupations	Creative
21-0000	Community and Social Service Occupations	Service
31-0000	Health-Care Support Occupations	Service
33-0000	Protective Service Occupations	Service
35-0000	Food Preparation and Serving Related Occupations	Service
37-0000	Building and Grounds Cleaning and Maintenance Occupations	Service
39-0000	Personal Care and Service Occupations	Service
41-0000	Sales and Related Occupations	Service
43-0000	Office and Administrative Support Occupations	Service
47-0000	Construction and Extraction Occupations	Working
49-0000	Installation, Maintenance, and Repair Occupations	Working
51-0000	Production Occupations	Working
53-0000	Transportation and Material Moving Occupations	Working

to CBSA boundaries. Employment by occupation (based on the categories delineated in the Occupational Employment Survey [OES]) was then aggregated to "creative," "service," and "working" class categories according to the framework provided by Florida (2012), as shown in Table 2.2.

Methods

To assess temporal retail trends, we used three methods: (1) a structural breaks analysis of the count of establishments by category (experience economy, basic needs, and general merchandise) and type (all independent and chain establishments across all three of these categories); (2) a visual exploration of box plots of the yearly percentage change in number of establishments by CBSA; and (3) Pearson's product-moment correlations (r) between the entire time series (by the same rank-order) by retail category and type.

The structural breaks approach starts with the assumption that the slope of a linear trend of the number of establishments (i.e., a linear regression of EST on YEAR) is not constant throughout the entire time period. Instead, several segments can be distinguished, each with its own coefficient. The transition from one segment to the next is termed a structural break. When one knows the position of the break a priori, then a test for structural stability is straightforward (the so-called Chow test). However, in practice, this is seldom known, nor is it known how many different segments are appropriate. A method to estimate both the number and the time for the break points from the data was suggested in the econometrics literature by Bai (1997) and Bai and Perron (2003), and implemented in the R software environment (see Zeileis et al., 2002, 2003).[4] With m break points, this results in $m + 1$ segments (starting with the first observation), each with its own regression coefficient. In our example, the dependent variable is the number of establishments of a given type (retail category, and chain-independent) regressed on the year. For each segment j, the regression equation is as follows:

$$EST_i = YEAR_i \beta_j + u_i \quad \left(i = i_{j-1} + 1, \ldots, i_j, \quad j = 1, \ldots, m+1 \right) \tag{1}$$

where j is the segment index and i corresponds to the years in the given segment j (i_j). The break points are estimated by means of an algorithm that essentially minimizes the residual sum of squares (RSS) of equation (1) over all possible combinations of m.

In addition to the aggregate structural breaks analysis that evaluates large-scale temporal trends, individual structural breaks analyses were run for each CBSA to determine regional variation in temporal retail trends. Once the structural breaks were determined, the average annual change in number of establishments over the period between the last empirically derived break and the last observation (2015) could be calculated for each CBSA:

$$\frac{EST_{C_{2015}}}{EST_{C_{BY}}}^{\frac{1}{COUNT(2015:BY)}} - 1 \tag{2}$$

where $EST_{C_{BY}}$ is the count of establishments of category/type C in the calculated final structural break year BY. This technique provides data-driven values for "decline" or "growth" over the recent period for each CBSA, which are then mapped to assess any spatial association between regions of decline or growth by retail category and type.

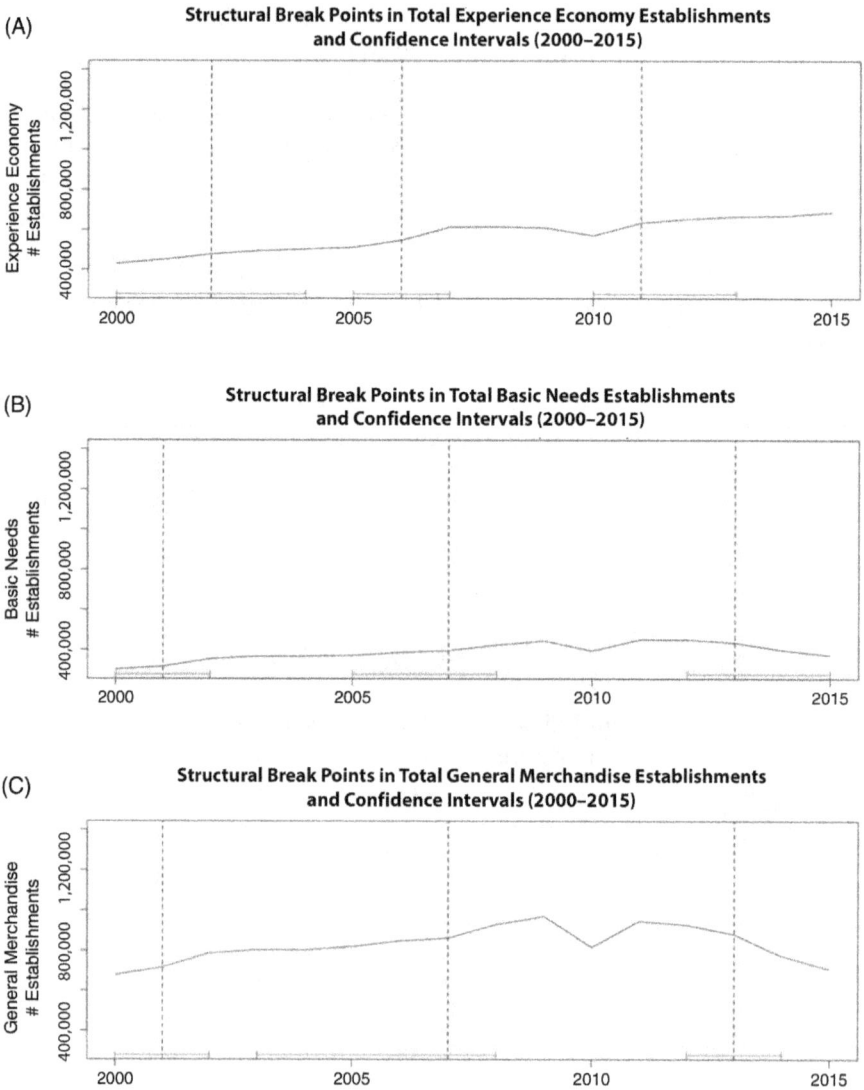

(A) Structural Break Points in Total Experience Economy Establishments and Confidence Intervals (2000–2015)

(B) Structural Break Points in Total Basic Needs Establishments and Confidence Intervals (2000–2015)

(C) Structural Break Points in Total General Merchandise Establishments and Confidence Intervals (2000–2015)

Figure 2.1. Graphs showing structural breaks for experience economy, basic needs, general merchandise, all independent, and all chain establishments, with confidence intervals, 2000–2015. (This figure continues on the facing page.)

(D)

Structural Break Points in Total Independent Establishments and Confidence Intervals (2000–2015)

(E)

Structural Break Points in Total Chain Establishments and Confidence Intervals (2000–2015)

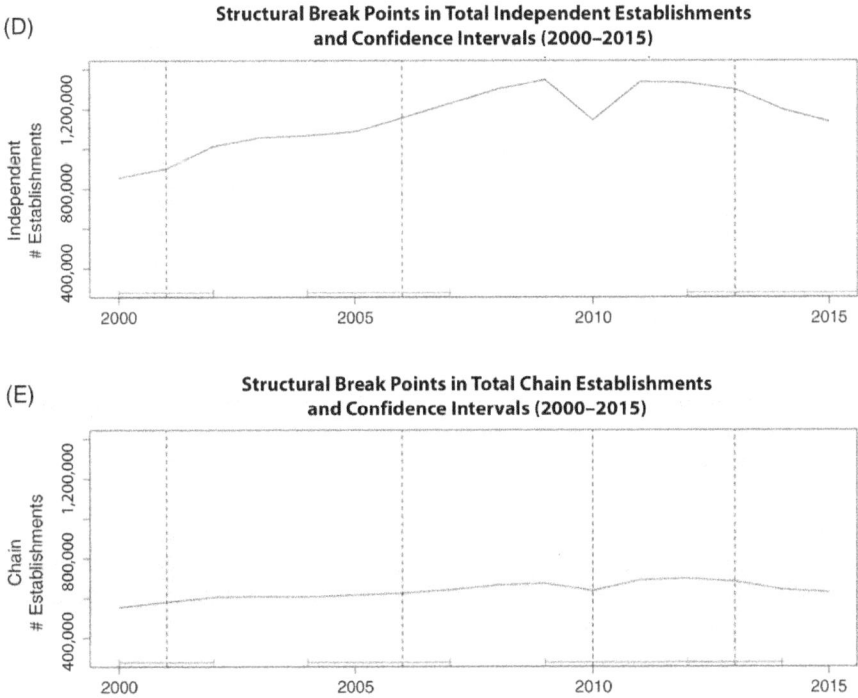

Correlations (r) between these individual "final break" average annual change values and each CBSA's percentage of working class (WC), service class (SC), and creative class (CC) occupations in 2015 are calculated to assess the relationship between occupational structure and recent retail trends. The average annual change values for each retail business type, occupational percentages, and 2015 population (POP) were also entered into a k-means clustering algorithm ($k = 5$), computed in GeoDa v.1.14 using a Z transformation and the KMeans++ initialization method (Anselin, 2018), to further explore the associations between these features and better contextualize the results as a whole.

Results

Our analysis begins with a look at temporal trends – and empirically derived structural breaks – for each of the three retail categories (experience economy, basic needs, and general merchandise) and two types (independent and chain) at the aggregate level. The graphs with total number of establishments by year, structural breaks, and confidence intervals for each category/type are shown in Figure 2.1.

Table 2.3. Correlations of yearly change (%)

	Correlation statistic (r)
Basic Needs and General Merchandise	0.88
Experience Economy and General Merchandise	0.59
Experience Economy and Basic Needs	0.54
Independent and Chain	0.77

Two interesting results stand out from the structural breaks analysis. First, experience economy businesses are the only retail category to experience growth over the most recent period (i.e., the time between the final structural break in the time series and 2015). Basic needs and general merchandise establishments have both experienced serious declines overall, with each category losing 14.2 per cent and 20 per cent of their stores, respectively, between 2013 and 2015. We observe a similar – although less drastic (because these figures include experience economy establishments) – decline over the most recent period when the data are broken down into independent and chain ownership types, with independent retailers losing 12.4 per cent of establishments and chain retailers losing 8 per cent between 2013 and 2015.

Table 2.3 shows the correlation for yearly percentage change in number of establishments (by CBSA) for each combination of categories and types.[5] Basic needs and general merchandise stores have generally followed quite similar trends over the time series, as have independent and chain businesses (though to a lesser extent). However, there is less similarity in time series trends between experience economy and general merchandise (0.59) and between experience economy and basic needs (0.54).

Second, despite the obvious visual appearance of the Recession dip in 2010 in the time series for each category/type, only chain retail stores (out of the five chosen ways to split the data) observed an empirical structural break at that point. This indicates that perhaps the Recession was not the primary or direct cause influencing the retail change that we observe over the most recent period; in fact, all categories/types experience a period of growth between the two break points that surround the Recession. It is possible, of course, that structural changes to the retail sector lag behind the macro-level shock of the Recession, but it appears from this analysis that the significant declines that we observe in the physical retail categories actually began to accelerate in 2013, which perhaps supports the hypothesis that changing technology and the rise of e-commerce have had more to do with recent retail decline than macro-level economic pressures (although the two could be related). Interestingly, this also matches the findings of the chapter by Anselin and Farah in this volume, which suggest that the observed downward trend in establishment birth and death rates predates the Recession and takes a steep downward turn more recently.

At the same time, there is significant heterogeneity among regions in these trends. Figure 2.2 highlights the distributional spread in yearly percentage change values between CBSAs for each delineated category and type. The median (central bar) of the distribution across CBSAs confirms the general pattern of growing experience economy and declining basic needs and general merchandise over the most recent period. However, there is a considerable spread in this distribution, which varies from year to year (e.g., in some years the box is compressed, suggesting less variation, but in others the range is quite large). For instance, the percentage change in general merchandise establishments between 2012 and 2013 ranges between 3.9 per cent growth (Sebastian-Vero Beach, FL) and 11.6 per cent decline (Racine, WI).

The box plots in Figure 2.2 indicate that variance in yearly percentage change values is generally very high during the Recession valley and bounce-back years of 2010–2011, perhaps demonstrating a heterogenous set of impacts and reactions to this macroeconomic shock across regions. Interestingly, however, in the post-Recession period, the variances tend to decrease, perhaps indicating that technological change and the rise of e-commerce have had a less heterogenous impact across regions, "compressing" the variance in retail outcomes in recent years.

The regional heterogeneity in trends in recent years is shown in Figure 2.3. These maps display the top (dark grey) and bottom (light grey) quintiles of average annual change in establishments by CBSA from the date of the uniquely calculated final structural break until 2015 by retail category. Thus these maps show us in broad strokes the regions that have been doing relatively well (dark grey) and poorly (light grey) in the post-Recession period for a given retail business type. The patterns are certainly mixed, but there does appear to be some spatial association between regions with high and low values.

For experience economy businesses, CBSAs that are doing relatively well (in the top quintile of recent average annual change) appear primarily in the South Atlantic, East South Central, West South Central, and Mountain US Census Divisions, that is, the "Sunbelt" areas that have seen the majority of population growth and domestic migration in recent years and whose labour forces (and economies) are often driven by service- and creative class employment. At the same time, CBSAs that

(A)

(B)

(C)

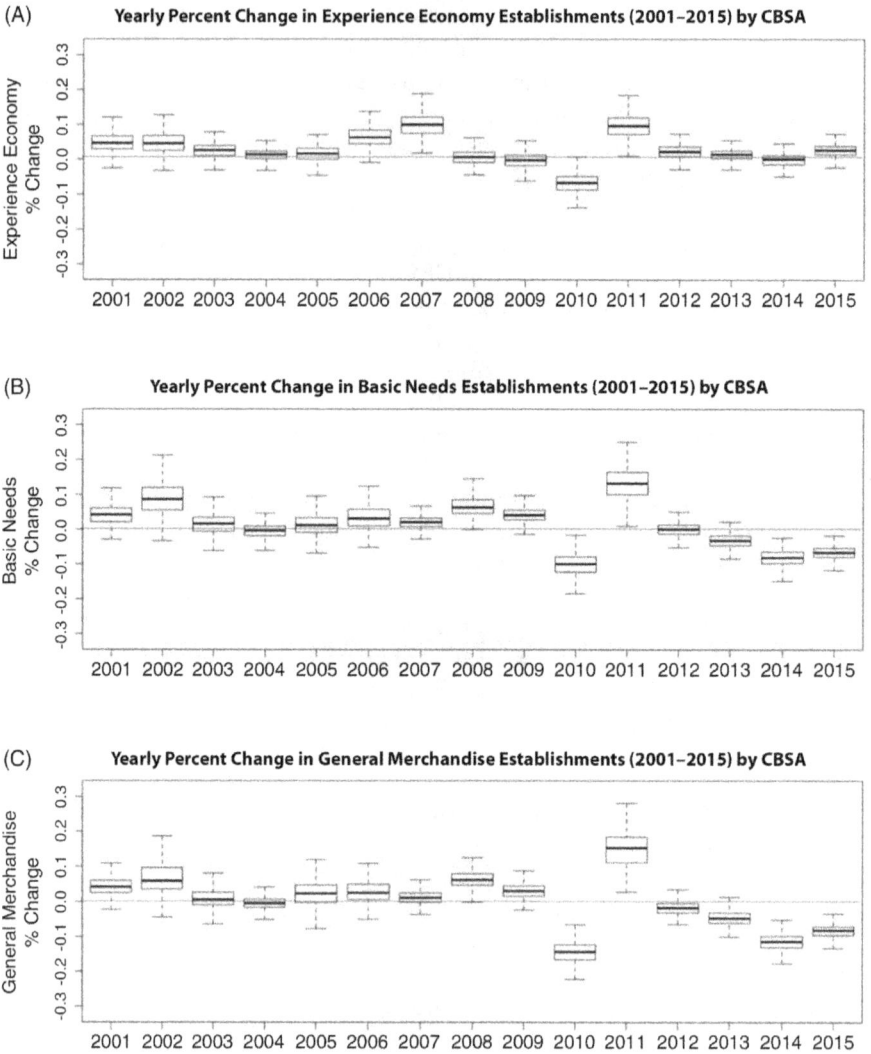

Figure 2.2. Box plots of yearly percentage change in establishments by CBSA for experience economy, basic needs, general merchandise, all independent, and all chain establishments, 2000–2015. (This figure continues on the facing page.)

(D)

Yearly Percent Change in Independent Establishments (2001–2015) by CBSA

(E)

Yearly Percent Change in Chain Establishments (2001–2015) by CBSA

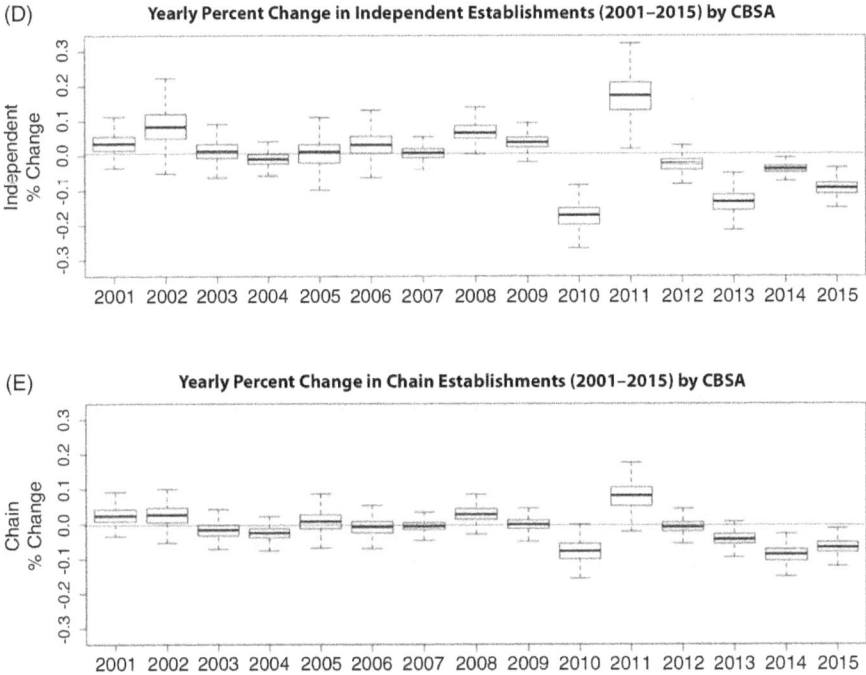

are doing poorly (in the bottom quintile of recent experience economy growth) – including some regions in which the number of experience economy businesses declined – show up generally in the Middle Atlantic and East North Central Census Divisions or "Rustbelt" areas that often have higher proportions of working class occupations and economies that were traditionally dependent on manufacturing.

Something of an inverse pattern is apparent for basic needs and general merchandise establishments: for basic needs, upper quintile regions tend to be located either in "Rustbelt" areas or in California and New England. These findings suggest that recent trends in basic needs businesses are positively related to both creative and working class occupations. Conversely, lower quintile regions tend to be those in the South Atlantic, East South Central, and Mountain Divisions, which we typically associate with service class economies (e.g., Atlanta–Sandy Springs–Roswell, GA). For general merchandise businesses the spatial pattern is more heterogeneous, but prominent working and service class metros

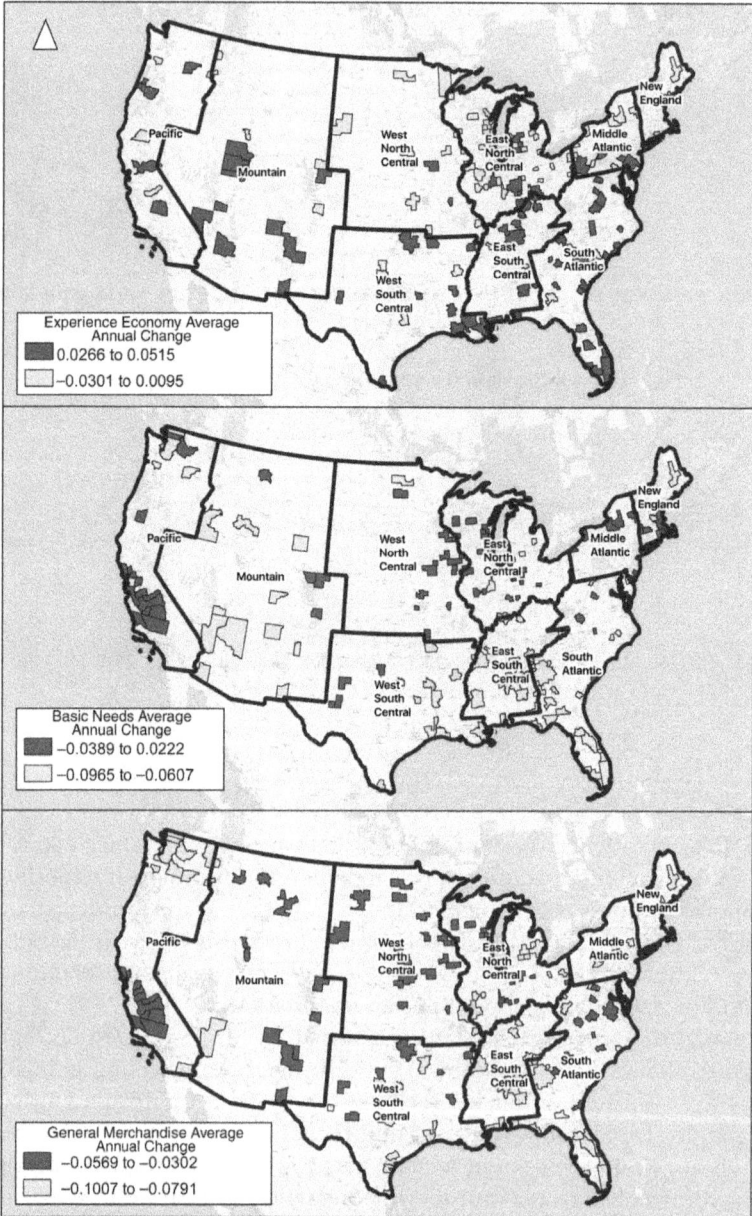

Figure 2.3. Top and bottom quintiles of average annual change in establishments between final structural break and 2015 for experience economy, basic needs, and general merchandise.

Table 2.4. Correlations between occupational structure and population in 2015 and retail trends by category and type

Correlation statistic (r)	Comparison	Description of trend
0.08	CC vs EE	More CC, increasing EE
0.07	CC vs BN	More CC, less decreasing/slightly increasing BN
0.10	CC vs GF	More CC, less decreasing GF
0.21	CC vs IND	More CC, less decreasing/slightly increasing IND
−0.03	CC vs CHN	More CC, decreasing CHN
0.09	SC vs EE	More SC, increasing EE
−0.14	SC vs BN	More SC, more decreasing BN
−0.11	SC vs GF	More SC, more decreasing GF
−0.03	SC vs IND	More SC, more decreasing IND
−0.17	SC vs CHN	More SC, more decreasing CHN
−0.17	WC vs EE	More WC, less increasing EE
0.02	WC vs BN	More WC, less decreasing BN
−0.03	WC vs GF	More WC, more decreasing GF
−0.17	WC vs IND	More WC, more decreasing IND
0.16	WC vs CHN	More WC, less decreasing CHN
0.09	POP vs EE	More POP, increasing EE
−0.03	POP vs BN	More POP, more decreasing BN
−0.10	POP vs GF	More POP, more decreasing GF
0.09	POP vs IND	More POP, less decreasing/slightly increasing IND
−0.13	POP vs CHN	More POP, more decreasing CHN

show up in the bottom quintile of recent change (e.g., Detroit, MI, and Atlanta, GA). Several CBSAs in California and the West North Central Division show up in the top quintile.

In the correlations between occupational structure and population in 2015 and average annual change in establishments over the most recent period by retail category and type,[6] many of the relationships observed in the spatial analysis are verified, although in most cases the correlations are relatively weak, as shown in Table 2.4. Regions with higher proportions of creative class occupations tend to have larger increases in experience economy businesses and lower decreases in basic needs and general merchandise businesses, lower decreases or even slight increases in independent businesses, and larger decreases for chain businesses. Regions with higher proportions of service class occupations tend to have larger increases in experience economy businesses, but they tend to have larger decreases in basic needs and general merchandise businesses, as well as for both chain and independent businesses. Working class regions have lower increases in experience economy businesses and larger decreases in general merchandise businesses, but lower decreases in basic needs. Interestingly, they also demonstrate larger decreases in independent businesses but lower decreases in chains. Larger CBSAs (in terms of population size in 2015) have a similar pattern of relationships with service class regions, with one important difference: they demonstrate a positive relationship with independently owned retailers.

To better characterize these trends, a k-means clustering algorithm with $k = 5$ clusters was computed using occupational structure, population, and post-break trend for each of the three retail establishments of interest as the input variables. The results of this analysis are shown in Table 2.5 and Figure 2.4. Cluster centres represent the central value (in attribute space) for each variable for each of the clusters; the distribution of these values across clusters are used to provide qualitative descriptions of the clusters. As the cluster analysis shows, these variables interact in interesting ways. Cluster 1 contains the largest CBSAs, which also tend to have high proportions of creative class occupations and high recent growth in experience economy businesses (and declines in other categories and types) – these are generally the largest, most diversified regions like New York, NY, Chicago, IL, and Los Angeles, CA. CBSAs in Cluster 2 also tend to have large populations and proportions of creative class employment, but have performed much better across all retail categories and types, in particular experience economy and independent businesses. This cluster includes some of the most prominent high technology entrepreneurial ecosystems, like Boston, MA, San Francisco, CA, and Austin, TX, and demonstrates the link between these creative regions and arrested decline for independent retailers.

However, Cluster 3 represents the best average performance across all physical retail categories and both ownership types. These (generally) smaller regions, which include a mix of institutionally anchored college towns like Lincoln, NE, and Manhattan, KS, and others with traditional main street urban design like Charleston, SC, have maintained the greatest physical retail resilience in the post-Recession period. Cluster 4, on the other hand, contains many service-class-oriented CBSAs with predominantly suburban development patterns that were particularly hard hit by the Recession, including Phoenix, AZ, Las Vegas, NV, Detroit, MI, and Memphis, TN; this cluster demonstrated the largest average declines in physical retail across the post-Recession period.

Table 2.5. Results of a *k*-means analysis including retail trend, occupational structure, and population. Qualitative cluster descriptions, example CBSAs, and cluster centres shown

						Cluster centres				
Clusters	Example CBSAs	Pop.	CC (%)	SC (%)	WC (%)	EE (%)	BN (%)	GF (%)	IND (%)	CHN (%)
1 Very large, creative class, experience economy growth + chain decline	New York; Los Angeles; Chicago; Houston; Washington, DC; Atlanta	8,967,640	32	49	19	2.0	-5.3	-7.5	-3.0	-3.1
2 Large, creative class, experience economy and independent superstars	Boston; San Francisco; Seattle; Minneapolis-St. Paul; Kansas City; Austin	1,203,270	34	47	19	2.1	-4.9	-6.6	-1.9	-2.5
3 Small physical retail superstars	Charleston; Lincoln; Lubbock; Cedar Rapids; Manhattan	383,734	27	50	22	1.9	-3.3	-5.5	-1.6	-1.3
4 Service class + physical retail decline	Phoenix; Detroit; Orlando; Las Vegas; Memphis; Springfield	470,969	26	53	20	1.7	-5.7	-7.6	-4.8	-3.0
5 Small, working class, independent decline + chain growth	Wichita; Toledo; Rockford; Green Bay; Dalton; Elkhart	260,954	24	46	29	1.0	-5.0	-7.0	-5.0	-1.9

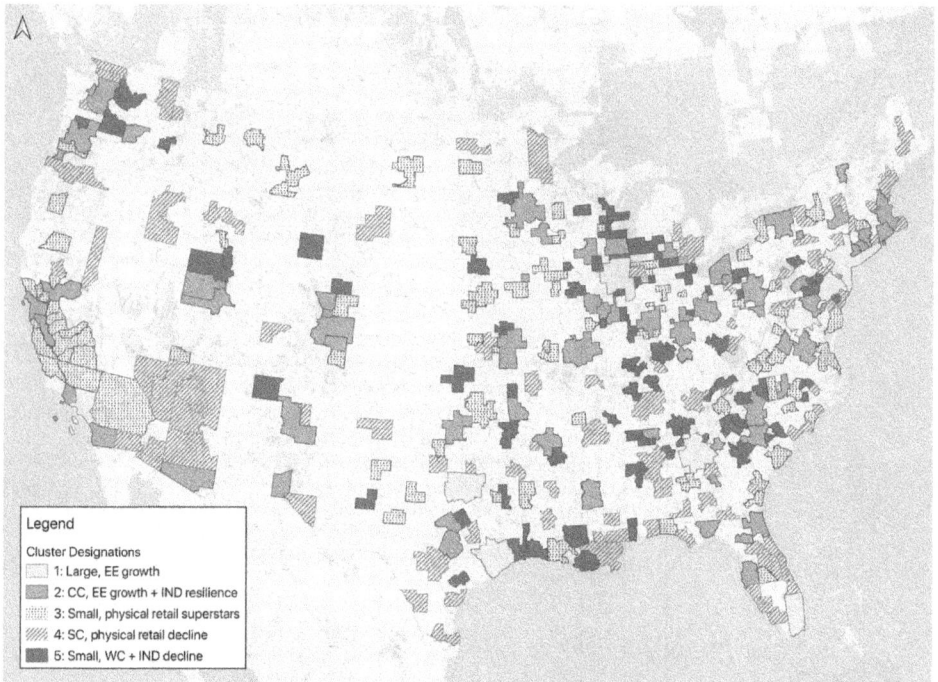

Figure 2.4. Spatial distribution of cluster types.

Finally, Cluster 5 contains small working-class metros like Toledo, OH, and Rockford, IL, and has demonstrated significant replacement of independently owned retailers with chains, while also showing the slowest growth in experience economy businesses and moderate decline in physical retailers.

While this analysis cannot lend insight into the direction of causality or the economic mechanisms at play, it does suggest that smaller, institutionally anchored regions have been more resilient to the recent downturn in physical retailing generally, while larger, more creative regions have fostered significant support for experience economy and independent retailers.

Discussion and Conclusion

Although additional research is needed to evaluate recent retailing trends and their economic impact, the results of the analysis presented here suggest several interesting findings. First, the empirical structural breaks analysis showed that a physical retail downturn has, in fact, occurred in recent years for basic needs and general merchandise businesses, as well as for independent and chain businesses. In terms of aggregate trends, general merchandise stores have declined slightly more than basic needs, losing 20 per cent and 14.2 per cent of their stores, respectively, from 2013 to 2015. Independent retailers have also declined more rapidly than their chain counterparts over this time period, at rates of 12.4 per cent and 8 per cent, respectively. The structural breaks analysis also suggests that, for most retail categories, the recent downturn did not start in earnest with the Recession, but rather began later, in 2013, perhaps suggesting that the recent retail downturn is due more directly to technological change and the rise of e-commerce than macroeconomic shocks.

Second, variations in these trends are expressed spatially, with "Sunbelt" areas in the Southeast and Mountain West seeing higher rates of experience economy growth and basic needs and general merchandise decline, while "Rustbelt" areas in the North Central and Midwest have generally seen lower rates of experience economy growth and mixed outcomes for basic needs and general merchandise retailers. These spatial trends become clearer when data on occupational structure are added to the analysis. Larger regions with higher proportions of workers in creative class occupations – such as Boston, MA, and Washington, DC – tend to have larger recent increases in experience economy establishments and lower decreases in basic needs, general merchandise, and independent establishments. Smaller regions with higher proportions of workers in working class occupations, like Toledo, OH, have seen a significant decline in independent retailers and moderately poor performance in physical retail more generally (with the exception of chain stores), while mid-sized regions with predominantly service-class-oriented economies (and often primarily automobile-oriented development patterns) that were hardest hit by the Recession – like Las Vegas, NV, and Orlando, FL – have seen the largest recent declines in physical retail. Interestingly, the most resilient regions for physical retail in the post-Recession period are smaller, institutionally anchored college towns that often have strong traditional main street design, like Lincoln, NE, and Charleston, SC. Although the direction of causality is not explored here, it seems reasonable to assume, on the basis of this analysis, that occupational-economic structure and physical main street design may both play significant roles in the growth of experience economy businesses and resilience to the recent retail downturn. Such regions might also have more ability to support independent retailers.

Of course, it is not yet possible to know what impact the COVID-19 pandemic and associated lockdowns, which have specifically targeted in-person activities – including both physical and experience-type retail – will have on these trends and associations. However, our findings suggest that efforts to foster sustainable patterns of development and investment in both declining, "Rustbelt" working class regions and auto-oriented "Sunbelt" regions – such as diversifying the economic base, enhancing the local entrepreneurial ecosystem, and investing in traditional main street design typologies – may also have positive effects for physical retail resilience.

NOTES

1 Based on the sample of 356 CBSAs used in this study; thus these figures exclude some small micropolitan areas and all non-metropolitan areas in the United States, based on occupational data availability throughout the entirety of the time series. Full list of CBSAs used available upon request.
2 The complete list includes membership clubs, sports centres, parks, theatres, events, museums, casino gambling, food service, accommodations, air travel, package tours, and foreign travel (Goldman et al., 2017).
3 We removed businesses categorized in NAICS codes 44–45 (retail trade) that are theoretically unrelated to the research interests of this chapter (e.g., motor vehicle and parts dealers).
4 We use the implementation in the R package strucchange (version 1.5–1).
5 Note: each observation in these correlations is a yearly percentage change value for a given retail category in a given CBSA. These observations are correlated by category/type in the matching rank-order over the entire time series (2001–2015).
6 The correlation and k-means analyses exclude the Urban Honolulu, HI, and Lafayette-West Lafayette, IN, CBSAs because data on occupational structure were unavailable.

REFERENCES

Anselin, L. (2018). Cluster analysis (2): Classic clustering methods. *GeoDa Workbook.* https://geodacenter.github.io/workbook/7b_clusters_2/lab7b .html
Bai, J. (1997). Estimation of a change point in multiple regression models. *Review of Economics and Statistics, 79,* 551–563. https://doi.org/10.1162 /003465397557132

Bai, J., & Perron, P. (2003). Computation and analysis of multiple structural change models. *Journal of Applied Econometrics, 18*, 1–22. https://doi.org /10.1002/jae.659

Baptista, R., & Karaöz, M. (2011). Turbulence in growing and declining industries. *Small Business Economics, 36*, 249–270.

Barata-Salgueiro, T., & Erkip, F. (2014). Retail planning and urban resilience: An introduction to the special issue. *Cities, 36*, 107–111. https://doi.org /10.1016/j.cities.2013.01.007

Bromley, R., & Thomas, C. (1993). The retail revolution, the carless shopper and disadvantage. *Transactions of the Institute of British Geographers, 18*, 222–236. https://doi.org/10.2307/622364

Bromley, R., & Thomas, C. (1995). Small town shopping centre decline: Dependence and inconvenience for the disadvantaged. *The International Review of Retail Distribution and Consumer Research, 5*, 433–456. https://doi.org /10.1080/09593969500000025

Coe, N. M., & Wrigley, N. (2007). Host economy impacts of transnational retail: The research agenda. *Journal of Economic Geography, 7*, 341–371. https://doi .org/10.1093/jeg/lbm012

Coresight Research. (2019, 12 April). Weekly US and UK store openings and closures tracker 2019, week 15: Fred's to shut 159 stores; in the UK, Debenhams enters administration. https://coresight.com/research/weekly -us-and-uk-store-openings-and-closures-tracker-2019-week-15-freds-to-shut-159 -stores-in-the-uk-debenhams-enters-administration/

Cowen. (2017). Retail's disruption yields opportunities: Store wars! Cowen and Company. https://www.cowen.com/reports/retaildisruption/

Dutta, N. (2018). Reports of retail's death are premature. *Bloomberg Opinion.* https://www.bloomberg.com/opinion/articles/2018-04-09/u-s-retail-is-alive -and-well-despite-the-apocalyptic-reports

Florida, R. 2012. *The rise of the creative class revisited.* Basic Books.

Fotopoulos, G., & Spence, N. (1998). Entry and exit from manufacturing industries: Symmetry, turbulence and simultaneity – some empirical evidence from Greek manufacturing industries, 1982–1988. *Applied Economics, 30*, 245–262.

Gardner, C., & Sheppard, J. (1989). *Consuming passion: The rise of retail culture.* Unwin Hyman.

Goldman, D., Marchessou, S., & Teichner, W. (2017). Cashing in on the US experience economy. McKinsey & Company: Private Equity & Principal Investors. https://www.mckinsey.com/industries/private-equity-and -principal-investors/our-insights/cashing-in-on-the-us-experience-economy

Guimaraes, P. P. C. (2018). The resilience of shopping centres: An analysis of retail resilience strategies in Lisbon, Portugal. *Moravian Geographical Reports, 26*, 160–172. https://doi.org/10.2478/mgr-2018-0013

Houston, D., & Eness, D. (2009). Thinking outside the box: A report on independent merchants and the New Orleans economy. The Urban Conservancy. https://www.independentwestand.org/wp-content/uploads/ThinkingOutsidetheBox_1.pdf

Jacobs, J. (1961). *Life and death of great American cities.* New York: Random House.

Jones, C., & Livingstone, N. (2018). The "online high street" or the high street online? The implications for the urban retail hierarchy. *International Review of Retail Distribution and Consumer Research, 28,* 47–63. https://doi.org/10.1080/09593969.2017.1393441

Klynveld Peat Marwick Goerdeler (KPMG). (2018). Global retail trends 2018. Global Consumer & Retail. https://assets.kpmg/content/dam/kpmg/xx/pdf/2018/03/global-retail-trends-2018.pdf

Montgomery, D. (2018). The 2018 retail apocalypse, in 6 charts and a map. CityLab. https://www.citylab.com/life/2018/12/2018-retail-apocalypse-online-shopping-charts-maps/579112/

Peterson, H. (2019). More than 7,100 stores are closing in 2019 as the retail apocalypse drags on: Here's the full list. Insider. https://www.businessinsider.com/stores-closing-in-2019-list-2019-3

Pine, B. J., & Gilmore, J. H. (1998). Welcome to the experience economy. *Harvard Business Review, 76.* https://hbr.org/1998/07/welcome-to-the-experience-economy

Retail Economics & Squire Patton Boggs. (2017). The retail experience economy: The behavioural revolution. https://www.squirepattonboggs.com/~/media/files/insights/publications/2017/06/the-retail-experience-economy-the-behavioural-revolution/the-retail-experience-economy-report.pdf

Singleton, A. D., Dolega, L., Riddlesden, D., & Longley, P. A. (2016). Measuring the spatial vulnerability of retail centres to online consumption through a framework of e-resilience. *Geoforum, 69,* 5–18. https://doi.org/10.1016/j.geoforum.2015.11.013

Smith, A., & Sparks, L. (2000). The independent small shop in Scotland: A discussion of roles and problems. *Scottish Geographical Journal, 116*(1), 41–58. https://doi.org/10.1080/00369220018737078

Talen, E., & Jeong, H. (2019). What is the value of "main street"? Framing and testing the arguments. *Cities,* 208–218.

Thompson, D. (2017). What in the world is causing the retail meltdown of 2017? *The Atlantic.* https://www.theatlantic.com/business/archive/2017/04/retail-meltdown-of-2017/522384/

US Census. (2020). Monthly Retain Trade: Definitions. https://www.census.gov/retail/definitions.html

US Department of Commerce. (2019). Quarterly retail e-commerce sales: 1st quarter 2019. *US Census Bureau News.* https://www2.census.gov/retail/releases/historical/ecomm/19q1.pdf

Wrigley, N., & Dolega, L. (2011). Resilience, fragility, and adaptation: New evidence on the performance of UK high streets during global economic crisis and its policy implications. *Environment and Planning A, 43,* 2337–2363. https://doi.org/10.1068/a44270

Wrigley, N., & Lambiri, D. (2015). British high streets: From crisis to recovery? Economic & Social Research Council. https://eprints.soton.ac.uk/375492/1/BRITISH%2520HIGH%2520STREETS_MARCH2015%2528V2%2529.pdf

Zeileis, A., Kleiber, C., Krämer, W., & Hornik, K. (2003). Testing and dating of structural changes in practice. *Computational Statistics and Data Analysis, 44,* 109–123. https://doi.org/10.1016/s0167-9473(03)00030-6

Zeileis, A., Leisch, F., Hornik, K., & Kleiber, C. (2002). Strucchange: An R package for testing for structural change in linear regression models. *Journal of Statistical Software, 7*(2), 1–38. https://doi.org/10.18637/jss.v007.i02

Business Profile: Scafuri Bakery

Chicago, Illinois

Scafuri Bakery was established in 1904 in the Little Italy neighbourhood, lying at the heart of Chicago. Luigi Scafuri immigrated from Calabria, Italy, in 1901 and decided to contribute to his neighbourhood by offering a slice of his native culture, a tradition that was continued by his daughter Annette. The bakery has gone through ups and downs over the last century – to the point of closing from 2007 to 2013. However, Michelle (Annette's great-niece) reopened Scafuri's doors and has continued her family's legacy. For Michelle, the bakery serves as a connection to the community, and the combination of traditional and contemporary tastes makes customers feel at home and makes all the hard work worthwhile. She mentions that even though most Italians have left the neighbourhood, during street festivals former residents visit Scafuri, with a longing to reconnect with the community and their Italian family traditions. Despite having a waiting line, today Michelle expresses her concern due to COVID-19. Many of the businesses surrounding the bakery are permanently closing, and she feels pressure to venture into innovative ideas like selling cookies, breads, and pastries in food trucks, adding more savoury elements to the menu, or offering delivery meals.

Figure B2.1. Michelle DiGiovanni, Scafuri Bakery. (Credit: Photo courtesy of the author.)

3 Commercial Gentrification: What Happens to Businesses and Services When the Neighbourhood Changes?

RACHEL MELTZER

Neighbourhoods Change

The idea that neighbourhoods are dynamic is not new. New residents move in, longstanding residents leave, and the circumstances for incumbent residents change. Gentrification is but a version of this – albeit a dramatic one with particular players. Whereas neighbourhood change is often gradual, gentrification – or the economic transitioning of low-income communities due to the entry of relatively affluent, educated, and typically white residents – can be accelerated and magnified. Accelerated, because of concentrated development and investment; and magnified, because the incoming residents are often economically and demographically distinct from those who have long occupied the neighbourhood.

Urban neighbourhoods are also characterized by their mixed use; they are places of production *and* consumption (Lees, 1994; Hamnett, 1991; Nichols Clark, 2003; Lloyd, 2010). Much of the commercial activity in a neighbourhood is tied to the preferences of the nearby residents and/or workers. Therefore, as these consumer bases shift in response to gradual neighbourhood change or more aggressive gentrification, it is not surprising that commercial activity also changes.

This chapter focuses on the challenges and opportunities that arise when neighbourhood change is dramatic and immediately path-shifting for the community, as with gentrification. Moreover, this chapter discusses the mechanisms behind the shifts in neighbourhood commercial services, and what they mean for local residents, workers, and business owners.

Much of the commercial change that accompanies gentrification is threatening, especially to incumbent residents in the neighbourhood. The affordability and nature of services can shift, and business owners may find it more challenging to meet the increasing costs and changing demands in the area. However, this is only one side of the story – gentrification, and the investment that often accompanies it, can bring into a neighbourhood services and economic opportunities that had stayed away. I weigh these different, but potentially co-existing, implications and discuss a public sector response that is moderating and inclusive.

What Is Commercial Gentrification?

Gentrification Defined

Gentrification, a phenomenon first documented in London in the 1960s by Ruth Glass (1964), has surged in cities across the United States. Places that were economically abandoned during the 1980s as middle- and upper-middle-class, white families left the cities to live in suburban communities thought to be safer and amenity-rich, are now confronting a demographic and economic reversal (Lees, 2000, 2016; Couture & Handbury, 2017). While urban neighbourhoods have exhibited signs of gentrification since the 1990s, dramatic surges in educated and white residents and rents started to emerge in the 2000s (Ellen & Ding, 2016; Hwang & Lin, 2016). One theory is that this urban resurgence is about jobs. Recent research, like that from Edlund et al. (2019) and Baum-Snow and Hartley (2015), documents how high-skilled jobs in particular have disproportionately returned to cities, in turn attracting residents back to the cities to be closer to those opportunities. These patterns have been starkest since 2000. However, more recent studies have shown that the return of younger, often childless residents to urban centres has been driven largely by the changing preferences for urban amenities (Carlino & Saiz, 2008; Cortright & Mahmoudi, 2016; Couture & Handbury, 2017). Crime has also plummeted in cities, attracting both single and family households (Ellen et al., 2017).

Gentrification is the classic "know it when you see it" phenomenon. Nevertheless, researchers have built a sizable body of work proposing and testing various statistical definitions of gentrification. Most studies prioritize the economic dimension of gentrification. That is, metrics rely on changes in either household income or housing costs (i.e., rents or prices; Gould Ellen & O'Regan, 2008; McKinnish et al., 2010; Meltzer & Schuetz, 2012; Meltzer, 2016; Meltzer & Ghorbani, 2017; Aron-Dine & Bunten, 2019). Conditional on being low-income, a neighbourhood is designated as gentrifying if, over a subsequent 5 to 10 years, the change in household income or housing costs exceeds that of the city or metro area overall. Aron-Dine and Bunten (2019) use both income and housing costs on an annual basis, specifically the gap between income and prices, to identify when a neighbourhood is gentrifying. Other studies solely implement socio-economic status or education, instead of economic dimensions (Hammel & Wyly, 1996; Vigdor et al., 2002; Freeman, 2005; Lester & Hartley, 2014; Waights, 2018). Rosenthal (2008) develops a model of house filtering, such that gentrification is more likely to take place in neighbourhoods with older housing stock that is ripe for redevelopment. Finally, Behrens et al. (2018) identify "pioneer" industries, or sectors that are usually found in affluent areas that are overrepresented in poor neighbourhoods, to predict a block's likelihood of gentrification. In general, the metrics correspond in terms of their designation; indeed changes across all of these dimensions tend to happen simultaneously in the context of gentrification.

Neighbourhood Commercial Services and Neighbourhood Change

Neighbourhood commercial businesses are subject to particular market dynamics. Their consumer base, mostly comprising nearby residents, is localized (DiPasquale & Wheaton, 1996); therefore necessary and frequently consumed goods and services are most in-demand. These goods and services are often perishable (like groceries) and contribute to everyday routines (like laundromats and drug stores; Stanback, 1981; Bingham & Zhang, 1997; Cortright & Mahmoudi, 2016; Meltzer & Capperis, 2017). They can also provide recreation (like restaurants and bars) and settings for social gathering (like barbershops, cafes, and bookstores; Oldenburg, 1999). These neighbourhood-based businesses are different from those in destination commercial corridors, which are not similarly tied to a localized consumer base. In some commercial areas, consumers will travel outside their residential community to attend theatre or purchase furniture and other high-cost durables. In contrast, the utility of neighbourhood services lies in their physical and economic accessibility and diversity.[1]

Thus, more than other commercial enterprises that rely on a broader consumer base, neighbourhood businesses are particularly vulnerable in the face of localized socio-economic transitions, like gentrification. New residents typically have a profile different from that of incumbent residents in low-income, gentrifying neighbourhoods, and they come with different tastes and preferences for services (Caulfield, 1994; Ley, 1996; Chapple & Jacobus, 2009). They can also have different abilities to pay. Businesses respond to these factors in deciding whether to stay open in the same location and whether to enter into the local market.

The goods and services of an existing business may not meet the needs of new residents – and, if the cycling in of new residents outnumbers the persistence of incumbent residents, an establishment's consumer base might dwindle. If rents go up as a result of increased demand for the area, the change in consumer characteristics might be even more threatening for maintaining a viable business. However, if the business can adjust its goods and services to meet the needs of the new local population, it might benefit from new, and perhaps bigger, spending power.

New businesses will decide to enter the market on the basis of an assessment of the costs of start-up and operation against the potential for revenues. Although new businesses will likely cater to the demands of the new set of residents, this depends on the nature and availability of information about local dynamics. Neighbourhoods in flux can be difficult to read – so businesses may refrain from or delay entry. On the other hand, if there are signals of viable markets, like a decline in crime, businesses may opt to locate in neighbourhoods that they had previously ruled out.

Similar to how residential gentrification is defined, commercial gentrification is the process through which incumbent businesses, typically serving the residential population that preceded the new "gentrifiers," are supplemented or replaced by new goods and services that tend to serve the new consumer base. Commercial gentrification is an economic shift – in the different kinds of transactions that will take place – but it is also a cultural shift. Since businesses that serve neighbourhood residents are places of social connection (Jacobs, 1961; Oldenburg, 1999), their replacement with commercial entities that cater to specific subsets of the population (as well as to shoppers drawn from outside the neighbourhood) are likely to have a dramatic effect on neighbourhood-based social interactions.

Commercial gentrification can be a product of market-driven shifts, such as the increase in millennial population in cities. However, it can also be triggered or accelerated by public interventions. For example, changes in the use of land, through rezonings or special district designations (Chapple et al., 2010; Shkuda, 2013; Yoon & Currid-Halkett, 2015; Ferm, 2016), can change the opportunity for and return on commercial investment. Transit-oriented development can shift the use and composition of an area dramatically – and is a prime location for commercial development (Schuetz, 2015; Chapple et al., 2017; Boarnet et al., 2018). The empirical evidence is inconclusive, however, on whether these interventions alone spawn commercial gentrification that would not have otherwise happened.

Documenting Commercial Gentrification: How Prevalent Is It?

Measuring Commercial Gentrification

There are several outcomes associated with commercial gentrification. Most obviously, gentrification can affect the viability of existing establishments. Therefore, measures of establishment counts and densities can indicate both the access to services for residents and the viability of operating in a particular location. Aggregate counts or densities, however, can obscure important dynamics. For example, they do not show gross flows of establishment entries or exits.

In addition to overall counts and densities, the types of services, and how they change, can be an important feature of commercial gentrification. Is the neighbourhood losing long-standing establishments that provided cultural and social assets? Is it gaining long-needed necessity goods and services, like supermarkets and drugstores, or gathering spots, like coffee shops and restaurants? And how are the goods and services priced – to whom are they financially accessible? It is important to look at not only the marginal establishment (i.e., one that is closing or entering); changes in the overall *mix* of goods and services also affect quality of life in a neighbourhood.

Finally, the organizational structure of the establishment matters. For example, are smaller and independently owned businesses being replaced by national or global chains? What are the demographic characteristics of the business owner, and how long has she or he been operating in the neighbourhood? The operator's ties to the neighbourhood can affect the quality and type of service and how the business's activities are integrated into the broader community.

Systematic documentation of commercial gentrification is limited by the scarcity of publicly available microdata on commercial activity over time. I've summarized the availability of data for the outcomes of interest in Table 3.1. Commercial gentrification, a neighbourhood-level phenomenon, needs to be observed at a small enough level that resident and consumer characteristics are credibly mapped onto local markets. In the next section, I use information from County Business Patterns (CBP) to document commercial change over time for zip codes in metros areas across the US.[2] ZIP codes, on average about 10,000 people, are not the ideal geography for measuring gentrification (about twice as large as the census tract, a common neighbourhood geography), but they are the level at which publicly available data are provided consistently over an extended period of time. The ZIP code, however, does represent a reasonably sized market for neighbourhood-based goods (Schuetz et al., 2012). Further, it provides enough variation within a metro area to

Table 3.1. Data sources for documenting commercial gentrification

Outcome	Data source	Availability*	Unit of observation**	Time period***
Net change in establishments	• US Economic Census	• Public	• ZIP codes • County	• County since 1964 • ZIP code since 1994
Gross change in establishments (entry, exit, stay)	• NETS • InfoUSA • Economic Census restricted data	• Proprietary (for purchase) • Public by application	• Estab./firm	• Depending on source, back to 1976
Type of services	• NETS • InfoUSA • US Economic Census • Economic Census restricted data	• Proprietary (for purchase) • Public and by application	• ZIP codes • County • Estab./firm	• Depending on source, back early 1980s
Products sold	• Nielson • Economic Census restricted data	• Proprietary (for purchase) • Public and by application	• Item purchased • Estab.	• Early 2000s • Depending on source, back to early 1980s
Prices of goods and services	• Nielson	• Proprietary (for purchase)	• Item purchased	• Early 2000s
Type of establishment: chain vs. independent	• NETS • InfoUSA • Economic Census restricted data	• Proprietary (for purchase) • Public and by application	• Estab./firm	• Depending on source, back to 1972
Owner characteristics	• Economic Census restricted data • Kauffman Firm Characteristics	• Public by application	• Estab./firm	• Mid-1970s • Kauffman: 2004–2011
Commercial rents	• CoStar • Economic Census restricted data	• Proprietary (for purchase) • Public by application	• Lease contract • Estab./firm	• Early 1990s
Reason for closure/entry	• Survey of business owners	• Ad hoc	• Business owner	• N/A
Social/cultural Identity of establishment	• Survey of consumers and business owners/ employees	• Ad hoc	• Business owner • Consumer	• N/A

* The proprietary datasets listed here are often prohibitively expensive for an independent researcher or small organization to purchase.

** All data are available on an annual basis except for those provided by Nielsen (transaction level) and CoStar (lease/contract level).

*** These dates will vary with the specific variables and geography of interest.

observe differences, if they exist, across areas that are gentrifying compared to those that are not. There are, on average, 20 ZIP codes per CBSA, or core-based statistical area, in my sample.[3]

Documenting a National Picture of Commercial Change

I rely on established measures of gentrification (discussed more generally above) to identify areas experiencing commercial gentrification. Specifically, I identify neighbourhoods with economic changes that exceed the pace of the surrounding metro area. I use changes in housing costs as the indicator of economic change for conceptual and practical reasons. Housing costs usually reflect other changes that accompany gentrification, such as demographic changes and investment surges. I do replicate my analysis with other metrics, such as those based on income, and the same patterns emerge. In addition, housing price indices (HPI) are publicly available at the ZIP level at annual intervals from the Federal Housing Finance Authority – equally rich data on household characteristics are not readily available. Following methods commonly used in the gentrification literature (for example, Ellen & O'Regan, 2008; Meltzer & Schuetz, 2012; Meltzer, 2016; Meltzer & Ghorbani, 2017), I first array all ZIPs and their HPIs across each CBSA at time t, and identify the ZIPs with HPIs that fall in the bottom 20th percentile of the ZIP-HPI distribution. This captures the relatively lower-valued, or more affordable, neighbourhoods of a particular metro area. For each time period $(t \in \{1990, 2000, 2011\})$ I construct a ratio of ZIP-level HPI to CBSA-level HPI to capture the price positioning of a particular neighbourhood (or ZIP) relative to its broader metro area context (or CBSA). The difference over time in these ratios should then capture the price appreciation of neighbourhoods relative to the price trends for the metro area more broadly. Then, I consider a ZIP, z, to be gentrifying over the period $\{t, t+1\}$ if:

(i) $HPI_{z,CBSA,t} < 20th\ percentile$ of the within-CBSA distribution; and

(ii) $\dfrac{HPI_{z,t+1}}{HPI_{CBSA,t+1}} - \dfrac{HPI_{z,t}}{HPI_{CBSA,t}} > 0$

Using this calculation method, between 1,640 and 2,256 ZIPs can be classified as gentrifying (depending on the time interval over which change is measured) across the entire sample of CBSAs. Further, between 175 and 230 ZIPs across a subsample of CBSAs experienced the steepest housing price appreciation.[4] I then track changes in commercial activity for 5-year intervals across gentrifying, non-gentrifying, and – to capture places that were already relatively affluent – middle/upper-income areas.

Retail
All CBSAs

Retail
High Price CBSAs

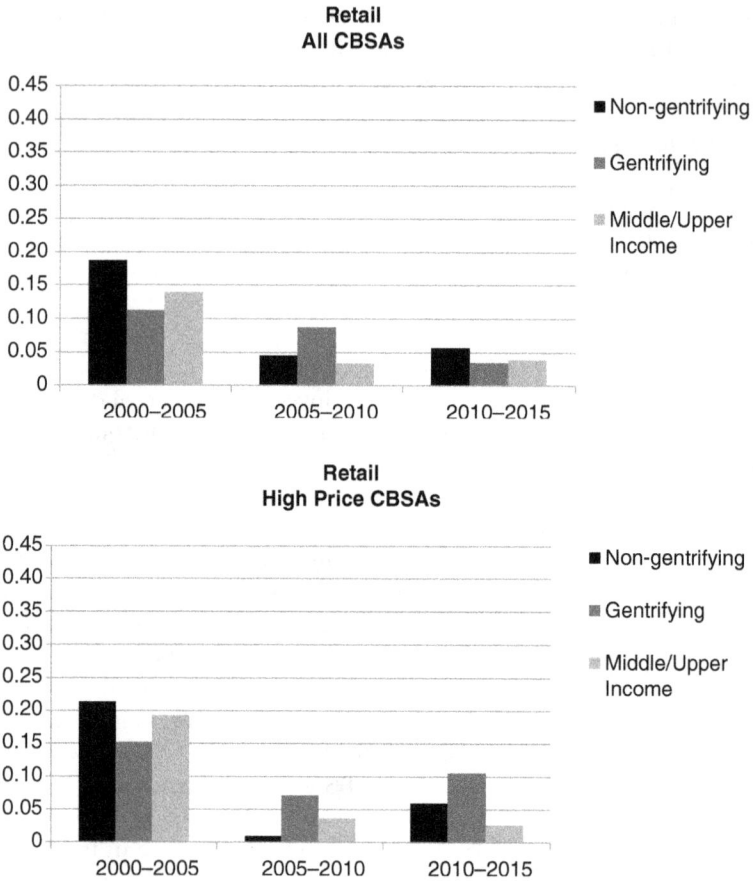

Figure 3.1. Percentage change in number of establishments, retail, all CBSAs and high price CBSAs. Gentrified ZIPs are identified for the 5-year interval preceding the displayed interval for commercial change; this is based on the assumption that commercial change will follow residential gentrification as indicated by relatively faster rising house prices.

I document establishments in three broad categories: *Retail*, which captures a range of neighbourhood and storefront goods and services, *FIRE* (finance, insurance, and real estate – professional services aligned with high-skilled jobs), and *Manufacturing* (goods producing).[5] The trends are displayed in Figures 3.1, 3.2, and 3.3.

Several observations can be made from these data. First, shifts in commercial activity in gentrifying neighbourhoods are most pronounced in metros areas that have experienced severe housing price appreciation.

**FIRE
All CBSAs**

**FIRE
High Price CBSAs**

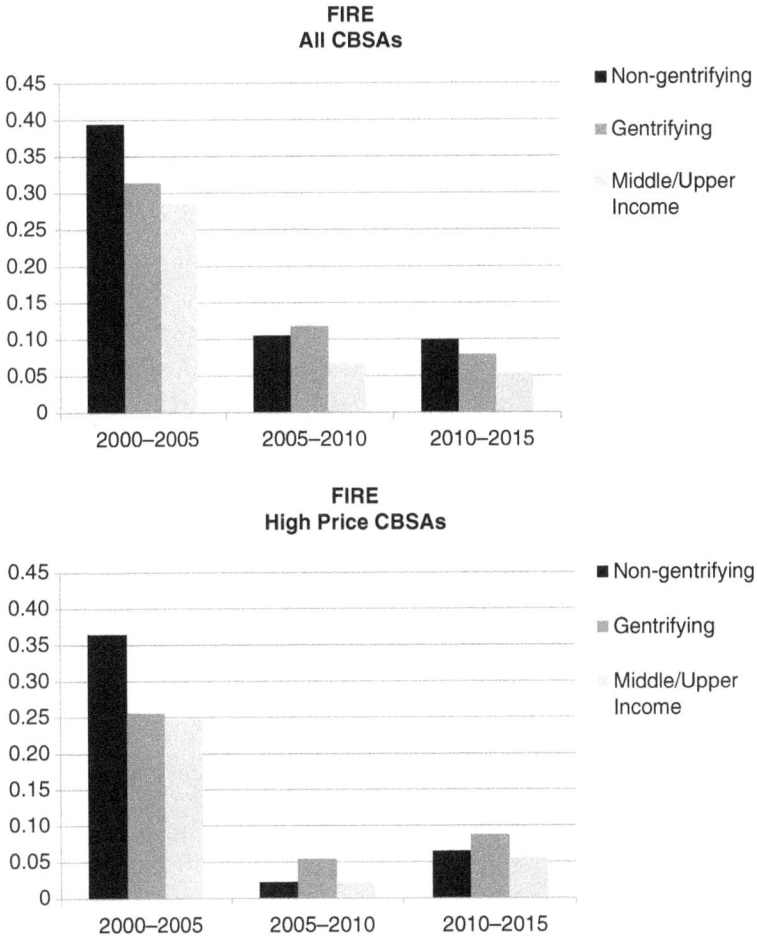

Figure 3.2. Percentage change in number of establishments, FIRE, all CBSAs and high price CBSA. Gentrified ZIPs are identified for the 5-year interval preceding the displayed interval for commercial change; this is based on the assumption that commercial change will follow residential gentrification as indicated by relatively faster rising house prices.

This confirms the expectation that gentrification is a phenomenon con-centrated among a subset of typically expensive and growing cities. Second, shifts have been most pronounced since the second half of the aughts.[6] This indicates that commercial gentrification is a more recent phenomenon than residential gentrification (which started to manifest in the late 1990s and early 2000s). These patterns are consistent with the expectation that commercial changes, especially among locally oriented enterprises, follow

Manufacturing
All CBSAs

Manufacturing
High Price CBSAs

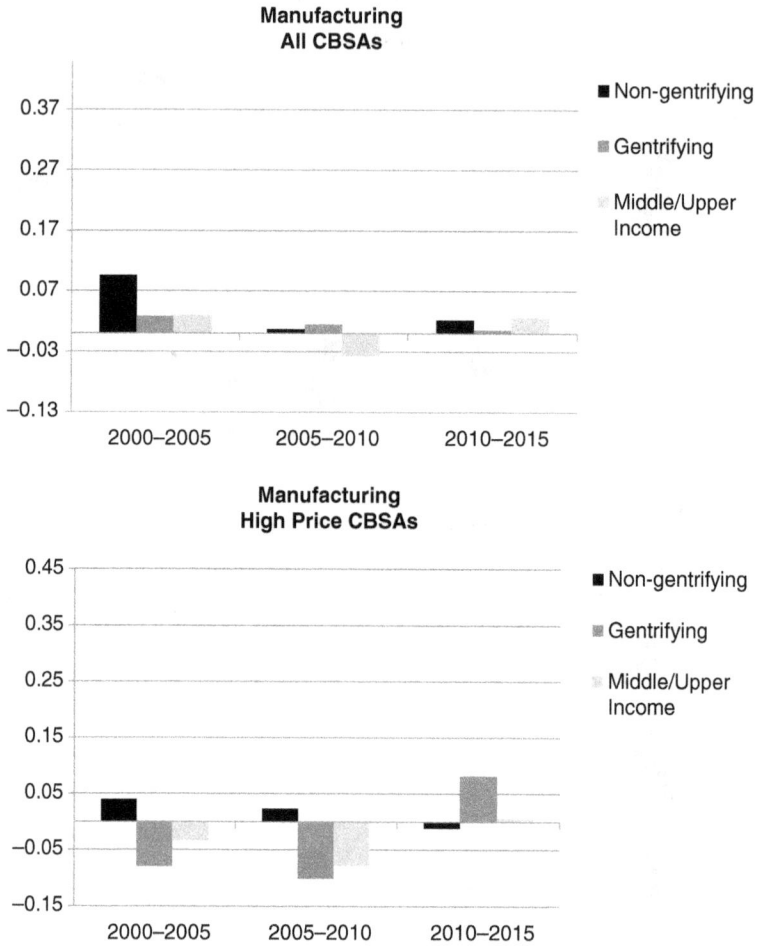

Figure 3.3. Percentage change in number of establishments, manufacturing, all CBSAs and high price CBSAs. Gentrified ZIPs are identified for the 5-year interval preceding the displayed interval for commercial change; this is based on the assumption that commercial change will follow residential gentrification as indicated by relatively faster rising house prices.

local demographic and economic shifts. Finally, in earlier periods, there is evidence of a compositional shift in gentrifying neighbourhoods, away from manufacturing and towards service-based sectors. This is consistent with the rezoning or repurposing of land in gentrifying neighbourhoods.

The aggregate data do not allow for more fine-grained analyses. For example, it is not possible to calculate the gross changes in establishments (i.e., entries and exits), nor document shifts in the composition of services. In the next section I summarize findings from other studies in order to get at these dimensions.

Commercial Gentrification: The Good and the Bad

While research on commercial gentrification is scarcer than that on residential gentrification, there are quantitative and qualitative empirical documentation of the phenomenon. Again, there is no public source of neighbourhood-level commercial activity with broad and deep coverage – therefore most of the studies have explored the phenomenon in particular cities or neighbourhood sites. I pull the findings together here to understand what gentrification means for the commercial landscape of neighbourhoods. The research confirms that commercial gentrification affects the economic well-being of both residents and entrepreneurs, as well as the general quality of life in the affected neighbourhoods.

Commercial Establishments and Services

How does gentrification affect the viability of commercial establishments? Does it change the prevalence or composition of commercial activity? Many expect that, as the result of dramatic shifts in consumers and rising rents, commercial establishments will shut down at an accelerated rate in gentrifying areas. However, the empirical evidence on this is mixed. First, there is consistent empirical documentation that the number of retail establishments increases in neighbourhoods undergoing economic upgrading (Chapple & Jacobus, 2009; Meltzer & Schuetz, 2012; Schuetz et al., 2012). This is a clear signal of increased commercial investment in areas undergoing demographic shifts towards more affluence and education – two features that businesses use to determine viable markets (Heather Arnold, this book).

These statistics, however, typically capture net changes in commercial activity. The number and type of establishments that enter or exit a neighbourhood under conditions of gentrification are actually the more important outcomes – that is, are retail gains occurring at the expense of certain establishments and services?[7] To investigate the nature of commercial flux, I conducted a citywide study of this question using microdata to track the rate of entry and exit of establishments at specific parcels. Using over two decades of data and definitions of gentrification similar to those specified earlier in this chapter to determine different degrees of neighbourhood change, I found no evidence of differential exit rates for businesses in gentrifying neighbourhoods compared to similar neighbourhoods that weren't gentrifying (see Figure 3.4).

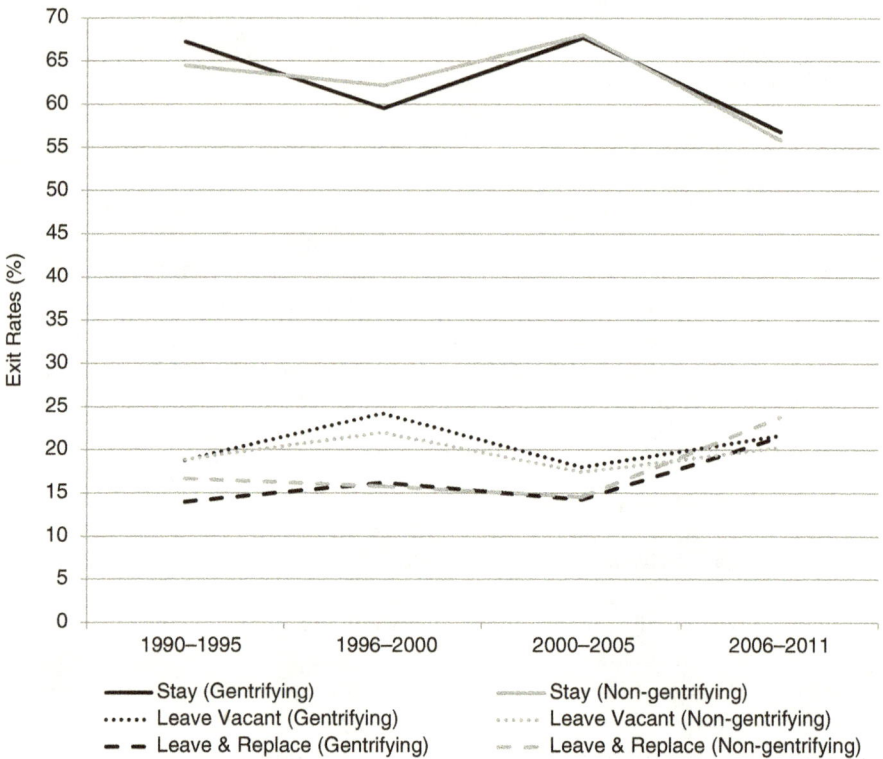

Figure 3.4. Establishment exit rates, gentrifying and non-gentrifying neighbourhoods, New York City. "Stay" designates the share of establishments that stay in place during the 5-year intervals. "Leave Vacant" designates the share of establishments that leave a commercial space that remains vacant after their exit through the end of the 5-year interval. "Leave & Replace" designates the share of establishments that leave a commercial space that is reoccupied by another establishment after their exit. (Source: Meltzer [2016].)

An important qualification for these findings is that they represent the experience for the "average" commercial location across the entire city – therefore, there could be places with both higher and lower rates of exit. A richer set of neighbourhood drill-down analyses confirmed this point, showing varied rates of establishment closures that correspond with different degrees and types of socio-economic shifts. For example, in East Harlem, an area of New York City that experienced some of the most dramatic demographic shifts during the 2000s, establishment retention rates declined with gentrification, and new establishments opened up with a range of services, many of which could serve local residents. This was not the case in gentrifying areas with less transformative demographic and economic changes, which also saw declines in retention rates, but a smaller influx of new establishments and services. Figures 3.5 (a–d) compare the changes in services across East Harlem and Sunset Park, the latter of which experienced less dramatic economic and demographic shifts during the 1990s and 2000s. East Harlem saw bigger net positive changes in services that are often viewed as necessary and desirable in the local neighbourhood, like drugstores and grocers.

In a recent nationwide analysis, Somashekhar (2020) also documents heterogeneity in retail responses across neighbourhoods. He looks specifically at the race of the newcomers and finds that retail development was significantly slower in areas gentrified by Black rather than white individuals. Therefore, while gentrifying neighbourhoods may see gains in retail activity, the benefits are not evenly distributed. On the other hand, there is no evidence that establishments are systematically displaced in areas undergoing gentrification.

What about those establishments that do exit in the face of gentrification – do certain kinds of businesses disproportionately bear this burden? When comparing closure rates across gentrifying and non-gentrifying neighbourhoods, I find that older, long-standing establishments are no more likely to close than newer ones. In addition, evidence that chains are replacing stand-alone businesses in gentrifying neighbourhoods is also weak.[8] However, Sutton (2010) finds that minority-owned businesses can suffer disproportionately more in the context of gentrification, since lower-income, predominantly Black or Latino communities tend to gentrify more than those that are predominantly white (see also Ong et al., 2014, for research on Asian-owned establishments).

Employment

Employment is another relevant outcome in the context of commercial gentrification. Increased commercial investment may generate more employment opportunities – for both incumbent and new residents. The kinds of jobs can also transition. Several studies document a sectoral shift in gentrification, suggesting that employment opportunities would change in a similar way. For example, Lester and Hartley (2014) find that jobs in restaurants and retail replace those in manufacturing at a faster rate in gentrifying neighbourhoods. Curran (2007) also finds what she terms "industrial displacement," whereby blue-collar work has

(A) **East Harlem: Drug Stores**

(B) **Sunset Park: Drug Stores**

(C) **East Harlem: Grocery Stores**

(D) **Sunset Park: Grocery Stores**

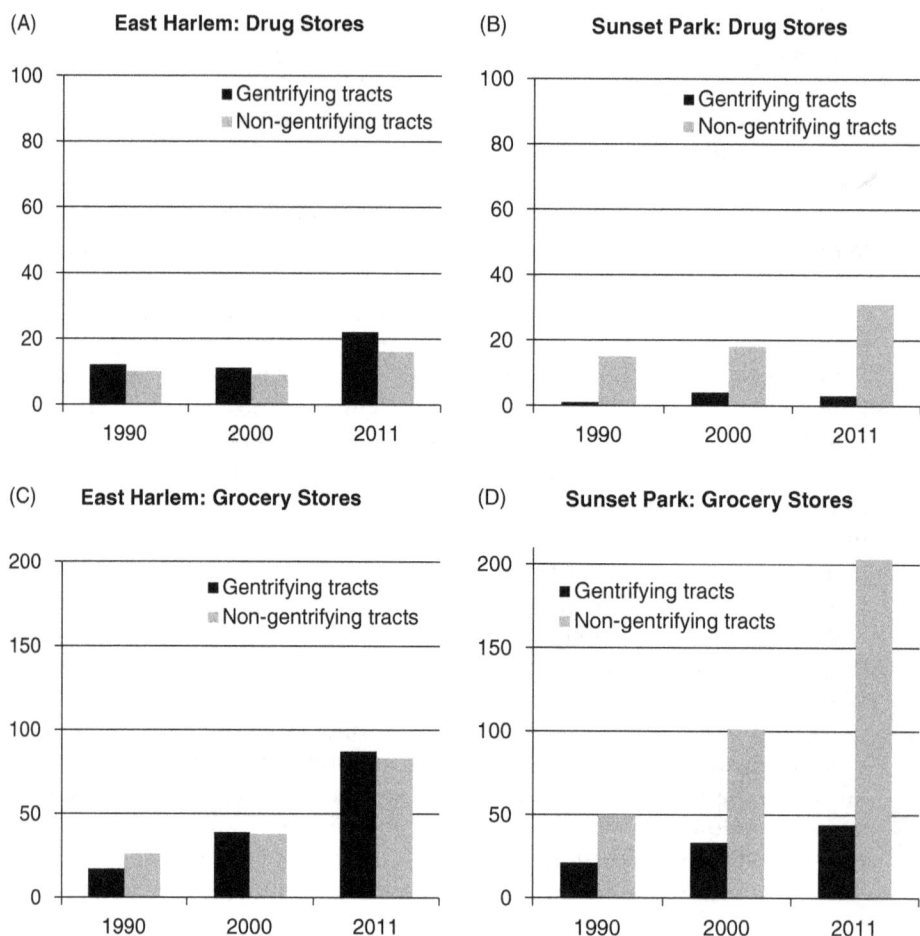

Figure 3.5. Changes in services, gentrifying and non-gentrifying tracts, East Harlem and Sunset Park.

been devalued and pushed into the informal sector. These findings raise concerns about the quality of new jobs, which, for lower-skilled individuals, would presumably be concentrated in lower-paying retail and service sectors.

The employment outcomes from gentrification, however, are particularly challenging to pinpoint, because the impetus for gentrification is often a movement towards jobs in the central city (often by well-educated, high-skilled, and white in-movers; Kolko, 2009; Baum-Snow & Hartley,

2015). Therefore, the simultaneity of the commercial and residential shifts, and the lack of data that follows both individuals and their employment locations, make it hard to disentangle how much gentrification is driving where the jobs are and who gets them.

For example, what about job prospects for existing (usually lower-income) residents in the gentrifying neighbourhoods? The results are mixed (Meltzer & Ghorbani, 2017). Job losses for incumbent residents are localized: even while jobs overall increase, incumbent residents experience a 63 per cent loss in employment in gentrifying neighbourhoods (compared to similar neighbourhoods that did not gentrify during the 2000s in the New York metro area). They lose jobs in both the service and goods-producing sectors, but universally on the lower-to-moderate end of the wage spectrum. There is also some evidence that these losses are compensated by gains farther away from their home census tract, mostly in goods-producing and lower-wage positions. While the numbers are small, there is also suggestive evidence that, outside of New York City, any job gains for local residents are driven by businesses that stay in place. This finding could be explained by the tendency of local businesses to have community ties, including local hires.

Commercial Culture

Much of the value from commercial activity in neighbourhoods is non-transactional. Jane Jacobs (1961) documented "eyes on the street" – the idea that commercial activity has an important role to play in the vitality and safety of neighbourhood streets. That is, beyond the economic value of purchasing a good or service, commercial uses can also generate meaning through the spaces and opportunities for social interactions (Crewe & Lowe, 1995; Patch, 2008; Zukin et al., 2009; Zukin, 2018). The emergence of cafes in gentrifying neighbourhoods epitomizes this cultural shift – their spaces are often designed to serve and attract the new, affluent class of residents (Bridge & Dowling, 2001; Zukin et al., 2009). These outcomes are harder to measure and unsurprisingly have been less studied. Bridge and Dowling (2001; p.95) assert that patterns of consumption are "classed and raced" (see also Deener, 2007; and Parker in this book). Therefore, when the composition of the neighbourhood's residents and consumers shifts, they observe that "the commodities sold, style, architecture and naming strategies of stores located in the retail spaces of these gentrified neighborhoods confirm the consumption practices of the new middle class" (Bridge & Dowling, 2001, p. 105). In turn, "retail upscaling" can reduce consumption opportunities for incumbent residents (Bridge & Dowling, 2001; Zukin et al., 2009).

Monroe Sullivan and Shaw (2011) illuminate the racial aspect of gentrification and how it interacts with the changing commercial scene. Through in-depth interviews with Black and white residents in a Portland, OR, neighbourhood, they find distinct perceptions of the commercial changes across these two groups. Black residents feel that new businesses don't cater to them and feel uncomfortable in white-owned businesses. However, they do value the new services offered – meaning that the appreciation is more transactional than cultural (this mix of resentment and optimism is consistent with Dastrup et al., 2015; and Freeman, 2006). White residents, on the other hand, appreciate the general growth in retail and view it as a positive cultural shift, even if the services do not appeal to them.

In aggregate, these findings suggest that some gentrifying neighbourhoods gain new services that can potentially improve quality of life for residents. The research to date, however, does not document the types or price-points of the new goods and services, and it is possible that they primarily serve the newer, more affluent consumer base (an expectation confirmed by the qualitative studies discussed above). This raises questions about the inclusiveness of service benefits from gentrification.

Commercial activity also contributes to the neighbourhood streetscape; vacant storefronts can change the feel of an area (and potentially threaten the viability of remaining stores that rely on the agglomerative benefits from retail clustering). Gentrification can induce a higher commercial vacancy rate for various reasons. Rising residential rents can spill over into commercial spaces, such that landlords expect much more than what was paid before. This means that they might hold out for chain establishments that can pay higher rents. Spaces can sit vacant as the local market adjusts to these elevated asking prices from landlords – even while business owners and entrepreneurs do not have complete information about the nature of the still-transitioning market. In addition, there is typically a spate of investment and construction in gentrifying neighbourhoods: commercial spaces can sit vacant while landowners assemble and vacate buildings in preparation for the renovations, and newly built units can also sit vacant as they wait to be occupied. In my study of New York City neighbourhoods (Meltzer, 2016), I find that storefronts in gentrifying neighbourhoods are more likely to sit vacant for longer than those in similar areas that are not gentrifying. Empty storefronts, especially those that languish, can blight the street and, in excess, threaten the vibrancy and safety of the community.

What We Still Don't Know

Commercial gentrification is particularly difficult to pin down. In addition to the data limitations discussed above, urban commercial markets are in many ways more opaque and "stickier" than residential ones. For example, commercial leases can extend up to 10 years. This means that it can take several years, at least, to observe gentrification-induced shifts in commercial activity. As long as the business can maintain operations to sustain its current lower rent, the establishment may stay open years after the neighbourhood has demographically transitioned. This is very different from the 1-to-2-year cycle of residential leases.

In addition, even if commercial transitions can be observed, the price and type of good or service consumed in new establishments is nearly impossible to observe in a systematic and nuanced way. This, again, is very different from residential units, which can easily be differentiated by features like age or size. Much of what we still don't know about commercial gentrification relates to *how* consumption has changed – who do the new businesses serve, to whom are they accessible, and what service and cultural amenities have disappeared? To credibly document these changes we would need to know the content and prices of the goods and services (over time), and the characteristics of the people who do or do not consume them (and why).

Finally, much of commercial gentrification has coincided with another transformative phenomenon: online commerce. The rise of Amazon and other online alternatives has tested the sustainability of brick-and-mortar establishments. While e-commerce might benefit some small establishments that are able to take advantage of new marketing opportunities, it can also threaten the viability of "offline" goods and services, changing how in-person goods and services are provided (see chapters by Daniel and Hernandez, Jones and Mack in this book).

Managing Commercial Change

Many local governments are faced with a phenomenon that stands to change the nature of urban neighbourhoods, and, in turn, the cities where they reside.[9] How can cities manage the neighbourhood commercial shifts that accompany gentrification, while maintaining the diversity and localness of consumption and production in urban neighbourhoods? While this chapter focused on gentrification, the lessons learned and remedies discussed below can easily pertain to many kinds of extreme threats to local neighbourhood economies. These can be driven as much by forces of gentrification as by disruptions from climate-related disasters, like extreme flooding, and health-related pandemics, like COVID-19. In all these cases, local retailers are left to navigate uncertain and unexpected shifts in their communities and business models. Managing commercial change means arming, usually small, commercial establishments with tools for resiliency.

Helping Businesses Adjust to Changes

Managing gentrification, from a business perspective, is challenging as a result of the dynamic and often obscure nature of localized neighbourhood conditions. The pace of change is also accelerated, making it difficult to respond effectively to economic and demographic shifts. These features can make it harder for existing businesses to stay open and can influence the kinds of businesses that do or do not enter the neighbourhood. The solution, in part, lies in arming local businesses with better information – on how the local market is changing, and what adjustments businesses need to make in response. This may be in the form of market research, technology support (see Mack in this book), subsidies, and business development services that help them serve a changing and perhaps more diverse clientele.[10]

A handful of cities have floated the idea of commercial rent control – not unlike what has been done to protect the affordability of residential markets. However, this approach can become detrimental to both business owners and neighbourhoods. As discussed throughout this chapter, the nature of neighbourhood commercial markets is unique. The behaviour of commercial actors – in opening a business in a neighbourhood, for example – is distinct from that of residents. It is difficult for neighbourhood-based enterprises to survive without the patronage of a local consumer base. In the face of gentrification, the best support a business can receive is resources to meet changing demand. While rent subsidies would help weather the transition in the short term (while the rents adjust and the business itself reads and responds to local market changes), it is hard to justify subsidizing a business via rent control in the long run if it does not adapt to the shifting conditions nearby.

Better long-term strategies include developing affordable and perhaps shared workspaces that are flexible and provide some of the support services described above (Ferm, 2016). The subsidization of commercial co-ops or the purchase of commercial properties by non-profit organizations can generate permanent commercial stock that remains affordable and accessible to business owners who otherwise face barriers to entry (e.g., minority-owned businesses). For example, local governments can set up acquisition funds that help non-profit organizations purchase properties that they could not otherwise afford. By helping these mission-based developers compete with private organizations that have more capital at their disposal, the local government ensures that properties are owned and managed by entities that would prioritize the tenancy of small and independently owned business.

Maintaining a Diverse and Vital Retail Landscape

The development of urban neighbourhoods comprising disparate property owners and interests is particularly hard to coordinate (compared to planned communities or malls with single developers or landlords, for example). Yet the public sector can play an important coordinating role in neighbourhood management and development. First, zoning is a powerful tool that cities can use to curate neighbourhood economic activity, at least to some degree. While zoning is not useful for dictating the ownership structure of businesses (for example, encouraging minority-owned enterprises), it can discourage uses that do not contribute to neighbourhood services or create barriers for small, independently owned businesses to thrive. For example, zoning has been used to direct certain kinds of commercial activity (i.e., banks) to second-floor spaces, so that large swathes of ground floor frontage are not left inactive. Zoning can also affect the size of commercial spaces: if storefronts are required to be small, then larger (often chain) businesses will be less likely to occupy those spaces.[11]

Second, Business Improvement Districts (BIDs) are sub-municipal financing and governance tools that can be important mediators of commercial gentrification.[12] They are incorporated institutions with a set of powerful fiscal tools and a direct line to government officials, commercial brokers, neighbourhood businesses, and property owners. Their nexus of relationships and knowledge positions them to support businesses in their efforts to weather the challenges of gentrification and also recruit commercial tenants that meet the comprehensive and increasingly diverse needs of the community. BIDs can also provide detailed accounts of commercial activity to government officials who make decisions about more systematic interventions. They are an important collective voice for an otherwise disparate group of commercial, often neighbourhood-based enterprises.[13]

Collecting Micro-Data on Commercial Activity

These strategies are speculative – there is little research and few track records to project costs and benefits. A persistent theme in the state of knowledge on neighbourhood commercial markets is the lack of broad *and* deep data. Local governments are poised to compile rich administrative datasets of commercial activity – information that is currently held by private vendors and sold at a large premium. For example, cities have information on land use, the incorporation and ownership of establishments, and sales revenues. Cities facing increasingly higher rates of commercial vacancies have proposed registries to record (and move towards taxing) persistently empty storefronts (for example, New York City and San Francisco). These data efforts need to be ongoing and the results made public. It is very difficult to make neighbourhood-level decisions about investment, development, or political advocacy without information that extends across time and drills down to a fine-grained (i.e., block or parcel) level.

Neighbourhood-based commercial activity is uniquely attached to the local market. As such, its success is an integral part of the challenges and opportunities that accompany gentrification, a process of dramatic and localized economic and cultural transition. The goal for local governments and community actors is to ensure that these changes are inclusive and productive for as many businesses and neighbourhood patrons as possible.

Appendix to Chapter 3

Table A3.1. NAICS codes for commercial categories

Commercial category	NAICS codes
Retail	441110, 441120, 441210, 441221, 441229, 441310, 441320, 442110, 442210, 442291, 442299, 443111, 443112, 443120, 443130, 444110, 444120, 444130, 444190, 444210, 444220, 445110, 445120, 445210, 445220, 445230, 445291, 445292, 445299, 445310, 446110, 446130, 446191, 446199, 447110, 447190, 448110, 448120, 448130, 448140, 448150, 448190, 448210, 448310, 448320, 451110, 451120, 451130, 451140, 451211, 451212,4 52111, 452910, 452990, 453110, 453210, 453220, 453310, 453910, 453920, 453930, 453991, 453998, 722, 812111, 812112, 812113, 812310
FIRE	51, 52, 53, 54
Manufacturing	31, 32, 33

NOTES

1 There are also theories of "corporatized gentrification" (Hackworth, 2002) and "tourism gentrification" (Hackworth & Rekers, 2005; Burnett, 2014), which describe a process by which commercial investors transform a community to exploit existing ethnic cultures or destinations in order to commodify the products or experience (Chapple et al., 2017).

2 CBP data are available at https://www.census.gov/programs-surveys/cbp.html.

3 CBSAs are geographic areas defined by the Office of Management and
   Budget (OMB). They consist of one or more counties (or equivalents)
   anchored by an urban centre of at least 10,000 people, in addition to adjacent
   counties that are socio-economically tied to the urban centre by commuting.

4 I isolate a subsample of eight CBSAs that represent some of the metros
   areas with the most aggressive price appreciation as of 2015: Boston-
   Cambridge-Newton, Chicago-Naperville-Joliet, Los Angeles–Long Beach–
   Santa Ana, Miami–Fort Lauderdale–Miami Beach, New York–NJ–Long
   Island, San Francisco–Oakland–Fremont, Seattle-Tacoma-Bellevue, and
   Washington-Arlington-Alexandria. Most of the metros areas with excessively
   high price appreciation are located in the western region of the United
   States (especially California); I select a sample of cities that is dispersed
   geographically across the country.

5 See full list of NAICS codes for all three categories in the appendix to this
   chapter.

6 However, the differences in percentage change across gentrifying and non-
   gentrifying strata are not statistically significant for the later intervals.

7 I am careful to not use the word "displacement," since establishments can
   close for a range of reasons, some of which are voluntary. There is no study
   or dataset that can distinguish among the reasons for establishment exit
   from a neighbourhood.

8 The entry of chain establishments is not always bad, however. It's possible
   that these establishments provide a broader range of products, at lower
   price points, and provide better working conditions and wages than
   independently owned businesses (Chapple et al. 2017). Unfortunately, there
   is no documentation of these benefits in the context of gentrification.

9 It is important to note that gentrification is concentrated in a subset of US
   and global cities (Kalhoon 2018). However, lessons can be learned for cities
   of all economic positions, since the viability of urban retail is under threat
   from other pervasive market shocks, like e-commerce and automation
   (Thompson 2018).

10 Mehta (this book) provides empirical evidence to elucidate the strategies that
   storefront retailers use to survive in the age of e-commerce and big-box retailers,
   and Talen (this book) presents results from interviews with mom-and-pop stores
   in Chicago to understand the challenges they face in staying open. Both studies
   can also be instructive for businesses facing gentrification pressures.

11 Although many chain retailers are now adapting their store model to fit into
   smaller, urban spaces.

12 Business Improvement Districts (BIDs) collect an assessment from member
   property owners that, when pooled together across all members, fund
   supplementary services to the prescribed BID area. BIDs are enabled by the

state, administered by the local government, and operated as independent non-profit organizations.

13 My proposal is an optimistic one – indeed, BIDs can accelerate commercial gentrification if their priorities align with new commercial investors and tenants. There are examples of BIDs who have facilitated the commercial transition of neighbourhoods both alongside and in the absence of incumbent business interests.

REFERENCES

Aron-Dine, S., & Bunten, D. (2019). *Yes, gentrification displaces people* [Working paper].

Baum-Snow, N., & Hartley, D. (2015). *Gentrification and changes in the spatial structure of labor demand* [Unpublished paper].

Behrens, K., Boualam, B., Martin, J., & Mayneris, F. (2018). Gentrification and pioneer businesses. *The Review of Economics and Statistics*, 1–45.

Bingham, R. D., & Zhang, Z. (1997). Poverty and economic morphology of Ohio central-city neighborhoods. *Urban Affairs Review*, *32*(6), 766–796. https://doi.org/10.1177/107808749703200602

Boarnet, M. G., Bostic, R. W., Burinskiy, E., Rodnyansky, S., & Prohofsky, A. (2018). Gentrification near rail transit areas: A micro-data analysis of moves into Los Angeles Metro Rail Station areas [Unpublished paper].

Bridge, G., & Dowling, R. (2001). Microgeographies of retailing and gentrification. *Australian Geographer*, *32*(1), 93–107. https://doi.org/10.1080/00049180020036259

Burnett, K. (2014). Commodifying poverty: Gentrification and consumption in Vancouver's Downtown Eastside. *Urban Geography*, *35*(2), 157–176. https://doi.org/10.1080/02723638.2013.867669

Carlino, G. A., & Saiz, A. (2019). Beautiful city: Leisure amenities and urban growth. *Journal of Regional Science*, *59*(3), 369–408.

Caulfield, J. (1994). *City form and everyday life: Toronto's gentrification and critical social practice*. University of Toronto Press.

Chapple, K., Jackson, S., & Martin, A. J. (2010). Concentrating creativity: The planning of formal and informal arts districts. *City, Culture and Society*, *1*(4), 225–234. https://doi.org/10.1016/j.ccs.2011.01.007

Chapple, K., & Jacobus, R. (2009). Retail trade as a route to neighborhood revitalization. *Urban and Regional Policy and Its effects*, *2*, 19–68.

Chapple, K., Loukaitou-Sideris, A., Gonzalez, S. R., Kadin, D., & Poirier, J. (2017). *Transit-oriented development & commercial gentrification: Exploring the linkages* [Unpublished paper].

Cortright, J., & Mahmoudi, D. (2016). *City report: The storefront index*. City Observatory.

Couture, V., & Handbury, J. (2017). *Urban revival in America, 2000 to 2010* (No. w24084). National Bureau of Economic Research.

Crewe, L., & Lowe, M. (1995). Gap on the map? Towards a geography of consumption and identity. *Environment and Planning A, 27*(12), 1877–1898. https://doi.org/10.1068/a271877

Curran, W. (2007). "From the frying pan to the oven": Gentrification and the experience of industrial displacement in Williamsburg, Brooklyn. *Urban Studies, 44*(8), 1427–1440.

Dastrup, S., Ellen, I., Jefferson, A., Weselcouch, M., Schwartz, D., & Cuenca, K. (2015). *The effects of neighborhood change on New York City Housing Authority residents*. Abt Associates.

Deener, A. (2007). Commerce as the structure and symbol of neighborhood life: Reshaping the meaning of community in Venice, California. *City & Community, 6*(4), 291–314. https://doi.org/10.1111/j.1540-6040.2007.00229.x

DiPasquale, D., & Wheaton, W. C. (1996). *Urban economics and real estate markets, 23*(7). Prentice Hall.

Edlund, L., Machado, C., & Sviatschi, M. M. (2015). *Bright minds, big rent: Gentrification and the rising returns to skill* (No. w21729). National Bureau of Economic Research.

Ellen, I. G., & Ding, L. (2016). Guest editors' introduction: Advancing our understanding of gentrification. *Cityscape, 18*(3), 3–8.

Ellen, I. G., Horn, K. M., & Reed, D. (2019). Has falling crime invited gentrification?. *Journal of Housing Economics, 46*, 101636.

Ferm, J. (2016). Preventing the displacement of small businesses through commercial gentrification: Are affordable workspace policies the solution? *Planning Practice & Research, 31*(4), 402–419. https://doi.org/10.1080/02697459.2016.1198546

Freeman, L. (2005). Displacement or succession? Residential mobility in gentrifying neighborhoods. *Urban Affairs Review, 40*(4), 463–491. https://doi.org/10.1177/1078087404273341

Freeman, L. (2006). *There goes the hood: Views of gentrification from the ground up.* Temple University Press.

Glass, R. L. (1964). *London: Aspects of change* (Vol. 3). MacGibbon & Kee.

Gould Ellen, I., & O'Regan, K. (2008). Reversal of fortunes? Lower-income urban neighbourhoods in the US in the 1990s. *Urban Studies, 45*(4), 845–869. https://doi.org/10.1177/0042098007088471

Hackworth, J. (2002). Postrecession gentrification in New York city. *Urban Affairs Review, 37*(6), 815–843. https://doi.org/10.1177/107874037006003

Hackworth, J., & Rekers, J. (2005). Ethnic packaging and gentrification: The case of four neighborhoods in Toronto. *Urban Affairs Review, 41*(2), 211–236.

Hammel, D. J., & Wyly, E. K. (1996). A model for identifying gentrified areas with census data. *Urban Geography, 17*(3), 248–268. https://doi.org/10.2747/0272-3638.17.3.248

Hamnett, C. (1991). The blind men and the elephant: The explanation of gentrification. *Transactions of the Institute of British Geographers, 16*(2), 173–189. https://doi.org/10.2307/622612

Hwang, J., & Lin, J. (2016). What have we learned about the causes of recent gentrification? *Cityscape, 18*(3), 9–26. https://doi.org/10.21799/frbp.wp.2016.20

Jacobs, J. (1961). *The death and life of great American cities*. Vintage.

Kalhoon, I. (2018). In praise of gentrification. *The Economist*, 21 June.

Kolko, J. (2009). *Job location, neighborhood change, and gentrification* [Unpublished manuscript].

Lees, L. (1994). Rethinking gentrification: Beyond the positions of economics or culture. *Progress in Human Geography, 18*(2), 137–150. https://doi.org/10.1177/030913259401800201

Lees, L. (2000). A reappraisal of gentrification: Towards a "geography of gentrification." *Progress in Human Geography, 24*(3), 389–408. https://doi.org/10.1191/030913200701540483

Lees, L. (2016). Gentrification, race, and ethnicity: Towards a global research agenda? *City & Community, 15*(3), 208–214. https://doi.org/10.1111/cico.12185

Lester, T. W., & Hartley, D. A. (2014). The long term employment impacts of gentrification in the 1990s. *Regional Science and Urban Economics, 45*, 80–89. https://doi.org/10.1016/j.regsciurbeco.2014.01.003

Ley, D. (1996). *The new middle class and the remaking of the central city*. Oxford University Press.

Lloyd, R. (2010). *Neo-bohemia: Art and commerce in the postindustrial city*. Routledge.

McKinnish, T., Walsh, R., & White, T. K. (2010). Who gentrifies low-income neighborhoods? *Journal of Urban Economics, 67*(2), 180–193. https://doi.org/10.1016/j.jue.2009.08.003

Meltzer, R. (2016). Gentrification and small business: Threat or opportunity? *Cityscape, 18*(3), 57–86.

Meltzer, R., & Capperis, S. (2017). Neighbourhood differences in retail turnover: Evidence from New York City. *Urban Studies, 54*(13), 3022–3057. https://doi.org/10.1177/0042098016661268

Meltzer, R., & Ghorbani, P. (2017). Does gentrification increase employment opportunities in low-income neighborhoods? *Regional Science and Urban Economics, 66*, 52–73. https://doi.org/10.1016/j.regsciurbeco.2017.06.002

Meltzer, R., & Schuetz, J. (2012). Bodegas or bagel shops? Neighborhood differences in retail and household services. *Economic Development Quarterly, 26*(1), 73–94. https://doi.org/10.1177/0891242411430328

Monroe Sullivan, D., & Shaw, S. C. (2011). Retail gentrification and race: The case of Alberta Street in Portland, Oregon. *Urban Affairs Review, 47*(3), 413–432. https://doi.org/10.1177/1078087410393472

Nichols Clark, T. (2003). Urban amenities: Lakes, opera, and juice bars: do they drive development? In T. Nichols Clark (Ed.), *The city as an entertainment machine* (pp. 103–140). Emerald Group Publishing Limited.

Oldenburg, R. (1999). *The great good place: Cafes, coffee shops, bookstores, bars, hair salons, and other hangouts at the heart of a community.* Marlowe & Company.

Ong, P., Pech, C., & Ray, R. (2014). *TOD impacts on businesses in four Asian American neighborhoods.* UCLA Center for the Study of Inequality.

Patch, J. (2008). "Ladies and gentrification": New stores, residents, and relationships in neighborhood change. In J. N. DeSena & R. Hutchison (Eds.), *Gender in an urban world* (pp. 103–126). Emerald Group Publishing Limited.

Rosenthal, S. S. (2008). Old homes, externalities, and poor neighborhoods: A model of urban decline and renewal. *Journal of Urban Economics, 63*(3), 816–840. https://doi.org/10.1016/j.jue.2007.06.003

Schuetz, J. (2015). Do rail transit stations encourage neighbourhood retail activity? *Urban Studies, 52*(14), 2699–2723. https://doi.org/10.1177/0042098014549128

Schuetz, J., Kolko, J., & Meltzer, R. (2012). Are poor neighborhoods "retail deserts"? *Regional Science and Urban Economics, 42*(1–2), 269–285. https://doi.org/10.1016/j.regsciurbeco.2011.09.005

Shkuda, A. (2013). The art market, arts funding, and sweat equity: The origins of gentrified retail. *Journal of Urban History, 39*(4), 601–619. https://doi.org/10.1177/0096144212443134

Somashekhar, M. (2020). Racial inequality between gentrifiers: How the race of gentrifiers affects retail development in gentrifying neighborhoods. *City & Community, 19*(4), 811–844.

Stanback, T. M. (1981). *Services, the new economy* (Vol. 20). Allanheld, Osmun.

Sutton, S. A. (2010). Rethinking commercial revitalization: A neighborhood small business perspective. *Economic Development Quarterly, 24*(4), 352–371. https://doi.org/10.1177/0891242410370679

Thompson, D. (2018). What's really happening to retail? *CityLab,* 3 December.

Vigdor, J. L., Massey, D. S., & Rivlin, A. M. (2002). Does gentrification harm the poor? [with Comments]. *Brookings-Wharton papers on urban affairs,* 133–182.

Waights, Sevrin. (2018). Does gentrification displace poor households? An "'identification-via-interaction' approach" [Unpublished paper].

Yoon, H., & Currid-Halkett, E. (2015). Industrial gentrification in West Chelsea, New York: Who survived and who did not? Empirical evidence from discrete-time survival analysis. *Urban Studies, 52*(1), 20–49. https://doi.org/10.1177/0042098014536785

Zukin, S. (2018). *Point of purchase: How shopping changed American culture.* Routledge.

Zukin, S., Trujillo, V., Frase, P., Jackson, D., Recuber, T., & Walker, A. (2009). New retail capital and neighborhood change: Boutiques and gentrification in New York City. *City & Community, 8*(1), 47–64. https://doi.org/10.1111/j.1540-6040.2009.01269.x

Business Profile: Roots Café

Brooklyn, New York

Roots Café is located in the Greenwood Heights neighbourhood of Brooklyn, NY. Patricia and her family live in the neighbourhood and took over Roots Café from the previous owner, whom they met at their church (also in the neighbourhood). In addition to serving up coffee and made-to-order food, Roots is a community space that showcases local artists' work, sells baked goods from local entrepreneurs, hosts music and literature events for adults and children, and more recently transformed its unused seating area into a food pantry during the COVID-19 pandemic.

Figure B3.1. Roots Café. (Credit: Photo courtesy of Rachel Meltzer.)

PART TWO

The Case of E-Commerce

4 Bricks and Clicks

Introduction

The internet is the general-purpose technology behind numerous innovations that changed how we live, work, and play (Bresnahan & Tratjenberg, 1995). People now operate at the intersection of the physical and the virtual. Nowhere is this truer than in e-commerce, where search, purchase, and pick-up options have segmented the shopping experience temporally and physically. People can search and buy products online and have them shipped directly to their front door. They can also search and buy online and then pick up items in the store. In just two decades, e-commerce became a trillion-dollar enterprise. In 2018, consumers around the world purchased goods worth almost $3 online – a 23 per cent increase from 2017 (Young, 2019). E-commerce market share is increasing globally, not only because of the COVID-19 pandemic, but because of larger technological shifts.

This segmentation in shopping is the latest in a long line of changes to the retail industry, which poses yet another set of challenges and opportunities for small independent retailers. The fate of mom-and-pop stores is increasingly unsure in the wake of these sweeping changes to the retail industry, which sets up as a veritable David and Goliath battle of small independent entities against online-only juggernauts like Amazon. To understand these changes in retailing, this chapter will trace the rise of e-commerce from its inauspicious beginning in the mid-1990s to the present. It will discuss the impact of these changes on the retail industry, particularly for small independent retailers. The chapter will review how these small businesses can survive in the digital age by leveraging their competitive advantages, while adopting the technological advantages of e-commerce in omnichannel online strategies, also known as "bricks and clicks."

Rise of E-Commerce

E-commerce is a process innovation (Burt & Sparks, 2003) that hinged on related innovations such as online payment systems, shipping, tracking, and logistics. It represents the most recent evolution in the history of retail. As discussed in the introduction of this book, shopping used to take place at smaller, independent retailers, evolving to shopping experiences via department stores in the late 1890s (Howard, 2015) and enclosed shopping malls in the 1950s (Binnie, 2018). Price-conscious superstores like Walmart first opened in 1962 (LeCavalier, 2016), followed by "category killer" stores of the 1980s specializing in particular types of products such as Toys "R" Us and Home Depot that were able to outcompete competitors on the basis of price, product variety, and massive square footage (Lal & Alvarez, 2011). In the 1990s, just as the malling of America reached its apogee (Howard, 2015; Behr, 2017; Kestenbaum, 2017), e-commerce emerged. In fact, the electronic purchase of a compact disc featuring the musician Sting in 1994 marked the birth of online shopping (Tuttle, 2014).

Throughout the 1990s, several internet-related innovations made e-commerce increasingly feasible. The invention of web browsers such as Mosaic and Netscape in the early part of the 1990s made the internet easier to explore and navigate. They also made it easier to find online retailers. Google, the search engine that ultimately trumped all competitors, was incorporated in 1998 (Redding, 2018). Interestingly, initial entrants into online retail in this decade were early successes who remain leaders in e-commerce to this day. Amazon incorporated as an online book retailer in 1994, before expanding into other product lines to become the dominant force in e-commerce today, selling everything from clothes to jewellery to lawnmowers (Stone, 2013). The online auction site eBay, founded in 1998, remains a top e-commerce destination today.

Bolstered by early success stories of online-only or dot.com companies and the soaring valuations of initial public offerings (IPOs), the second half of the 1990s coincides with the dot.com bubble as stark reminders of the early limits to e-commerce growth. At this time, online businesses selling everything from delivery services to pet food were founded in the hope of cashing in on a big IPO payday. In 2000, this bubble expanded to its limit and the IPO market crashed in 2001 when people began to question the growth prospects and associated valuations of these virtual corporations (Geier, 2015). Two of the more spectacular failures of this dot.com era are the online grocer Webvan.com and the online pet supply company Pets.com (Lanxon, 2009).

Although many companies did not survive the dot.com crash in 2001, online retailing would continue growing in the following decade. Several internet-oriented innovations throughout the 2000s would cement online platforms as a retail staple. In 2003, Apple launched its iTunes store, which revolutionized how people purchase music (Chen, 2010). Instead of buying a physical CD with multiple songs, consumers could now purchase individual songs without ever acquiring a physical good. The following year, in 2004, Google went public (Redding, 2018), and in 2005, three employees of PayPal launched YouTube (Fitzpatrick, 2018). Also in 2005, the National Retail Federation coined the term Cyber Monday, which is the Monday after Black Friday, designed to encourage online shopping in preparation for the Christmas holiday season (Davis, 2005; Nowak, 2021.). As of 2017, consumers spent $6.59 billion on Cyber Monday – and Black Friday sales are increasingly moving online as well (Adobe, 2017).

Other innovations in the 2000s that changed how people purchase and use products and services include Amazon's launch of the Kindle in 2007, which enabled people to purchase digital books and read them on a tablet (Stone, 2013). Apple also launched its smartphone, called the iPhone, in 2007 (Carey, 2018). While it was not the first smartphone launched, this phone, in combination with the launch of Apple's app store in 2010 (Apple, 2018), changed phone use from voice calling only to multimedia use that enabled people to send text messages, browse the Web, watch videos, and shop online. In many ways, it wrapped the innovations of the prior two decades into one device and disentangled people from their desktop devices, a trend that has strongly influenced e-commerce ever since.

Since the purchase of Sting's first compact disc in 1994, e-commerce has grown tremendously. The range of products and services available online has expanded from the 2000s, boosting the e-commerce market share of total retail sales (Figure 4.1). While e-commerce accounted for less than 1 per cent of all American retail sales when counts began in the fourth quarter of 1999, the tally in the first quarter of 2019 put e-commerce market share at over 10 per cent (US Census Bureau, 2019).

Impacts of E-Commerce

The online revolution in retail has been the subject of substantial research, which has examined impacts that range from travel behaviour (Weltevreden & Rietbergen, 2009) to commercial real estate sales (Zhang et al., 2016). Explanatory studies identified the elements through which e-commerce has affected the retail industry, which include product characteristics, consumer shopping behaviour, culture, and geography.

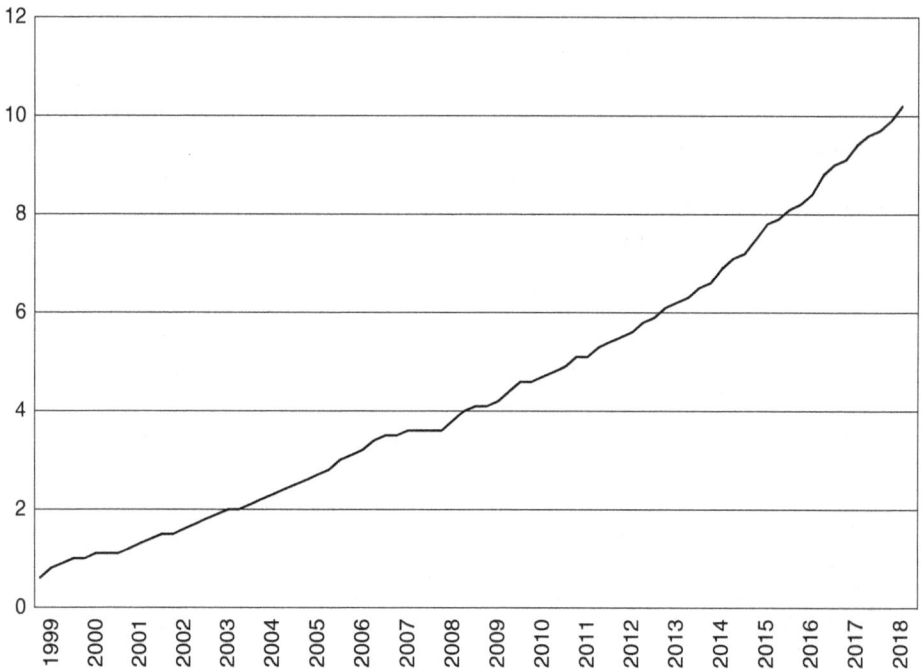

Figure 4.1. E-commerce market share (percentage of total retail sales) in the United States. (Source: Data from Retail Indicators Branch, US Census Bureau.)

The extent that e-commerce is a viable substitute for physical retail depends on product characteristics that include physical size, weight, perishability, and fashionability (Burt & Sparks, 2003). In understanding the implication of e-commerce for traditional forms of retailing, Stern (1999) suggests three types of sectors: high-, moderate-, and low-impact sectors. High-impact sectors are anticipated to be affected heavily by online retailers – or already have seen these impacts. Perhaps the earliest and most obvious impact of e-commerce has become visible in the music, video, and book industries. The ability to purchase songs one at a time and the subsequent innovation in music-streaming entities such as Spotify changed how people consume music. Instead of cassette tapes or compact discs, much of the music consumed today is from online resources. Another visible example is the video retail industry. People no longer rent movies from physical stores. Instead, most people watch movies from the on-demand channel from their cable provider or stream videos from companies including Hulu and Netflix. This change in how movies are consumed meant that the former industry leader, Blockbuster video, filed for bankruptcy in 2010 (Satell, 2014). The online bookstore Amazon and the popularity of e-books like Kindle have similarly decimated book retailers – between 1995 and 2000 alone, the number of independent bookstores declined by 43 per cent (Nobel, 2017).

In other product categories the impact of e-commerce is lower, although the online market share of fashion items is growing rapidly (Keller et al., 2014). Even the relatively low adoption of online grocery deliveries is set to grow as grocery store chains experiment with home-delivery strategies for both perishable and non-perishable goods (Binnie, 2018). As explained in the following chapter by Colin Jones, market shares for presumably low-impact goods also have plenty of room to grow.

In combination with the type of product purchased, the impact of e-commerce also depends on the type of shopping trip. Categorizing shopping visits according to their stage of the purchasing process (e.g., product awareness, product search, and actual purchase) can help us understand which retail activities are place-bound and which are prone to move online (Couclelis, 2004; Verhoef et al., 2015). Furthermore, consumers make shopping trips for several different reasons (e.g., essential, purposive, leisure, convenience, experimental), and retailers need to customize shopping experiences to differentiate them from the virtual shopping experience (Burt & Sparks, 2003). For example, retailers might choose to play up the social or experiential aspects of the shopping experience, which are not possible to achieve to the same degree in online environments (Burt & Sparks, 2003).

Geographic location matters in the impact of e-commerce. Despite early claims that the internet would render distances irrelevant, consumers continue to value proximity and the quality of physical environments in their shopping choices (Couclelis, 2004). Several studies found that the location of retailers and their consumers has implications for e-commerce impacts (Farag et al., 2006). Burt and Sparks (2003) suggest centrally located physical retailers will be less affected by e-commerce than are less centrally located retailers. More recent work by Singleton et al. (2016) in the United Kingdom finds that mid-sized retail centres are more vulnerable to e-commerce than large centres that provide a destination shopping experience or small retailers that serve a convenience function – which is corroborated by the findings in the following chapter. On the other hand, the geographic location of consumers may also affect the vulnerability of physical retailers to e-commerce. For example, customers travelling further distances to shop may be more likely to substitute e-commerce for in-store purchases (Weltevreden & Rietbergen, 2009).

From a global perspective, other factors that may explain the impact of e-commerce on retailers are culture and politics. A study of demand for commercial real estate in China suggests this cultural aspect is important to consider (Zhang et al., 2016). As is the case in other countries, Chinese commercial properties dealing in specific types of goods such as clothes, shoes, and electronics are more likely to be affected by e-commerce (Zhang et al., 2016). However, shopping malls continue to thrive as popular Chinese destinations (Zhang et al., 2016). This is a distinct difference from American cities, where shopping malls are on the decline (Tomasic, 2019).

Hybrid Retail: Brick and Clicks

Along a continuum with pure online e-tailers on one end and store-only brick-and-mortar businesses on the other, hybrid retailers (multichannel or omnichannel retailers) fall somewhere in the middle. The phrase "bricks and clicks" is often used to refer to these hybrid retail strategies. Formally defined, "bricks and clicks" refers to using e-commerce in combination with bricks-and-mortar operations (Stojković et al., 2016). Hybrid strategies are varied and popular with big-box retailers including Target and Walmart, which sell products and fulfil orders both online and in the store. Many retailers also allow customers to order online and pick up orders in the store, or return online orders in the store. More recently, grocery stores in particular have capitalized on the online order and kerbside pickup method for delivering goods to customers. It involves the time-saving and convenient online channel for ordering products and circumnavigates the inconvenience of shipping highly perishable goods.

There is evidence that this hybrid strategy is growing in popularity. Studies find that e-commerce is responsible for the growth in bricks-and-clicks retailing, and that companies using a multichannel strategy are more successful (Stojković et al., 2016). Online and physical retailing are not polar opposites, and several studies have found complementarities between online and traditional shopping (Cao et al., 2010), which manifest themselves in several ways. For example, the internet provides the ability to search for and view items online and then purchase those items in stores (Hernández et al., 2001) – known as "webrooming," as opposed to "showrooming" items in stores before buying them online (Verhoef et al., 2007; Kang, 2018). E-commerce can also create additional demand beyond store purchases. Weltevreden and Rietbergen (2009) found that e-commerce can foster online purchases for physical retailers that would not otherwise have been made in store. Conversely, opening physical stores can provide up to 20 per cent more sales for online retailers in defined markets (Pauwels & Neslin, 2015). For these reasons, studies are recommending a mixed retail strategy that combines the advantages of online and in-store shopping (Weltevreden, 2007). Maintaining a physical presence in an increasingly virtual world also serves several functions for big name stores. Beyond the purpose of browsing, other logics for a physical presence include conveying trust to customers (Benedicktus et al., 2010), reinforcing brand power, and creating an experience for customers.

Small Independent Retailers

Although the retail environment is evolving rapidly in the digital age, this does not necessarily spell doom for small, independent retailers. Information from the American Booksellers Association highlights 35 per cent growth in independent booksellers between 2009 and 2015 (Nobel, 2017). There is also evidence that online environments have helped independent entities. Etsy, the online retailer for vintage, handmade products, is a visible success story; it connects customers with smaller-scale artisans making a variety of goods from clothes to toys. An economic impact study of this online retailer noted that sellers on this site have contributed a combined $4.7 billion to the US economy (Guta, 2018). In 2017 alone, over 300,000 small businesses opened a virtual storefront, and half of the items sold on Amazon's Marketplace are from small and medium-sized businesses (BusinessWire, 2018). The success of Etsy and Amazon Marketplace speaks to the benefits that the internet can offer very small retailers and producers, such as a reduction of overhead from rent, staffing, utilities, and other operations costs that come with bricks-and-mortar stores.

Small retailers or mom-and-pop stores have several advantages, even in the digital age. They can fulfil consumer demand for unique artisanal products and offer unique shopping environments for customers (Goldman & Solomon, 2015). Many people also want to support smaller retailers over global retail chains, a movement coined "localism" (Andres Coca-Stefaniak et al., 2010). However, consumers are using mobile phones and smart technologies to shop (Reinartz et al., 2019). These trends, in combination with the complementarities between online and bricks-and-mortar operations suggests that small independent retailers must seriously reconsider a bricks-and-clicks only retailing strategy. Being online would not only bolster the visibility of small businesses, but also provide consumers with the shopping flexibility desired in the present retail environment. That said, implementing a hybrid strategy may not be easy for small business owners. Studies show that small businesses (defined as businesses with fewer than 10 employees) are less likely to adopt information and communications technologies than are slightly larger businesses with between 10 and 49 employees (Alonso-Almeid & Llach, 2013).

There is a variety of reasons small and medium-sized enterprises (SMEs) do not adopt e-commerce. Aside from small business owners' lack of time, Simpson and Docherty (2004) offer a range of reasons why SMEs do not adopt e-commerce: lack of skills and knowledge, company age, costs, and lack of resources and qualified employees. Thus, it is important to note several nuances about small businesses when considering their hybrid bricks-and-clicks retailing strategy. Differences in business characteristics are critical considerations for business owners looking to implement e-commerce, but also for entities that provide small business support and for craft training programs that will help small businesses become successful in the digital economy.

Prior work finds owner characteristics are important to e-commerce adoption. For example, an Indonesian study found younger, more educated owners were more willing to adopt e-commerce (Astuti & Nasution, 2014). This same study also found men were more likely than women to adopt e-commerce. In terms of business characteristics, industry membership affects the propensity to adopt e-commerce (Simpson & Docherty, 2004). As mentioned previously, the type of product sold by the business, as well as business size affects online strategy and success (Weltevreden et al., 2005). Sales volume should be evaluated, as e-commerce may not be the best strategy for businesses with low sales volumes (Manning, 2016). Older SMEs are less likely to adopt e-commerce (Simpson & Docherty, 2004). Finally, the demography of customers is important in e-commerce operations. For example, an anticipated trend in e-commerce is to offer online assistants to help older customers shop online (Stone, 2018).

If small businesses decide to implement e-commerce, they face critical considerations. Managing delivery and returns is critical to maintaining profits; their spiralling costs has prompted many small businesses to abandon e-commerce (Manning, 2016). Failure to maintain cyber-security for business records and customers shopping online can lead to a host of problems for businesses, ranging from financial losses to lost customer information to reputation damage (Rahman & Lackey, 2013). Customer expectations are also important to keep in mind, particularly when designing company websites. Table 4.1 compares the outcomes of a survey conducted by the North American Retail Hardware Association of 400 independent home improvement retailers and consumer information from the Farnsworth Group, respectively, asking retailers and consumers what

Table 4.1. Difference in retailer and consumer expectations of retail websites

Retailer	Consumer
Basic information about the business	Product pricing
Service offerings	Product availability in stores
Products and department offerings	Easy to find product information
Contact information	Easy to search for information
Ability to buy products online and shipped to store	Review and recommendations

Source: Adapted from Klein & Taber (2018).

they expect from retail websites. It highlights a mismatch in what retailers perceive to be customer wants and needs from their websites and the actual needs of customers. Given this mismatch, as retailers develop their strategies to remain competitive in the digital age, they need to survey customers about their needs and structure websites accordingly.

The Future of E-Commerce

In their quest for online presence and success, small retailers should stay aware of trends in e-commerce. First, it is important to note that the bricks-and-mortar business model is not dead, especially in hybrid formats with e-commerce. In fact, many retail forecasters indicate an increase in bricks-and-clicks business models that blend the best of virtual and physical shopping experiences. Forecasters also predict the continued decline of big-box retailers (Stone, 2018) and the entry of more independent retailers into the market. After all, they have distinct advantages over chains, such as store ambiance, a strong customer service image, and convenience (Barber & Tietje, 2004; Goodman & Remaud, 2015). While these are promising indicators, independent retailers need to be more reflective about their market niches and more strategic about reaching consumers in a full-fledged digital age. This is particularly true as innovation accelerates and larger retailers use new technology to reduce the competitive advantages of independents.

For example, chain and online retailers increasingly understand the value of strategies driven by data and technology to streamline and personalize the shopping experience for customers, to compete with the customer-centric focus of independent competitors (Barseghian, 2018). There are several methods for personalizing the shopping experience, which is considered a top priority by retail chains and online retailers. At the core of this effort is the ability to gather customer data, using loyalty programs, registered gift cards, online purchase tracking behaviour, and the Internet of Things (Barseghian, 2018). Artificial intelligence helps find patterns in these data and is reshaping the customer experience to provide personalized product recommendations, remember customer preferences, and maintain inventory levels for popular items (Morgan, 2019).

Retailers are also leveraging technology to enhance online shopping experiences. One of these technologies is augmented reality (Poushneh & Vasquez-Parraga, 2017), which superimposes digital images and graphics so customers can preview and understand how items will appear prior to purchase (Vilner, 2018). For example, British clothing company Top-Shop uses augmented reality so customers can try on clothing virtually and preview it on themselves prior to making purchases (Barseghian, 2018). Projections also indicate retailers will use artificial intelligence to recognize and link customer faces with past purchasing activity (Vilner, 2018). Online retailers are also using chatbots to answer customer questions online in an effort to streamline and personalize the virtual shopping experience (Chung et al., 2018). These innovations are seen by retail scholars and professionals as key trends and priorities for growth (Grewal et al., 2017).

Independent retailers will need to stay ahead of these trends, leveraging their unique advantages in customer service, convenience, and ambiance through a hybrid physical and online presence. This presence needs to be visible, easy to navigate, useful, and streamlined to accept a growing array of payment options, from credit cards to ApplePay, Venmo, PayPal, and potentially even cryptocurrencies (e.g., Bitcoin). These payments can increasingly become integrated through social media platforms, which is a growing channel to engage customers (Jones et al., 2015). In creating these hybrid strategies, independents need to make their websites and payment systems mobile friendly to take advantage of the rapid growth in mobile shopping "mcommerce" and the utility of mobile apps to streamline the customer experience.

The Future of Small Independent Retailers

These trends suggest small independent retailers must have an online presence to remain competitive in the digital era. While smaller retailers may not have the same level of sophistication in digital interactions with customers as the result of financial, organizational and technological restraints, there are strategies for small retailers to enhance their online capacity.

One way for smaller enterprises to grow their online presence is to use the infrastructure of e-commerce platforms such as Shopify,

Bigcommerce, Wix, or Squarespace to set up an e-commerce-enabled website. These platforms offer a range of services for e-tailers including inventory methods, flexible payment options, shipping, and gift-card capability. Prices for the platforms vary, based on complexity and diversity of services needed. Another strategy for establishing an online presence is to list items with established e-tailers (Amazon, Etsy or eBay), to offer online sales with relatively low barriers to entry.

Offline, small retailers can also be competitive by meeting growing consumer demand for local alternatives to global retailers, customized shopping experiences, and superior customer service. There is evidence that people want to support small businesses (Andres Coca-Stefaniak et al., 2010) and that small retailers have an advantage in providing customized shopping experiences with superior customer service. In fact, independent bookstores have survived in the age of Amazon because they have focused on community connections and are able to provide personalized shopping experiences for consumers because of knowledge of their buying preferences, and by hosting events (e.g., book signings, story time, and readings groups) that bring people into the store (Nobel, 2017). From this perspective, they function as third places (Oldenburg, 1989) where people can explore product options, relax, and socialize (Barseghian, 2018). They can also capitalize on the buy online pick up in store (BOPIS) model to increase customer traffic.

Small independent retailers need to leverage the growing array of e-commerce options to improve marketing, sales, and consumer interaction. As they may not be able to compete with the digital prowess of larger retailers, small independent retailers will need to think strategically about their market niche and customer profile. They will also need to consider how they differ from larger retailers and how they can exploit these differences to foster a competitive advantage.

REFERENCES

Adobe. (2017, 27 November). Adobe data shows Cyber Monday is largest online sales day in history with $6.59 billion [Press release]. https://rb.gy/eggfqt

Alonso-Almeida, M. D. M., & Llach, J. (2013). Adoption and use of technology in small business environments. *The Service Industries Journal, 33*(15–16), 1456–1472. https://doi.org/10.1080/02642069.2011.634904

Andres Coca-Stefaniak, J., Parker, C., & Rees, P. (2010). Localisation as a marketing strategy for small retailers. *International Journal of Retail & Distribution Management, 38*(9), 677–697. https://doi.org/10.1108/09590551011062439

Apple. (2018, 5 July). The App Store turns 10. https://www.apple.com/za /newsroom/2018/07/app-store-turns-10/

Astuti, N. C., & Nasution, R. A. (2014). Technology readiness and e-commerce adoption among entrepreneurs of SMEs in Bandung City, Indonesia. *Gadjah Mada International Journal of Business, 16*(1), 69–88. https://doi .org/10.22146/gamaijb.5468

Barber, C. S., & Tietje, B. C. (2004). A distribution services approach for developing effective competitive strategies against "big box" retailers. *Journal of Retailing and Consumer services, 11*(2), 95–107. https://doi.org/10.1016 /s0969-6989(03)00009-2

Barseghian, A. (2018). *Localmotion: How technology is personalizing the global marketplace.* Lioncrest Publications.

Behr, A. (2017, 20 September). When malls were the disruptors of retail. *Chain Store Age.* https://chainstoreage.com/real-estate/when-malls-were-disruptors-retail

Benedicktus, R. L., Brady, M. K., Darke, P. R., & Voorhees, C. M. (2010). Conveying trustworthiness to online consumers: Reactions to consensus, physical store presence, brand familiarity, and generalized suspicion. *Journal of Retailing, 86*(4), 322–335. https://doi.org/10.1016/j.jretai.2010.04.002

Binnie, L. (2018). *The future of omni-channel retail: Predictions in the age of Amazon.* Emerald Lake Books.

Bresnahan, T. F., & Trajtenberg, M. (1995). General purpose technologies "Engines of growth"? *Journal of Econometrics, 65*(1), 83–108. https://doi .org/10.1016/0304-4076(94)01598-t

Burt, S., & Sparks, L. (2003). E-commerce and the retail process: A review. *Journal of Retailing and Consumer Services, 10*(5), 275–286. https://doi .org/10.1016/s0969-6989(02)00062-0

BusinessWire. (2018, 10 January). Shopping local on Amazon: More than 300,000 US-based small and medium-Ssized businesses started selling on Amazon in 2017. https://www.businesswire.com/news/home/20180110005439/en /Shopping-Local-Amazon-300000-U.S.-based-Small-Medium-Sized

Cao, X., Douma, F., & Cleaveland, F. (2010, 1 January). The influence of e-shopping on shopping travel: Evidence from twin cities. *Transportation Research Record: Journal of the Transportation Research Board, 2157*(1), 147–154. https://doi.org/10.3141/2157-18

Carey, C. (2018, 26 November). The evolution of the iPhone: Every model from 2007–2018. *iPhone +iPad Life Magazine.* https://www.iphonelife.com /content/evolution-iphone-every-model-2007-2016

Chen, B. X. (2010, 28 April). April 28, 2003: Apple opens iTunes Store. *Wired.* https://www.wired.com/2010/04/0428itunes-music-store-opens/

Chung, M., Ko, E., Joung, H., & Kim, S. J. (2018). Chatbot e-service and customer satisfaction regarding luxury brands. *Journal of Business Research, 117*, 587–595. https://doi.org/10.1016/j.jbusres.2018.10.004

Couclelis, H. (2004). Pizza over the internet: e-commerce, the fragmentation of activity and the tyranny of the region. *Entrepreneurship & Regional Development, 16*(1), 41–54. https://doi.org/10.1080/0898562042000205027

Davis, E. (2005). Cyber Monday quickly becoming one of the biggest online shopping days of the year [Press release].

Farag, S., Weltevreden, J., Van Rietbergen, T., Dijst, M., & van Oort, F. (2006). E-shopping in the Netherlands: Does geography matter? *Environment and Planning B: Planning and Design, 33*(1), 59–74. https://doi.org/10.1068/b31083

Fitzpatrick, L. (2018, 31 May). Brief history YouTube. *Time.* http://content.time.com/time/magazine/article/0,9171,1990787,00.html

Geier, B. (2015, 12 March). What did we learn from the dotcom stock bubble of 2000? *Time.* https://time.com/3741681/2000-dotcom-stock-bust/

Goldman, D., & Solomon, K. (2015, 30 March). Mom-and-pops are cool again. Chain Store Age. https://chainstoreage.com/real-estate/mom-and-pops-are-cool-again

Goodman, S., & Remaud, H. (2015). Store choice: How understanding consumer choice of "where" to shop may assist the small retailer. *Journal of Retailing and Consumer Services, 23*, 118–124. https://doi.org/10.1016/j.jretconser.2014.12.008

Grewal, D., Roggeveen, A. L., & Nordfält, J. (2017). The future of retailing. *Journal of Retailing, 93*(1), 1–6. https://doi.org/10.1016/j.jretai.2016.12.008

Guta, M. (2018, 25 November). Etsy study says small businesses selling on the platform contribute $4.7B to the US economy. Small Business Trends. https://smallbiztrends.com/2018/11/2018-etsy-statistics.html 1/6

Hernandez, T., Gomez-Insausti, R., & Biasiotto, M. (2001). Non-store retailing and shopping centre vitality. *Journal of Shopping Centre Research, 8*, 58–81.

Howard, V. (2015). *From Main Street to mall: The rise and fall of the American department store.* University of Pennsylvania Press.

Jones, N., Borgman, R., & Ulusoy, E. (2015). Impact of social media on small businesses. *Journal of Small Business and Enterprise Development, 22*(4), 611–632. https://doi.org/10.1108/jsbed-09-2013-0133

Kang, J. Y. M. (2018). Showrooming, webrooming, and user-generated content creation in the omnichannel era. *Journal of Internet Commerce, 17*(2), 145–169. https://doi.org/10.1080/15332861.2018.1433907

Keller, C., Magnus, K., Hedrich, S., Nava, P., & Tochtermann, T. (2014, 1 September). Succeeding in Succeeding in tomorrow's global fashion market. McKinsey & Company. https://www.mckinsey.com/business-functions/marketing-and-sales/our-insights/succeeding-in-tomorrows-global-fashion-market#

Kestenbaum, R. (2017, 7 April). Why so many stores are closing now. Forbes. https://www.forbes.com/sites/richardkestenbaum/2017/04/07/why-so-many-stores-are-closing-now/#6443d25c4159

Klein, K., & Taber, T. (2018, March). Competing online in an Amazon world. *Hardware Retailing.*

Lal, R., & Alvarez, J.B. (2011, 10 October). Retailing revolution: Category killers on the brink. *Harvard Business Review.* https://hbswk.hbs.edu/item/retailing-revolution-category-killers-on-the-brink

Lanxon, N. (2009, 18 November). The greatest defunct Web sites and dotcom disasters. *CNET.* https://www.cnet.com/news/the-greatest-defunct-web-sites-and-dotcom-disasters/

LeCavalier, J. (2016). *The rule of logistics: Walmart and the architecture of fulfillment.* University of Minnesota Press.

Manning, E. (2016, 15 December). Why retailers stop selling online: The hidden cost of e-commerce. *The Guardian.* https://www.theguardian.com/small-business-network/2016/dec/15/hidden-cost-e-commerce-online-shopping-entrepreneurs

Morgan, B. (2019, 4 March). The 20 best examples of using artificial intelligence for retail experiences. Forbes. https://www.forbes.com/sites/blakemorgan/2019/03/04/the-20-best-examples-of-using-artificial-intelligence-for-retail-experiences/#1ea5d8f84466

Nobel, C. (2017, 26 November). How independent bookstores thrived in spite of Amazon.com. Harvard Business Working Knowledge. https://hbswk.hbs.edu/item/why-independent-bookstores-haved-thrived-in-spite-of-amazon-com

Nowak, C. (2021, 26 October). Finally! Here's how Cyber Monday even became a thing. *Reader's Digest.* https://www.rd.com/culture/history-of-cyber-monday/

Oldenburg, R. (1989). *The great good place.* Paragon House.

Pauwels, K., and Neslin, S. A. (2015). Building with bricks and mortar: The revenue impact of opening physical stores in a multichannel environment. *Journal of Retailing, 91*(2), 182–197. https://doi.org/10.1016/j.jretai.2015.02.001

Poushneh, A., & Vasquez-Parraga, A. Z. (2017). Discernible impact of augmented reality on retail customer's experience, satisfaction and willingness to buy. *Journal of Retailing and Consumer Services, 34,* 229–234. https://doi.org/10.1016/j.jretconser.2016.10.005

Rahman, S., & Lackey, R. (2013). E-commerce systems security for small businesses. *International Journal of Network Security & Its Applications (IJNSA), 5*(2). https://doi.org/10.5121/ijnsa.2013.5215

Redding, A. C. (2018). *Google it: A history of Google.* MacMillan Publishing.

Reinartz, W., Wiegand, N., & Imschloss, M. (2019). The impact of digital transformation on the retailing value chain. *International Journal of Research in Marketing, 36*(3), 350–366.

Satell, G. (2014, 5, September). A look back at why Blockbuster really failed and why it didn't have to. Forbes. https://www.forbes.com/sites

/gregsatell/2014/09/05/a-look-back-at-why-blockbuster-really-failed-and
-why-it-didnt-have-to/#36fc5b501d64

Simpson, M., & Docherty, A. J. (2004). E-commerce adoption support and advice for UK SMEs. *Journal of Small Business and Enterprise Development, 11*(3), 315–328. https://doi.org/10.1108/14626000410551573

Singleton, A. D., Dolega, L., Riddlesden, D., & Longley, P. A. (2016). Measuring the spatial vulnerability of retail centres to online consumption through a framework of e-resilience. *Geoforum, 69*, 5–18. https://doi.org/10.1016/j.geoforum.2015.11.013

Stern, N. Z. (1999). The impact of the internet on retailing. *International Trends in Retailing, 16*(2), 71–87.

Stojković, D., Lovreta, S., & Bogetić, Z. (2016). Multichannel strategy: The dominant approach in Modern retailing. *Economic Annals, 61*(209): 105–127. https://doi.org/10.2298/eka1609105s

Stone, B. L. (2013). *The everything store: Jeff Bezos and the age of Amazon.* Little, Brown and Company.

Stone, M. (2018, 12 November). 5 trends that will drive online shopping in 2019. Forbes. https://www.forbes.com/sites/braintree/2018/11/12/5-trends-that-will-drive-online-shopping-in-2019/#1e2587a384ac

Tomasic, M. (2019, 23 March). Shopping mall decline: What malls are doing to stay open. *US News & World Report.* https://www.usnews.com/news/best-states/pennsylvania/articles/2019-03-23/shopping-mall-decline-what-malls-are-doing-to-stay-open

Tuttle, B. (2014, 15 August). 8 amazing things people said when online shopping was born 20 years ago. Money. https://money.com/online-shopping-history-anniversary/

US Census Bureau. (2019). Retail Indicators Branch. https://www.census.gov/retail/index.html

Verhoef, P. C., Kannan, P. K., & Inman, J. J. (2015). From multi-channel retailing to omni-channel retailing: Introduction to the special issue on multi-channel retailing. *Journal of Retailing, 91*(2), 174–181. https://doi.org/10.1016/j.jretai.2015.02.005

Verhoef, P. C., Neslin, S. A., & Vroomen, B. (2007). Multichannel customer management: Understanding the research-shopper phenomenon. *International Journal of Research in Marketing, 24*(2), 129–148. https://doi.org/10.1016/j.ijresmar.2006.11.002

Vilner, Y. (2018, 19 September). 2019 will be a revolutionary year for online shopping. Entrepreneur. https://www.entrepreneur.com/article/319339

Weltevreden, J. W. (2007). Substitution or complementarity? How the internet changes city centre shopping. *Journal of Retailing and consumer Services, 14*(3), 192–207. https://doi.org/10.1016/j.jretconser.2006.09.001

Weltevreden, J. W. J., Atzema, O. A. L. C., & Boschma, R. A. (2005). The adoption of the internet by retailers: A new typology of strategies. *Journal of Urban Technology, 12*(3), 59–87. https://doi.org/10.1080/10630730500417281

Weltevreden, J. W., & van Rietbergen, T. (2009). The implications of e-shopping for in-store shopping at various shopping locations in the Netherlands. *Environment and Planning B: Planning and Design, 36*(2), 279–299. https://doi.org/10.1068/b34011t

Young, J. (2019, 21 January). Global ecommerce sales grow 18% in 2018. Digital Commerce 360. https://www.digitalcommerce360.com/article/global-ecommerce-sales/

Zhang, D., Zhu, P., & Ye, Y. (2016). The effects of e-commerce on the demand for commercial real estate. *Cities, 51*, 106–120. https://doi.org/10.1016/j.cities.2015.11.012

5 The Changing Demand for Urban Retail Space: Evidence from Canada

CHRISTOPHER DANIEL AND TONY HERNANDEZ

Introduction

The demand and supply of space for retail and consumer service activities has been a long-standing area of study amongst academics and practitioners (Proudfoot, 1937; Berry, 1963; Simmons, 1966; Jones & Simmons, 1993; Birkin et al., 2002, 2019). Physical space needs for retail and service companies have changed in parallel with the evolution of business models and technologies such as automobiles and the internet. The locational imprint of such change is highly visible in the urban landscape, as the legacy of space needs of the past interacts with the space needs of the present day and foreseeable future. This transformation of urban retail space needs is negotiated and often contested amongst a multitude of stakeholders, including retailers, real estate developers, investors, urban planners, local communities, municipalities, and of course the consumer. This negotiation and contest has intensified in recent years with the rise of e-commerce.

However, the debate on the impacts of e-commerce on the retail landscape is hardly new. Early forecasts on the risks this new retail innovation posed for traditional retailers, and the resultant space needs of retailers, date back to the rise of e-commerce in the mid- to late 1990s (Maruca, 1999; Kotha, 1998). Sir Richard Greenbury, at that time chairman and chief executive of the UK High Street retailer Marks and Spencer, noted that opinions about the impact of e-commerce on physical retail space needs were quite varied and there was little agreement.

> One school of thought maintains that E-commerce will affect all retailers and all types of products, that the Internet will change the face of retailing thoroughly and permanently. Another school believes the Internet is not as much an issue for bricks-and-mortar retailers as it is for direct mail retailers,

that it will prove easier for consumers to switch from catalogue shopping to on-line shopping, but that most people who currently enjoy shopping face-to-face will not find the new medium particularly attractive. (Maruca, 1999, p. 162)

While some persisted in the belief of continued vibrancy of physical retailers, others already foresaw an eventual retail environment where e-commerce businesses would dominate and eliminate most traditional physical retailers because of the advantages e-commerce held relating to their lack of need for physical space. These advantages included lower overhead and stock management costs for sellers (Chen & Leteney 2000), lower search costs for consumers, better price competition for consumers, better price discovery mechanisms for sellers, and more varied types of intermediary choices (Bakos, 2001). This upheaval of the retail industry was also expected to have knock-on effects for the shopping centre industry where the viability of shopping centres as retail destinations was called into question. Despite the bursting of the dot.com bubble in the late 1990s and subsequent tech market turmoil, many continued to forecast that e-commerce was likely to be a major disruptive force over the coming decades.

More than two decades on from the early crystal ball–gazing days of e-commerce impacts, the very same debates over space continue (Yeates & Hernandez, 2018a, 2018b; ICSC, 2017; Zhang et al., 2016). However, the e-commerce landscape has significantly changed. Behemoth pure-play online retailers such as Amazon and Alibaba have grown rapidly, along with several other e-commerce retail platform providers such as eBay and Shopify. Major retail chains and retail brands have ubiquitously adopted e-commerce and associated omnichannel business models throughout the world (Mason, 2019; Binnie, 2018).

In 2019, e-commerce retail sales in Canada were estimated to total US$44 billion (approximately 10 per cent of total non-automotive sales), with projected sales increasing to over US$55 billion by 2023 (Statista, 2019). The penetration of e-commerce by retail sector varies widely, as categories like fashion have a much greater penetration than other categories like grocery (JCWG, 2018).[1] Given the growth of e-commerce, the prevailing news media coverage has been sensationalized under "the retail apocalypse" or "death of the main street or shopping mall" headlines (Barrabi, 2019; Bhattarai, 2019; Petro, 2019; Sanburn, 2017; Townsend et al., 2017). In essence, the "retail apocalypse" idea predicts that in the future virtually all retail sales transactions will transition online rather than in-store, resulting in widespread and persistently high vacancy rates across the commercial real estate industry. However, there is no consensus on the future as reported by the media, with the majority of academic debate focusing on the United States as opposed to Canada (Baird, 2018; Matthews, 2018, Sagan, 2018; Shaw, 2018; Kurutz, 2017).

The dire headlines reflect the fact that the retail industry has indeed seen widespread disruption with the growing adoption of e-commerce and associated omnichannel activities by consumers in Canada (Mason, 2019; Binnie, 2018; Dart, 2017; Levy et al., 2017; Stephens, 2013, 2017; Murray & Hernandez, 2016a; Treadgold & Reynolds, 2016; Yeates & Hernandez, 2013, 2016; Lewis & Dart, 2014). As a result, retailers and consumer service firms have been reassessing the space needs of their store networks and, in turn, shopping centre landlords, owners, commercial real estate developers, and investors have had to re-examine the scale, function, and form of the commercial spaces within their property portfolios (McLean, 2018; Sekus, 2018; Schnurr, 2018; Israelson, 2017; ICSC, 2016).

The increasing pace of retail transformation, with shorter innovation cycles, places an even greater urgency on a broader range of private and public sector stakeholders to understand the impact of e-commerce on the viability of existing physical retail stores and the retail space needs of the future. The uncertainty of the impact of e-commerce on physical retail space needs, coupled with the lengthy planning and investment horizons associated with physical stores, have created heightened business risk. There are increasing demands for data-driven metrics and objective analysis of the nature and extent of the retail evolution in order to support business, investment, and policy decision-making (Gibbs, 2012; Lorinc, 2017, 2018; Spivak, 2018; Tate & Patel, 2018).

This chapter focuses on the changing retail space needs in urban Canada, using the Toronto region as a case study. The chapter is divided into three sections. First, a historical overview is provided of the Canadian retail landscape from pre–Second World War to the present day. Second, a series of forecasts for future retail space needs in the Toronto region is presented based on Delphi surveys of key public and private sector stakeholders. Finally, the implications for the future of urban retail are discussed and areas for further research identified.

The Continual Retail Apocalypse: The Evolution of the Canadian Retail Landscape

Retailers are some of the most responsive elements in the urban landscape (Yeates, 2011; Simmons & Kamikihara, 2009; Jones & Hernandez, 2005; Yeates, 1998, 2000; Bromley & Thomas, 1993), as they experience great volatility (Ceh et al., 2018; Hernandez, 2012; Simmons & Hernandez, 2004a, 2004b; Hernandez, 2003). Minor shifts in the income, demographics, or competitive characteristics of an area will lead to rapid changes in form and structure of the retail environment. The development of systematic classifications of retail and consumer service structures can be traced back to the pioneering works of Proudfoot (1937), Berry (1963), Simmons (1966), and Jones & Simmons (1993). Retailers are broadly located in either planned or unplanned environments. The planned environment comprises a broad range of managed shopping centres, whereas unplanned environments primarily include free-standing locations and commercial strips such as main streets.

More than any other retail environment, the shopping mall has become a functional symbol of mass consumerism – akin to capitalist cathedrals to which shopping pilgrims travel far and wide to pay homage (Stokan, 2005). The centrally owned and managed mall is an entrenched part of our consumer society, serving as the backdrop of Hollywood movies such as *Dawn of the Dead* and *Mall Cop*, and a topic of popular fiction such as J.G. Ballard's *Kingdom Come* (Ballard, 2006; Morrison, 2012, ICSC, 2014; Lambert, 2008; Cohen, 2002; Kowinski, 1985). As the "Scientist of Shopping" Paco Underhill notes the allure of shopping malls in North America: "[We] have a love/hate relationship with the mall. On the one hand, we claim to loathe its homogeneity … on the other hand, we return there time and again – to shop, to dine, for entertainment, to people-watch or even just to pass the time" (Underhill, 2004, cover). Shopping centres play an important role in the provision of and access to retail products and services. In 2019, it was estimated that there were over 5,000 shopping centres across Canada, accounting for more than 600 million square feet of retail space – twice the 300 million square feet of Canada's street front and freestanding retail space (CSCA, 2019). The size and type of these planned centres range in Canada from major destination and tourist shopping centres, such as The Eaton Centre in downtown Toronto, Ontario, and the super-sized West Edmonton Mall in the Alberta suburbs, down to the ubiquitous local community and neighbourhood malls that serve day-to-day needs. The shopping centre landscape has evolved to meet the changing needs of consumers in where they live, work, and play (Lau & Hernandez, 2014; Hernandez & Du, 2009; Hernandez et al., 2008).

However, the shopping mall is only part of a longer continuum of retail transformation, as the Canadian retail landscape has experienced several transformations over the last century. In the pre-war city of low consumer mobility and car ownership, many Canadians shopped daily

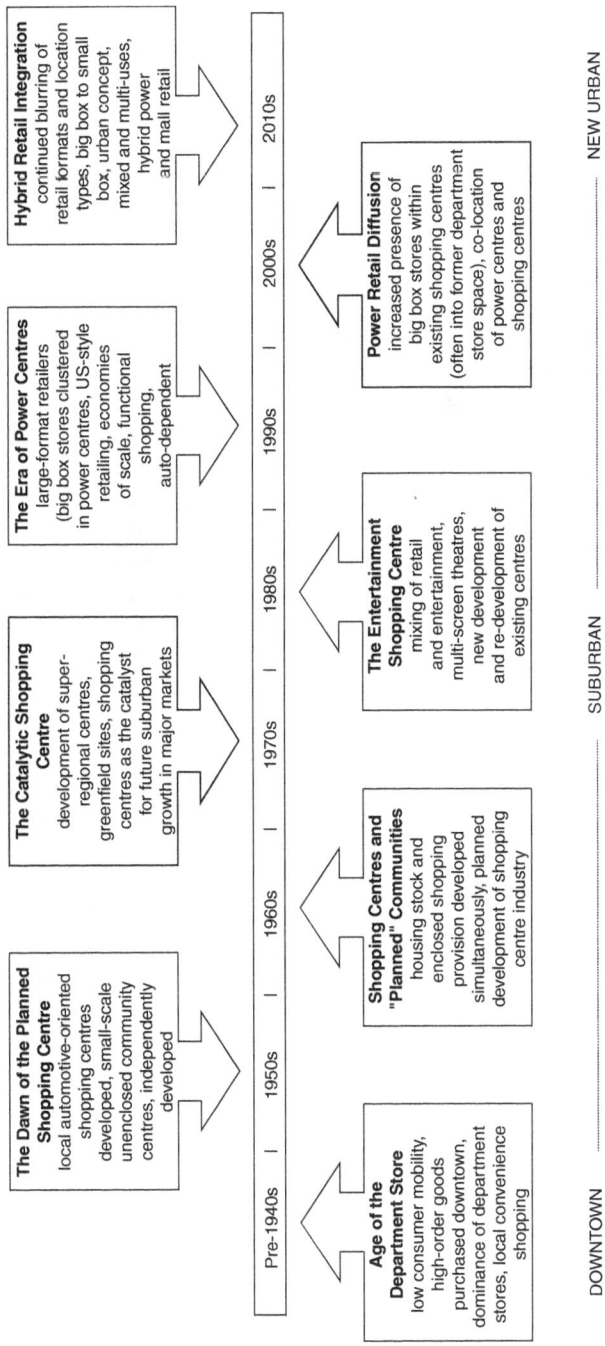

| Pre-1940s | 1950s | 1960s | 1970s | 1980s | 1990s | 2000s | 2010s |

The Dawn of the Planned Shopping Centre
local automotive-oriented shopping centres developed, small-scale unenclosed community centres, independently developed

The Catalytic Shopping Centre
development of super-regional centres, greenfield sites, shopping centres as the catalyst for future suburban growth in major markets

The Era of Power Centres
large-format retailers (big box stores clustered in power centres, US-style retailing, economies of scale, functional shopping, auto-dependent

Hybrid Retail Integration
continued blurring of retail formats and location types, big box to small box, urban concept, mixed and multi-uses, hybrid power and mall retail

Age of the Department Store
low consumer mobility, high-order goods purchased downtown, dominance of department stores, local convenience shopping

Shopping Centres and "Planned" Communities
housing stock and enclosed shopping provision developed simultaneously, planned development of shopping centre industry

The Entertainment Shopping Centre
mixing of retail and entertainment, multi-screen theatres, new development and re-development of existing centres

Power Retail Diffusion
increased presence of big box stores within existing shopping centres (often into former department store space), co-location of power centres and shopping centres

DOWNTOWN ———————— SUBURBAN ———————— NEW URBAN

Figure 5.1. Waves of retail change.

for food and visited downtown for durable goods. Most shops were located along streets, either in neighbourhood strips or in downtown clusters, anchored by department stores. This pattern began to change by the 1950s; for the next 40 years, the automobile and suburbanization were the major forces that shaped commercial structure, evolving from initial small car-oriented retail plazas to full-scale planned shopping centres. The 1960s saw a synchronization of shopping centre development and suburban housing stock, as malls became the centre of "planned" communities – a divergence from relatively asynchronous suburban development in the United States. The early 1970s saw a gradual shift to larger "catalytic" shopping centres, and by the 1980s there was a shift to the revitalization of existing retail properties and a growing focus on entertainment to spur consumption. During the 1990s, virtually no traditional enclosed shopping centre growth took place in Canada. Instead, large format big-box retailers expanded their store networks into "power centres" across metropolitan Canada (Simmons & Hernandez, 2008). The rapid growth of these power centres and their locational preference for clustering in suburban and exurban locations was a major contributor to urban sprawl (Murray & Hernandez, 2016b; Hernandez et al., 2013), as retail real estate development took place in parallel with new residential subdivision expansion along the urban fringe (Buliung & Hernandez, 2009, 2013). Towards the end of the 1990s and into the first decade of the millennium a small number of lifestyle centres (Hernandez, 2008), outlet centres (Hernandez & Murray, 2015), and ethnic malls (Wang & Hernandez, 2017, 2018; Zhuang, 2016; Wang & Zhong, 2013; Preston & Lo, 2000; Wang, 1999; Qadeer, 1998) were built in the suburbs of major markets. During the mid-aughts, retail refocused on the urban experience in densifying downtown and inner-urban areas, along with denser New Urbanist–inspired retail redevelopment in the suburbs. In recent years, widespread concerns over e-commerce disruption and the associated re-evaluation of retail space needs – amplified by major anchor store closures (Emmons & Hernandez, 2017) – prompted a new wave of retail transformation with emphasis on redeveloping former single-use properties to mixed-use centres. The functional line between traditional street-front retail and the planned shopping centre industry is becoming increasingly blurred, as shopping centres are being redeveloped with urban streetscapes, and traditional urban streetscapes are transitioning into more elaborately designed and planned retail environments. However, it is largely unknown whether these physical redevelopments are sufficient to mitigate or even reverse the continued rise of e-commerce.

Forecasting Commercial Space Demand in the Toronto Region

Given the apparent lack of consensus on what the impacts of e-commerce will be on traditional brick and mortar retailing, how can we anticipate and plan vibrant commercial spaces for our future cities? Without reliable forecasts, the effects of e-commerce will be unknown, unpredictable, and unmitigated. Will e-commerce and its associated technologies have the devastating impact predicted by the retail apocalypse doomsayers, or are the traditionalists right and do we have nothing to fear but fear itself?

A common answer these questions may come from a round table of industry experts, which would arrive at a consensus about the future impacts of e-commerce. As noted in the chapter by Rosa Danenberg, this consensus may become significantly affected by social dynamics (Krueger & Casey, 2014). In the private sector, forecasting the impact of disruptive technologies is regularly carried out by groups of company leadership or expert panels (often drawing from external consultants), enabling businesses to pre-emptively adapt to future conditions. The methodology commonly adopted in the business community to avoid the distortions of social dynamics is referred to as "estimate, talk, estimate," or the Delphi method. This method surveys participants through quantitative questions (for example, asking for a forecast); asks them to qualitatively explain their quantitative answers; and subsequently collates and shares the qualitative and quantitative answers with all participants. The survey is repeated after this feedback loop, allowing participants to see how their quantitative and qualitative inputs compare to those of the rest of the group, and to react to group opinions in subsequent rounds. Incorporating the forecasts and opinions of their peers increases the likelihood that the responses take into account considerations, information, and expertise that may not have been available to each individual participant.

This section presents the outcomes of a Delphi survey of retail transformation for the Greater Toronto, Canada region, known as the Greater Golden Horseshoe Area (GGHA), asking industry experts to forecast the impact of e-commerce on physical retail space needs. It is important to forecast trends of e-commerce impact on retail at this local scale, as e-commerce adoption is significantly affected by culture and lifestyle at varying spatial scales (Hallikainen & Laukkanen, 2018). Uniquely to the GGHA, the provincial government has implemented urban growth containment policies since 2006, which prioritize the development of relatively dense mixed-use developments (Government of Ontario, 2005). For many such developments, the inclusion of commercial space is mandated by urban planners primarily as a way to animate public space rather than serve a local demand for commercial amenities. These policies could disrupt the present commercial space balance in the region, adding to the importance of forecasting future commercial space needs in the region.

The first round of the Delphi survey was conducted from 6 April to 31 May 2018 and the second round from 7 June to 20 July 2018. Thirty-one industry practitioners and consultants drawn from urban planning, commercial real estate development, and retail management, with experience working in the GGHA, were identified and invited to participate. Participants were asked to forecast the online market share, size of sales floor space per capita, and number of stores per capita for nine key merchandise categories that correspond to North American Industrial Classification System (NAICS) three-digit codes. Responses were aggregated using a 20 per cent trimmed average in order to prevent strategic manipulation by the participants (Balinksi & Laraki, 2011).

E-Commerce Space Impacts

The final round of the Delphi survey resulted in 18 responses from 31 experts invited to participate. The aggregated results of the three main Delphi survey questions are shown in figures 5.2, 5.3, and 5.4. The collective expert evidence presented in the preceding research indicates that demand for commercial space in the GGHA will not collapse in the future because of e-commerce, as is generally predicted in the media. Instead, participants believe that e-commerce will create a decline in retail floor space and number of establishments per capita, but the extent of this slight estimated decline is far smaller than the trends this same group forecasted for the growth of e-commerce. While participants forecast an average growth of e-commerce sales of approximately 137 per cent by 2041 for the nine commercial categories, they expected the square footage per capita to decline by an average of just 16 per cent, and stores per capita by approximately 12 per cent.

While these forecasts might appear for some to represent a collapsing of commercial space demand, this is not so when they are offset with forecasted population growth. The Ontario Ministry of Municipal Affairs and Housing estimates that the population of the GGHA will grow by almost 50 per cent from 9.2 million people in 2016 to approximately 13.5 million people by 2041. When this growth is taken into account, total demand for commercial square footage is expected to actually increase by 21 per cent to 214.65 million square feet. Stores would increase by 32 per cent to approximately 56,700. These numbers demonstrate that even in the face of decreasing demand for retail space, the current supply will remain viable as long as the development of new supply is not excessive and current spaces are sufficiently accessible to new residents. Instead

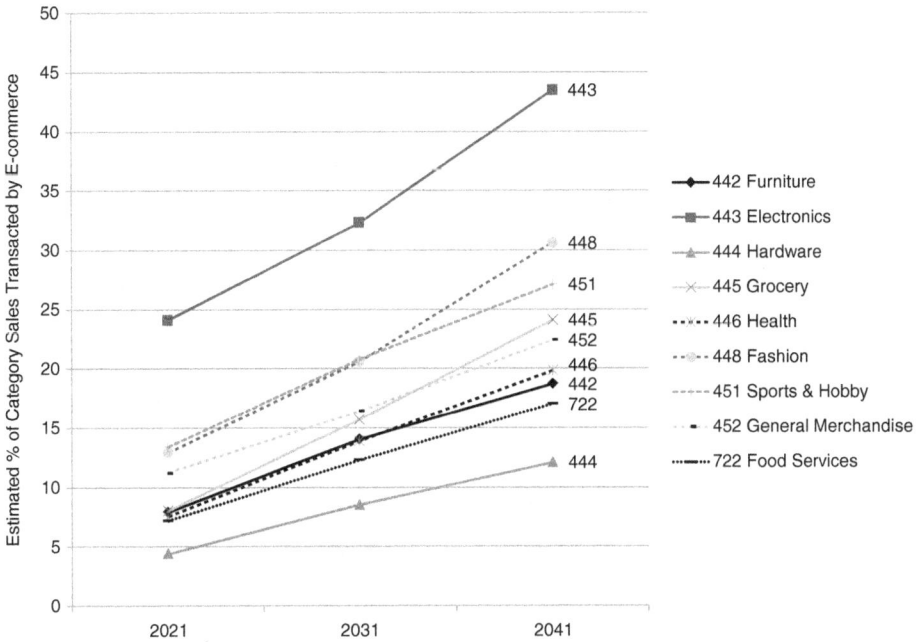

Figure 5.2. Percentage of purchases ordered online, by store type, 2021, 2031, and 2041 (estimates).

of showing an industry in complete distress, these numbers seem to represent an industry that is adapting to a new way of doing business. The numbers also forecast a decrease in average store size by 8 per cent from 4,100 square feet per store to 3,700 square feet. While there is significant variation within the nine commercial categories, this decrease suggests that e-commerce may be creating changes in the functionality of our commercial space. What forces might be driving such a change? Decreasing store sizes must be more complicated than a simple reduction of in-store spending or lower stock levels on the shelves, since the forecasted growth of e-commerce is not equally reflected in the forecasted demand decline for retail space. Instead, changing store sizes could be related to other emerging factors, such as the present levels of e-commerce penetration in Canada and the changing role of retail stores.

For many years, Canada was seen as a laggard in the adoption of e-commerce. Current Canadian e-commerce penetration rates have grown to 10 per cent of non-automotive sales – slightly below the level of the

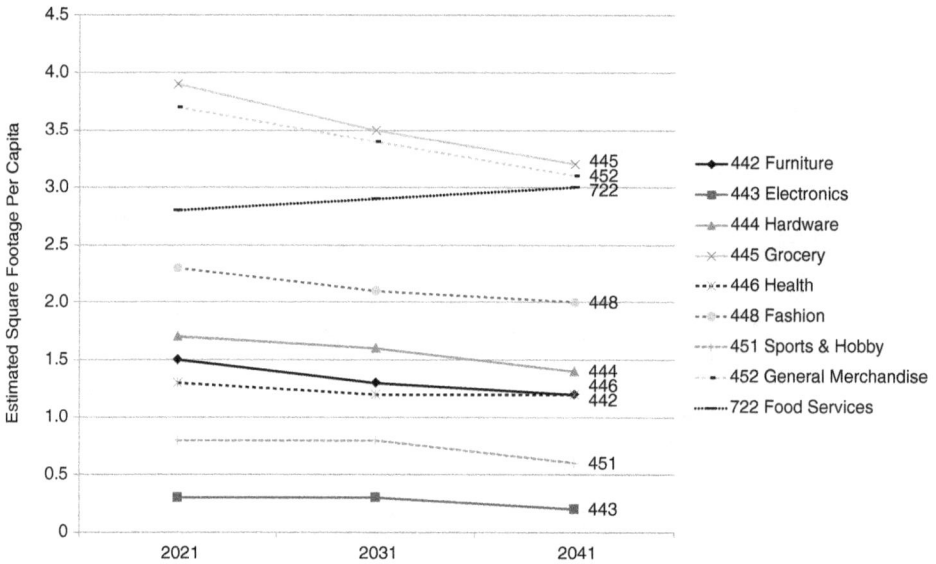

Figure 5.3. Square feet per capita, by store type, 2021, 2031, and 2041 (estimates).

United States and many European markets. This middle-of-the-pack penetration rate suggests that Canada's retail sector has undergone some of
the transformation fuelled by e-commerce, but has not reached maturity.
This may explain the expected pattern of retail transformation in the
Delphi survey. As e-commerce market penetration in Canada catches up
to that of more mature markets, Canadian retail space dynamics can be
expected to change in a similar pattern. Countries with similar urban
environments but higher rates of e-commerce penetration than Canada,
such as the United States or the United Kingdom, could be seen as examples of what is yet to come. As demonstrated in the chapter by Arnold
in the United States, and by Jones in the United Kingdom, e-commerce
growth often leads to smaller store sizes. While the retail environment
in these countries is comparable in many ways, there is additional complexity in the Canadian environment. In Canada, traditional commercial
streetscapes in major urban centres have aged to reach the end of their
development life cycle. They are increasingly becoming the targets of
government-promoted larger-scale mixed-use redevelopment. The redevelopment of these ageing commercial strips is increasing population
density and street-level vibrancy, although evidence suggests that the

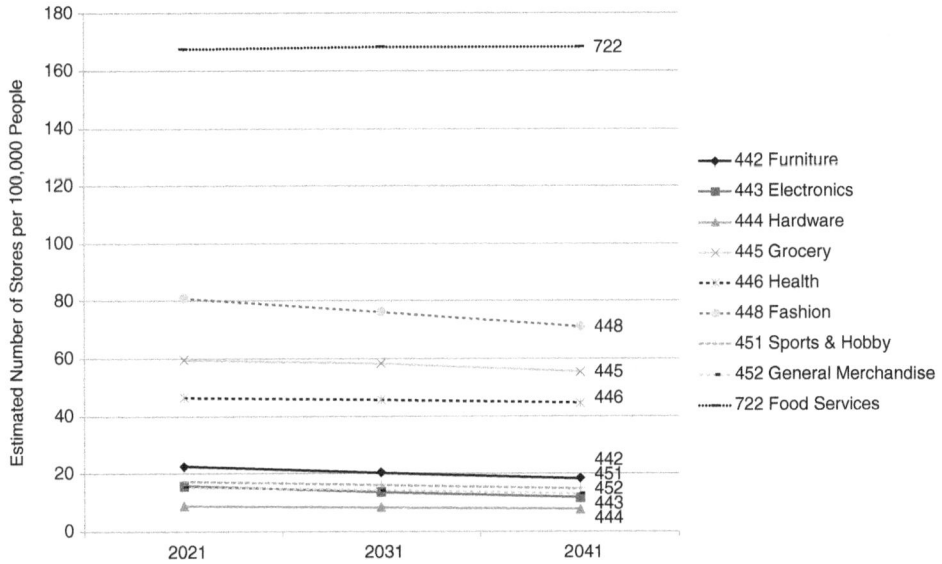

Figure 5.4. Change in the number of stores per 100,000 people, by store type, 2021, 2031, and 2041 (estimates).

functionality of these reinvented legacy spaces is shifting from retail to foodservice and personal service firms (Yeates et al., 2015a). This shift is corroborated by our Delphi panel and may have an impact on the number of stores and their dimensions.

The final factor that may explain the Delphi survey results that project a small increase in commercial square footage with smaller average store sizes is the changing role of physical stores in the customer "journey" to make a purchase in the e-commerce and omnichannel era. Before the rise of e-commerce, the typical path to purchase would linearly take consumers through a narrowing funnel of brand awareness, brand familiarity, purchase consideration, and then a final choice – ideally resulting in customer loyalty if they were happy with the process and product (Murray & Hernandez, 2016a). In this model, brand awareness was driven by traditional paid advertising, bringing consumers to brick and mortar stores to interact with and ultimately acquire products. With the rise of e-commerce and omnichannel shopping, many of these in-store functions are no longer necessary. Consumers increasingly build brand awareness through online research and evaluation and are able to decide on the transaction mode, payment type, delivery, and shipping – all without ever entering a store (Edelman & Singer, 2015). In the contemporary customer journey, stores have shifted from traditional transactional environments to experiential places where consumers can interact with brands and merchandise. In other words, the omnichannel era disrupts the traditional linear path to purchase and conceptualizes customer loyalty more like a cyclical loop or journey (Edelman & Singer, 2015).

In this omnichannel "journey" scenario, customers begin with brand awareness through an initial set of physical or digital touch points, from online advertisements to storefronts as billboards. Customers then move into an evaluation phase where they conduct research to determine which brands and products they will consider more seriously, using tools like Web searches, curated products shown on retailers' websites, but also show-rooming products in physical stores. Once research has narrowed their search to a particular product, customers will decide to purchase, as well as the transaction mode, payment type, and shipping and delivery modes. After the purchase, customers who saw their expectations for their product met and enjoy its use will advocate for its brand or even bond with it. With positive brand attachment, customers effectively enter a loyalty loop where positive experiences with a brand reduce the set of brands they consider when they initiate another purchase journey.

The role of physical retail space completely shifts in this new customer journey. In the traditional linear path to purchase, most commercial space essentially functioned as a transactional space, a means of physical interaction to actually acquire products. Big-box stores with rows of product available are the ultimate example of this transactional commercial space. In an omnichannel environment, customers can conduct research and transactions online rather than in-store, leaving the only exclusive function of commercial space as the ability to physically interact with and experience products, essentially an experiential space rather than a transactional space. Successful retailers will understand that their stores are no longer a place to buy and have become instead a place to be, with experiential product displays that maximize the emotional responses of consumers to the brands they interact with. Furthermore, they will understand that the shifting role of physical stores requires a far smaller footprint, validating the predictions of the Delphi experts.

The COVID-19 pandemic that swept across the globe in 2020 will drastically influence the future of e-commerce growth and its impact on physical store space predictions. As the Delphi panel survey has been completed prior to the pandemic, the impacts of COVID have not been incorporated into the expert predictions. According to Statistics Canada, the proportion of sales generated through e-commerce increased in every commercial category during the first wave of the pandemic in the spring of 2020. In general, Canadian e-commerce sales market share skyrocketed from 3.8 per cent in April 2019 to 11.4 per cent in April of 2020. This increase in e-commerce has been coupled with overall declines in total retail sales and has fuelled a similar narrative to the retail apocalypse predictions of previous years. However, industry observers predict that many customers will return to bricks and mortar after a couple years of uncertainty (e.g., Toneguzzi, 2020). As more consumers experience e-commerce platforms during the COVID-19 pandemic, likely the e-commerce growth trends predicted by the Delphi experts will be accelerated. However, a consumer shift toward digital safety and convenience does not necessarily indicate their preferred method of shopping after the pandemic concludes. An extended period of uncertainty fuelled by the pandemic and any post-pandemic economic impacts likely will see the demise of many operators who cannot provide an effective e-commerce experience. However, many physical spaces they leave behind may be taken up by more innovative retailers who are able to build and balance an engaging omnichannel commercial environment for their customers.

The shifting role of retail stores from transactional to experiential spaces will require a similar change in current retail research. Transactional commercial spaces have provided businesses with an excellent set of performance metrics like sales per square foot or comparable-store sales that made it much easier to understand how to value commercial space. In an experiential environment, these metrics no longer apply – as explained in detail in Heather Arnold's chapter in this volume. Additional research is needed in order to understand how consumers interact with and value experiential spaces – yielding new metrics that define and predict the viability of retail spaces and locations. Although expert opinion is a valuable starting point on the journey to understanding this professional and scholarly paradigm shift, consumer behaviour and changing lifestyle preferences will ultimately determine what vibrant commercial space will look like in the future, and where it will be located.

NOTE

1 Note that Canadian consumer-based estimates of retail e-commerce are significantly greater than those produced by Statistics Canada. This national statistical agency in Canada only asks domestic retailers to report their e-retail sales. However, it has been estimated than over half of e-retail sales in Canada are from retailers outside of Canada, primarily from US-based retailers (Statista, 2019; Daniel & Hernandez 2018; Hernandez & Emmons, 2018). This, in turn, raises issues about currency fluctuations, duty, trade tariffs, and shipping costs

that can significantly affect the value of e-retail sales. As a result, it is difficult to compare e-retail sales between countries (Yeates & Hernandez, 2016).

REFERENCES

Baird, N. (2018, 12 March). Retail apocalypse explained: Retailers have not actually embraced digital transformation. Forbes.

Bakos, Y. (2001). The emerging landscape for retail e-commerce. *Journal of Economic Perspectives, 15*(1), 69–80. https://doi.org/10.1257/jep.15.1.69

Balinski, M., & Laraki, R. (2011). *Majority judgment.* MIT Press.

Ballard, J. G. (2006). *Kingdom come.* Harper Perennial.

Barrabi, T. (2019, 29 May). Retail apocalypse: These big retailers closing stores, filing for bankruptcy. Yahoo Finance. https://finance.yahoo.com/news/retail-apocalypse-23-big-retailers-004654125.html

Berry, B. J. L. (1963). *Commercial structure and commercial blight* (Research Paper No. 85). University of Chicago, Department of Geography.

Bhattarai, A. (2019, 10 April). "Retail apocalypse" now: Analysts say 75,000 more US stores could be doomed. *Washington Post.*

Binnie, L. (2018). *The future of omni-channel retail.* Emerald Lake Books.

Birkin, M., Clarke, G., & Clarke, M. (2002). *Retail geography and intelligent network planning.* Wiley.

Birkin, M., Clarke, G., & Clarke, M. (2019). *Retail location planning in an era of multi-channel growth.* Routledge.

Bromley, R., & Thomas, C. (1993). *Retail change: Contemporary issues.* UCL Press Ltd.

Buliung, R., & Hernandez, T. (2009). *Power retail, growth management, and consumer travel behaviour in the Greater Toronto Area.* Neptis Foundation.

Buliung, R., & Hernandez, T. (2013). Retail development in Urban Canada: Exploring the changing retail landscape of the Greater Toronto Area: 1996–2005. *International Journal of Applied Geospatial Research, 4*(1), 32–48. https://doi.org/10.4018/jagr.2013010103

Ceh, B., Hernandez, T., & Kressell, D. (2018). Implications of supersizing a box store: Localized effects in the inner city. *Papers in Applied Geography, 4*(2), 157–174. https://doi.org/10.1080/23754931.2017.1399159

Chen, S., & Leteney, F. (2000). Get real! Managing the next stage of internet retail. *European Management Journal, 18*(5), 519–528. https://doi.org/10.1016/s0263-2373(00)00041-4

Cohen, N. E. (2002). *America's marketplace: The history of shopping centers.* ICSC, Greenwich Publishing Group.

CSCA. (2019). Canadian Shopping Centre Database, Centre for the Study of Commercial Activity, Ryerson University.

Daniel, C., & Hernandez, T. (2018). *CSCA Retail 100.* CSCA Research Report, Ryerson University.

Dart, M. (2017). *Retail's seismic shift*. St. Martin's Press.

Edelman, D. C., & Singer, M. (2015). Competing on customer journeys. *Harvard Business Review, 93*(11), 88–100.

Emmons, M., & Hernandez, T. (2017). *The absorption of Target's former store portfolio in Canada*. CSCA Research Insight, Centre for the Study of Commercial Activity, Ryerson University.

Gibbs, R. (2012). *Principles of urban retail planning and development*. Wiley.

Government of Ontario. (2005). Places to Grow Act. Government of Ontario Service Ontario eLaws. http://www.e-laws.gov.on.ca/html/statutes/english/elaws_statutes_05p13_e.htm

Hallikainen, H., & Laukkanen, T. (2018). National culture and consumer trust in e-commerce. *International Journal of Information Management, 38*(1), 97–106. https://doi.org/10.1016/j.ijinfomgt.2017.07.002

Hernandez, T. (2003). The impact of big box internationalization on a national market: A case study of Home Depot in Canada. *The International Review of Retail, Distribution and Consumer Research, 13*(1), 77–98. https://doi.org/10.1080/0959396032000065364

Hernandez, T. (2008). The future of the lifestyle concept in Canadian shopping centres. *International Council of Shopping Centers Research Review, 15*(1), 23–26.

Hernandez, T. (2012). *Power retail vacancy*. CSCA Research 2012-03, Centre for the Study of Commercial Activity, Ryerson University.

Hernandez, T., & Du, P. (2009). *Tracking the evolution of the Canadian mall*. CSCA Research Report, 2009-07, Centre for the Study of Commercial Activity, Ryerson University.

Hernandez, T., Erguden, T., & Murray, A. (2013). *Power retail growth in Canada and the GTA*. CSCA Research Report 2013-01, Centre for the Study of Commercial Activity, Ryerson University.

Hernandez, T., Lorch, B., & Du, P. (2008). *The changing face of the mall*. CSCA Research Letter, 2008-05, Centre for the Study of Commercial Activity, Ryerson University.

Hernandez, T., & Murray, A. (2015). Factory outlet centers in Canada: Trends and prospects. *Retail Property Insights, 22*(1), 7–13.

ICSC. (2014). *Shopping centers: America's first and foremost marketplace*. International Council of Shopping Centres.

ICSC. (2016). *Exploring new leasing models in an omni-channel world*. International Council of Shopping Centres.

ICSC. (2017). *The halo effect*. International Council of Shopping Centers.

Israelson, J. (2017, 28 November). Regional malls evolve into town centres. *The Globe and Mail*.

JCWG. (2018). *eTail Report*. JC Williams Group.

Jones, K., & Hernandez, T. (2005). Dynamics of the Canadian retail environment. In T. Bunting and P. Fillion (Eds.), *Canadian cities in transition* (pp. 404–422). Oxford University Press.

Jones, K., & Simmons, J. (1993). *Location, location, location.* Nelson Publishing.

Kotha, S. (1998). Competing on the internet: The case of Amazon.com. *European Management Journal, 16*(2), 212–222. https://doi.org/10.1016/s0263-2373(97)00089-3

Kowinski, W. S. (1985). *The Malling of America.* W. Morrow Publishers.

Krueger, R. A., & Casey, M. A. (2014). *Focus groups: A practical guide for applied research.* Sage Publications.

Kurutz, S. (2017, 2 July). An ode to the shopping mall. *New York Times.*

Lambert, J. (2008). Healthy growth in the Canadian shopping center industry. *ICSC Research Review, 15*(2), 15–18.

Lau, S., & Hernandez, T. (2014). *The evolution of major shopping centres in Canada: 1996–2013.* CSCA Research Report, Centre for the Study of Commercial Activity, Ryerson University.

Levy, M., Weitz, B., Watson, D., & Madore, M. (2017). *Retailing management.* McGraw-Hill.

Lewis, R., & Dart, M. (2014). *The new rules of retail: Competing in the world's toughest marketplace.* St. Martin's Press.

Lorinc, J. (2017, 9 November). Dead space, *The Globe and Mail.*

Lorinc, J. (2018, 25 January). Home sweet mall, *The Globe and Mail.*

Maruca, R. F. (1999). Retailing: Confronting the challenges that face bricks-and-mortar stores. *Harvard Business Review, 77*(4), 159–168.

Mason, T. (2019). *Omnichannel retail.* Kogan Page.

Matthews, M. (2018). No sign of the retail apocalypse. National Retail Federation.

McLean, S. (2018, 13 June). More malls set to become mixed-use developments. *Property Biz News.*

Morrison, E. (2012, 25 April). Ewan Morrison's top 10 books about shopping malls. *The Guardian.*

Murray, A., & Hernandez, T. (2016a). *The Canadian omni-channel retail landscape.* CSCA Research Insight, Centre for the Study of Commercial Activity, Ryerson University.

Murray, A., & Hernandez, T. (2016b). *Market thresholds of major retail chains in Canada.* CSCA Research Insight, Centre for the Study of Commercial Activity, Ryerson University.

Petro, G. (2019, 21 June). The bogus "retail apocalypse" looks more like a renaissance. *Forbes.*

Preston, V., & Lo, L. (2000). Asian theme malls in suburban Toronto: Land use conflict in Richmond Hill. *Canadian Geographer, 44*(2), 182–190. https://doi.org/10.1111/j.1541-0064.2000.tb00701.x

Proudfoot, M. J. (1937). City retail structure. *Economic Geography, 13*, 425–28. https://doi.org/10.2307/141589

Qadeer, M. (1998). *Ethnic malls and plazas: Chinese commercial developments in Scarborough, Ontario* [Working paper]. Joint Center of Excellence for Research on Immigration and Settlement (CERIS).

Sagan, A. (2018, 11 March). Online shopping not the only factor behind Canada's dying malls. The Canadian Press.

Sanburn, J. (2017, 20 July). Why the death of malls is about more than shopping. *Time.*

Schnurr, J. (2018, 28 March). Ottawa's Planning Committee approves plan to redevelop Westgate Shopping Centre. CTV News Ottawa.

Sekus, T. (2018, 20 January). Retrofitting suburbia: Old shopping malls can be saved by their parking lots, CBC News.

Shaw, H. (2018, 12 January). Transformation over turmoil: Why reports of traditional retail's imminent death may be premature. *Financial Post.*

Simmons, J. (1966). *Toronto's changing retail complex* (Research Paper No. 104). University of Chicago, Department of Geography.

Simmons, J., & Hernandez, T. (2004a). *Power retailing: Close to saturation.* CSCA Research Report 2004-08, Centre for the Study of Commercial Activity, Ryerson University.

Simmons, J., & Hernandez, T. (2004b). *Power retailing in Canada major urban markets.* CSCA Research Report 2004-09, Centre for the Study of Commercial Activity, Ryerson University.

Simmons, J., & Hernandez, T. (2008). *Defining power retailing.* CSCA Research Report, 2008-09, Centre for the Study of Commercial Activity, Ryerson University.

Simmons, J., & Kamikihara, S. (2009). *Commercial activities in Canada.* CSCA Research Report, 2009-02, Centre for the Study of Commercial Activity, Ryerson University.

Spivak, J. (2018, July). *Retail realities,* American Planning Association.

Statista. (2019). *E-commerce in Canada.* Statista Digital Market Outlook.

Stephens, D. (2013). *The retail revival: Reimagining business for the new age of consumerism.* Wiley.

Stephens, D. (2017). *Re-engineering retail.* Figure 1 Publishing.

Stokan, A. J. (2005). *Naked consumption: Retail trends uncovered.* Anthony Russell Inc.

Tate, J., & Patel, S. (2018). Challenge of retail in mixed-use buildings [Special section]. *Ontario Planning Journal, 33*(3), 7–8.

Toneguzzi, M. (2020, 29 July). Retail e-commerce explodes in Canada amid COVID-19 pandemic. *Retail Insider.* https://www.retail-insider.com/retail-insider/2020/7/retail-e-commerce-explodes-in-canada-amid-covid-19-pandemic

Townsend M., Surane, J., Orr, E., & Cannon, C. (2017, 8 November). America's "retail apocalypse" is really just beginning. *Bloomberg.*

Treadgold, A., & Reynolds, J. (2016). *Navigating the new retail landscape.* Oxford University Press.

Underhill, P. (1999). *Why we buy: The science of shopping.* Simon & Schuster.

Underhill, P. (2004). *Call of the mall.* Simon & Schuster.

Wang, S. (1999). Chinese commercial activity in the Toronto CMA: New development patterns and impacts. *The Canadian Geographer, 43*(1), 19–35. https://doi.org/10.1111/j.1541-0064.1999.tb01358.x

Wang, S., & Hernandez, T. (2017). *Reconceptualizing ethnic retailing.* CSCA Research Insight, Centre for Study of Commercial Activity, Ryerson University.

Wang, S., & Hernandez, T. (2018). Contemporary ethnic retailing: An expanded framework. *Canadian Ethnic Studies, 50*(1), 37–68. https://doi .org/10.1353/ces.2018.0003

Wang, S., & Zhong, J. (2013). *Delineating ethnoburbs in Metropolitan Toronto* (CERIS Working Paper No. 100). The Ontario Metropolis Centre.

Yeates, M. (1998). *The North American city* (5th ed.). Longman.

Yeates, M. (2000). *The GTA @ Y2K: The dynamics of change in the commercial structure of the Greater Toronto Area.* Centre for the Study of Commercial Activity, Ryerson University.

Yeates, M. (2011). *Charting the GTA.* CSCA Research Report 2011-06, Centre for the Study of Commercial Activity, Ryerson University.

Yeates, M., & Hernandez, T. (2013). *E-retail in Canada: Performance meets "bricks and mortar."* CSCA Research Report 2013-10, Centre for the Study of Commercial Activity, Ryerson University.

Yeates, M., & Hernandez, T. (2016). *E-retail in Canada: Growth and the exchange rate.* CSCA Research Insight, Centre for the Study of Commercial Activity, Ryerson University.

Yeates, M., & Hernandez, T. (2018a). *Assessing shopping centre space needs in Canada.* CSCA Research Insight, Ryerson University, Centre for the Study of Commercial Activity, Ryerson University.

Yeates, M., & Hernandez, T. (2018b). *Shopping centre over-spacing in Canadian cities.* CSCA Research Insight, Centre for the Study of Commercial Activity, Ryerson University.

Yeates, M., Hernandez, T., & Murray, A. (2015a). *Commercial strips in the Greater Toronto Area.* CSCA Research Report, Ryerson University.

Zhang, D., Zhu, P., & Ye, Y. (2016). The effects of e-commerce on the demand for commercial real estate. *Cities, 51,* 106–120. https://doi.org/10.1016/j .cities.2015.11.012

Zhuang, Z. C. (2016). Planning for diversity in a suburban retrofit context: The case of ethnic shopping malls in the Toronto area. In R. Thomas (Ed.), *Planning Canada: A case study approach* (pp. 134–142). Oxford University Press.

6 Online Sales and the British Urban Retail Hierarchy

COLIN JONES

Introduction

The expansion of online sales has a significant impact on the demand for and nature of physical stores. This chapter examines these effects, focusing on the impact of online sales on retail locations and store networks, and hence the urban retail hierarchy. The chapter focuses on the United Kingdom, where the decentralization of retailing has been more muted and more recent than in the United States, although more pronounced than in other European nations. Britain's relatively weak decentralization can be linked to the strong tradition of planning policies that have sought to maintain the existing retail hierarchy, especially town centre shops. Yet compared to the United States, British retail has a much higher penetration of online sales, which is one of the highest in the world.

The year 2018 saw a wave of major British retail brand closures/financial difficulties. While in some cases, brands that have failed to interface with the internet have gone under, this is not the only explanation for the financial woes of retailers. An important influence on recent retailing patterns has been the global financial crisis and subsequent government austerity policies that have stifled real incomes and consumer spending in shops (ONS, 2019a, 2019b). These problems have been compounded, as noted by several authors in this volume, by the 2020 COVID-19 crisis that has accelerated the virtualization of retailing.

The main objective of this chapter is to explore the influence of online retailing in the decision-making of leading British retailers, specifically to evaluate the effects on the physical hierarchy of retail locations and clusters. It begins by explaining the regulatory and planning framework for retailing, connecting with wider social and economic trends in the United Kingdom, and their impacts on the contemporary British retail hierarchy. The following sections examine the growth of online sales and the response of physical retailers to this growth, concluding with a wider perspective on contemporary British retail dynamics.

Regulation

In common with other European peer nations, retail policies in the United Kingdom have followed a relatively restrictive approach (Evers, 2002). A key regulatory change for retailing in the United Kingdom was the banning of resale price maintenance (RPM) in 1964, whereby manufacturers required that retailers charge a specified price for a good. While the scale of RPM was not all encompassing, its abolition supported the rise of large retailers who could sell goods more cheaply than small independent retailers could, which especially affected the electronics segment. Simultaneous to the abolition of RPM, the growth of pre-packaged goods of uniform quality, especially in the grocery sector, and the increasing adoption of self-service set the stage for the emergence of economies of scale that propelled discounters and superstores.

Up until this point, the spatial retail pattern of British cities and towns had remained remarkably unchanged for decades. It comprised a clear hierarchical structure of historical town and city central retail and business centres and organically grown suburban shopping streets along arterial routes, all colloquially known as High Streets. In addition, there were scattered neighbourhood shops. Rather than focus on mounting challenges to this retail hierarchy, British planners in the 1950s and 1960s were occupied mostly with alleviating the post-war housing shortages by creating new towns and overspill estates. Urban planning policies in the late 1960s switched their focus to urban regeneration, including the demolition of slums, followed by the rehabilitation of the housing stock and promoting reuse of brownfield land, much of it from deindustrialization. Even during the 1970s, the interface between planning and retailing received little attention, except for public concerns that the massive urban slum clearance programs were destroying independent shops in neighbourhoods. Planned shopping centres began to anchor new neighbourhoods and towns from the 1950s onward, but these were diminutive compared to the size and ubiquity of the American regional mall.

All this changed in the 1980s when the Conservative government led by Mrs Thatcher heralded the arrival of a shift away from planning constraints, which gave spatial market forces full throttle, and in particular enabled the decentralization of retailing. Nevertheless, subsequent national retail planning guidance in the United Kingdom, with the exception of this significant "liberal" period during the 1980s and early 1990s, has sought to maintain the retail hierarchy status quo, especially protecting town centres by restricting development outside of settlements. Since 1994, applications for planning permissions for new out-of-town shopping centres must pass sequential tests to demonstrate that these developments would not affect the vitality and viability of existing centres (Jones, 2014; McArthur et al., 2016).

However, many cities have allowed for key breaches to this policy, which have been justified by the need to regenerate parts of these cities. In fact, many of the largest out-of-town centres in the United Kingdom have been built since the millennium through this justification. Guy (2010) argues that while it was ultimately unable to stem the decentralization tide, the restrictive planning regime was able to slow it; however, arguably it has created a haphazard pattern of shopping centres. Nevertheless, restrictive policy has likely contributed to a lower than average retail sales floor space per capita in the United Kingdom compared with other developed countries, and in particular half that in the United States (CBRE, 2019).

Changing Retail Context

The transformation of retail patterns from High Streets and neighbourhood corner stores from the 1970s onward was of course not only the result of regulatory dynamics – it responded to technological and organizational shifts in retailing as well. Most notably, automobiles began to significantly affect British shopping patterns and retail development from this decade onward. This process occurred much later than in the United States, where the first suburban covered shopping mall opened in 1956 (Eppli & Benjamin, 1994; Hardwick, 2010). The removal of RPM that enabled large retailers to exploit economies of scale paved the way for a similar trend in the United Kingdom (Dawson, 2004). The first signs could be seen as large supermarkets left town centres, subsequently followed by new out-of-town retail formats such as big-box stores (known in the United Kingdom as retail warehouses), a trend that culminated in fully fledged regional centres (Schiller, 1986). Initial retail warehouse development began in the first half of the 1980s on industrial estates – sometimes in converted units – but by the end of that decade warehouses were being purpose built together to form retail parks (known in the United States as power centres). These retail parks have grown to generate more than 4 million square feet of new space per annum on average in the 13 years up to 2006, before the real estate market downturn brought a rapid contraction (CBRE, 2014a). There has also been a historically high numerical expansion of out-of-town shopping centres despite increasingly strict planning constraints on such development since 1994 (CBRE, 2014b).

Furthermore, the nature of these out-of-town developments has increasingly encroached High Street territory. Early retail warehouse developments focused on the sale of bulky non-food goods such as furniture, carpets, and electrical appliances. Out-of-town locations for such goods had the advantage of more space and lower rents than the traditional High Street, as well as better automobile parking facilities. The focus on these retailers was at first deemed beneficial to High Streets that had been subject to significant historic supply constraints to expansion because of the existing urban infrastructure. However, the bulky goods stores that originally predominated in retail parks became outnumbered by High Street retailers in the late 1990s (Guy, 2000). Thomas et al. (2006) describe retail parks as effectively suburban High Streets with overlapping goods sold in town centres. This transformation also reflects the long-term trend toward chain retailers, and the rise of brand power (Durand & Wrigley, 2009). Chain retailers generally require larger shops, which has made retail parks more attractive locations for them.

There are also wider underlying influences on the changing physical pattern of retail centres. The long-term increase in real incomes for the five decades from the 1950s led to a demand for more shops that could not be met by existing town and city retail centres. At the same time, the population was suburbanizing, leading to a demand for shops near their homes. Suburbanization went hand-in-hand with greater automobile use, which enabled more flexibility and complexity of shopping patterns, including combining activities such as shopping on the way home from work (Ferguson & Woods, 2010). Historic shopping centres of small towns were particularly affected by the growing dominance of automobiles and the subsequent extension of motorway networks, as they suffered from "out shopping" because of the ease of travel to larger centres with greater choice. High streets also increasingly suffered from automobile parking constraints introduced by the planning system (Portas, 2011). The overall outcome of these forces has been a reframing of the shopping centre hierarchy that is part of a much wider restructuring of the spatial economy (Jones, 2017), together with internal substantial structural changes to retailing.

These trends are illustrated by Table 6.1, which records the change in the retail investment portfolios of financial institutions such as major life assurance companies. At the beginning of the 1980s, retail real estate portfolios were dominated by standard High Street shops, reflecting the nature of retailing at that time. Yet just over 30 years later, standard shops now represent less than a quarter of their investments, and vast numbers of these shops have been sold as obsolescent. Over this period, retail

Table 6.1. Breakdown of retail real estate portfolios of financial institutions, 1981, 2000, and 2014 (%)

Retail type	1981	2000	2014
Standard shop	58.9	31.0	24.2
Shopping centre/mall	38.6	41.3	37.1
Retail warehouse	2.4	27.7	38.7

Source: IPD (2014).

warehouses – which hardly existed in 1981 – have risen in significance to being the most popular form of investment, representing nearly 40 per cent of retail investments. Despite the increase in the number of planned shopping centres or malls, their share of total investment is broadly unchanged. Overall, these changes to investment and shopping centre patterns reflect the retail landscape prior to the arrival of the impact of online sales that the chapter now turns to.

Online Retailing and Its Effects on British Towns and Cities

The opportunities of the internet for retailing were first identified in the mid-1990s, with the establishment of Amazon and eBay as pure play (internet only) retailers; the growth of British online retailing accelerated in the second half of the aughts when broadband connections became widely available in homes. In 2006, over 57 per cent of households had internet access in the United Kingdom, but the largest annual percentage (11 per cent) increase in internet accessibility occurred during 2008–9 as broadband became the norm (ONS, 2014). This was the effective platform for the online sales revolution supported by improvements in Web connection technology and faster browsing.

Consumers have increasing opportunities to shop online. At the turn of the millennium, there were just four channels through which purchases could be made: physical stores, catalogues, telephone ordering, and online via desktop or laptop computers; nearly two-thirds of consumers shopped primarily through visiting shops or via catalogues. By 2013, three additional retail channels had become available to consumers, who could now also shop online via mobile phones and tablets, social media, and interactive TV. These changes have all contributed to online sales becoming an integrated component of retail sales, growing continuously since 2009 with similar quarterly trends to the wider retail market in purchasing peaks and troughs (ONS, 2019b).

Table 6.2. Online sales as a percentage of all sales broken down by sector

Year	Food	Clothing/ footwear	Household goods	Total non-food	Total
October 2008	2.0	4.1	4.5	4.4	5.2
October 2010	2.8	6.7	6.2	6.9	7.9
October 2012	3.1	9.9	5.8	7.7	9.3
October 2014	3.9	11.5	6.6	8.9	11.5
October 2016	5.1	13.6	10.7	11.5	15.7
October 2018	5.5	18.1	14.1	14.5	18.2
July 2019	5.5	19.3	14.4	15.8	19.9

Data source: ONS (2019b).

Table 6.2 charts the near quadrupling of online sales from 5.2 per cent of total retail sales in 2008 to 19.3 per cent in May 2019. The table hints that the rate of growth may be slowing as online market share may be reaching saturation (although these numbers are prior to the COVID-19 pandemic). The headline figure also perhaps slightly exaggerates the picture as online sales have captured catalogue sales. Nevertheless, a nearly 20 per cent online market share represents the highest share in Europe and almost double that of the United States. The table also illustrates wide differences in online market penetration between merchandise categories. The least intrusion has been into food sales, where the market share of online sales has been steady for a few years at just over 5 per cent. The sector where online sales has shown the most penetration is clothing and footwear, reaching 19 per cent of sales in 2019. An equivalent share can be seen for non-specialized stores such as department stores. Although not presented in Table 6.2, online sales are just over 18 per cent of the total for these stores. These figures hide some of the main effects on shopping centres, such as the severe loss of banks and other financial services, as well as travel agents (Teale, 2013). The closures of book and music shops were also part of the first wave of impacts of online sales, but these represented only a small percentage of shopping centre space.

Online purchases have become a normal part of shopping. However, physical shops are still important. Many British retailers have promoted a strategy known as "click and collect" whereby customers can order online and collect in a store. Consumers benefit from the choice provided by the internet and the convenience that they can pick the purchased item up at a shop with the certainty it is in stock. Retailers benefit from the possibility of further purchases when consumers collect online-bought items in a store. In this way, retailers with a visible and recognizable brand build on consumer trust and utilize store networks to attract online sales and permit easy return of unwanted goods. Many successful large retailers have a central strategy to integrate online and physical sales through multichannel or omnichannel shopping. The result is that almost 90 per cent of British retail sales still involve physical stores (CBRE, 2019), utilizing a variety of delivery and click-and-collect hybrids (Odell & Pickford, 2013).

As online strategies are very capital intensive, small retailers are at a disadvantage in developing such strategies, topic that is further discussed in the chapter by Elizabeth Mack. Instead, large chains and department stores selling fashion and other comparison shopping goods have been the primary drivers of click and collect strategies, backed by their brand recognition, extensive network of shops, and potential to access the necessary investment funds for development. Part of this investment is in adapting the physical retail space, including reformulating shops to act as "showrooms" so that some consumers can peruse products before buying online (Thompson, 2012). Furthermore, some fashion retailers seek to extend the purchasing experience, turning shopping into an "event" and entertainment (Dennis et al., 2002), creating a "destination store mind set" (Katros, 2000, p. 76). This perspective fits the broader shopping centre context, where success depends partly on ensuring appeal to the "leisure-tourist-shopper" rather than just the utilitarian shopper (Howard, 2007). In this context, the next section of this chapter focuses on the adaptive policies of individual fashion and department chain store retailers to the online revolution.

Online Retailing and Retail Real Estate

This section considers the connection between real estate portfolios and organizational strategies of three leading major omnichannel online shopping chains – Next, John Lewis, and Debenhams – based on an updated analysis originally reported in Jones and Livingstone (2018). As the largest fashion store chain in the United Kingdom, Next is omnipresent in many town and city centre malls, but predominantly in out-of-town locations in retail parks. John Lewis and Debenhams had the largest department store networks in the United Kingdom and acted as anchors in urban and suburban shopping centres. Each of these store chains inculcated multichannel and omnichannel shopping as central to their commercial strategy (Jones & Livingstone, 2015). The analysis is based on semi-structured interviews with either the property director or the property acquisitions manager of each company. In addition, the changing locations of leases held by Next in relation to the retail centre hierarchy have been tracked by using CoStar property market data.

The interviews with the three major retailers reveal that the growth in online shopping in the form of click and collect is becoming a key consideration in their real estate decision-making. All three retailers see responding to the rise of online shopping as an essential motive for adjusting their portfolios of stores. Online and physical retailing are seen as having a symbiotic relationship. However, their real estate holdings differ significantly, and the varied ways in which the retailers hold properties in their portfolio, combined with their retail strategies, are a key influence in how they integrate and deliver online retailing. Specifically, retailers' existing portfolio can prove a major constraint on adjusting to the online age, as these three retailers demonstrate.

Next is predominantly a short leasehold business, with a weighted average unexpired lease term of only 6 years across their whole portfolio. This gives them flexibility to adapt to the fluidity of retailing. Since 2003, two-thirds of their new store leases have been in retail parks, with Next often closing town centre branches to move to a retail park. Indeed, their latest preferred store format encompasses male and female fashion, home furnishings, and a coffee shop, which require the flexibility of large units in retail parks. This retail park presence is seen to be "vital" to their online success, in terms of being accessible to consumers. At the other extreme, John Lewis had a 65–70 per cent freehold or long leasehold retail real estate portfolio of 46 stores in 2016, with expansion plans moving to more flexible leases. It expected that the "estate will be 50–55 [per cent freehold or long leasehold], by 2020." The lack of real estate flexibility contributed in part to John Lewis's profits decrease by 45 per cent in 2019 (Jahshan, 2019). Debenhams noted that they open new stores as a "showcase," which it claims can generate an increase of online market penetration in that locality by 30 per cent. However, Debenhams also recognized that, unlike Next, the desired size of their department stores was now 30–40 per cent smaller than before the rise of online shopping, and stores thrive primarily but not exclusively in out-of-town locations such as retail parks. In 2016, Debenhams' property portfolio contained stores on restrictive longer lease terms, which meant it had approximately 1 million square feet of redundant space. This property baggage contributed to its financial problems in 2019 that saw it entered into a Company Voluntary Arrangement (CVA) insolvency process to restructure its business, including rent reductions and at least 22 store closures from its 166 total stores (Butler, 2019). Eventually the business completely failed.

The case studies illustrate that the growth of online sales has brought an imperative for retailers to change not only the amount and design of their physical stores but also their locations, particularly to decentralized locations with automobile parking. This restructuring has prompted department stores to move to smaller shops, while some fashion retailers are conversely seeking to increase their store size, simultaneously reducing the size of their networks. Notwithstanding the financial problems of Debenhams and the plunge in John Lewis's profits, the physical stores remain of considerable importance to the retailers. Retailing is now much more complicated than it was before the arrival of online sales, as online sales complement the in-store experience and enhances profitability. E-commerce is particularly *most* cost effective when customers collect goods from stores (LDC, 2019). The promotion of online sales has involved a reappraisal of the real estate strategies of the three retailers in shop size and location, as click and collect bolsters the continued growth of fashion and comparison goods stores in retail parks.

Socio-economic Forces behind Retail Change

While online shopping has changed British retailers significantly, it is not the only force to affect the viability of physical stores. There has been little growth in British wages since the global financial crisis of 2008, with average incomes in 2019 still below that in 2007. The years 2018 and 2019 have seen a tsunami of insolvency agreements in the retail sector, as major retailers find themselves in financial difficulties. As part of these CVA agreements, retailers renegotiate their leases with their landlords to reduce rents and accept store closures. The scale of rent reductions can range between 30 and 40 per cent, in cases even 50 per cent (Evans, 2018). Rent reductions are not only the outcome of insolvencies: with all its shops on short leases, Next reports an average annual reduction of rents in recent years of 29 per cent (Hipwell, 2019). These reductions also have more general long-term implications for shopping centres and investors, as leases come up for renewal.

In addition to stagnant real incomes and growing online sales competition, further woes were created by the referendum vote in 2016 to leave the European Union (known as Brexit), which has brought a significant devaluation in of the British pound, increasing the cost of imports. Most of Britain's clothing and other retail merchandise is imported, putting great strain on retailers' margins (CBRE, 2019). Structural changes in retail have therefore been accelerated by the cyclical position of the economy and the one-off effect of Brexit. Many leading shopping chains – certainly including fashion – have recently closed or are expected to close a significant proportion of their shops, many of which were and are anchor tenants of small town High Street shopping centres. As of

2019, the national retail vacancy rate was 13 per cent, but it was much higher in small towns in peripheral regions. Out-of-town centres such as retail parks have not been immune to the recent wave of shop closures, but the vacancy rates are much smaller and it has been easier to find replacement tenants (LDC, 2019; Savills, 2019).

The COVID-19 pandemic led to fashion and other non-essential stores being closed for over three months, forcing customers to buy these items online. The financial consequences have been an expansion of existing closure plans with all the main retail chains announcing a further reduction in their store networks. The pandemic also saw an increase in customers shopping for food in neighbourhood stores rather than larger supermarkets, but it is still unclear whether this marks a permanent change in shopping habits. Small stores have also turned to click and deliver as a means to help their businesses survive. Bars and restaurants, previously hailed as partial saviours of many British storefronts, reeled, at least in the short term, under the pressure of social distancing mandates and customer hesitance. Many closed and have not reopened.

Conclusions

The recent history of retailing in the United Kingdom has been dominated by stories of store closures and the rise of online sales. Especially department stores and fashion shops have seen their profits squeezed by stagnant real incomes and Brexit, but most notably the 19 per cent market penetration of online sales. Shopping chains have had to embrace the internet and there has been a swift learning curve. Online retailing has now become an essential element within major retailers' business and therefore real estate strategies, especially as click and collect has become part of the English retail vocabulary as a central strategy for promoting online sales. Through click and collect, physical retailers can maximize the potential of their store network, and a high proportion of other types of online sales still involve an actual store.

Click and collect benefits from easy automobile access, so its overall impact on the urban retail hierarchy has been and will likely continue to decentralize retail. Despite planners' attempts to maintain town centre and suburban High Streets, the automobile has dispersed retail over the past decades, prompting the growth of out-of-town shopping malls and retail parks. Many large sub-regional town and city shopping cores have survived this decentralizing tendency, but especially small towns have been in decline, also partly caused by long distance "out-shopping" to still-vibrant regional centres. Shoppers have either been attracted to the utilitarian retail parks or to town and city centres offering a "day out" experience.

The problems of many traditional centres has been worsened by the internet, not just due to the contraction of retail chains but also to the abandonment of these centres by services such as banks and travel agents, with stark consequences for local footfall. Small and medium sized towns decimated by closures do not have immediate prospects of revival, especially with the imminent loss of many of their traditional anchor fashion and department stores – if still present. However, falling retail rents may provide some salvation. While for the foreseeable future, new small town retail developments will be at best muted because of low capital values diminishing viability, these low values could draw new (start-up) retailers and stimulate the prospect for the redevelopment of parts of these centres into a more mixed use future.

The trends outlined in this chapter demonstrate a long-term structural retail change, prompted by automobile and online technology, and accelerated by Brexit uncertainty, the pandemic, and the lingering fallout of the global financial crisis of 2008. However, the statistics in this chapter suggest that the period of rapid online sales expansion may be coming to an end, as market share is stabilizing. Nevertheless, significant reductions in retail space by major retailers continue to be announced, adding to significant closures that have already occurred. This could be the finale of the retail shake-out that began at the turn of the century. Even if it is not, there is a symbiotic relationship between physical stores and online sales, through online order fulfilment, click and collect, and experiential retail strategies. The COVID-19 crisis may only cement this symbiosis.

While the physical store is sure to survive, the confidence of financial institutions in British retail real estate investment has been challenged, even severely damaged. Future investment in retail real estate is likely to be more cautious, and disinvestment is inevitable. The real estate investment portfolios of British financial institutions are increasingly likely to be focused on regional city shopping centres, the largest out-of-town malls and retail parks. The combination of the restructuring of multiple retailers' store networks and investment portfolios will see an acceleration of the reframing of the hierarchy of retail centres away from traditional High Streets. Accelerated by the COVID-19 crisis, the decline of High Street retail creates an urgent mandate for urban planners and policymakers to counter this process or adapt to it.

REFERENCES

Butler, S. (2019, 9 May). Debenhams wins green light for restructuring plan. *Guardian*. https://www.theguardian.com/business/2019/may/09/debenhams-wins-green-light-for-restructuring-plan

CB Richard Ellis (CBRE). (2014a). *UK retail warehouse parks in the pipeline: Market view.* CBRE.

CB Richard Ellis (CBRE). (2014b). *UK shopping centres in the pipeline: Market view,* CBRE.

CB Richard Ellis (CBRE). (2019). *Weathering the perfect storm in UK retail real estate.* CBRE.

Dawson, J. (2004). *Retail change in Britain during 30 years: The strategic use of economies of scale and scope.* (Research Papers in Retailing 0402). Centre for the Study of Retailing in Scotland, University of Edinburgh.

Dennis, C., Harris, I., & Sandhu, B. (2002). From bricks to clicks: Understanding the e-consumer. *Qualitative Market Research, 5*(4): 281–290. https://doi.org/10.1108/13522750210443236

Durand, C., & Wrigley, N. (2009). Institutional and economic determinants of transnational retailer expansion and performance: A comparative analysis of Wal-Mart and Carrefour. *Environment and Planning A, 41*(7), 1534–1555. https://doi.org/10.1068/a4137

Eppli, M. J., & Benjamin, J. D. (1994). The evolution of shopping center research: A review and analysis. *Journal of Real Estate Research, 9*(1): 5–32. https://doi.org/10.1080/10835547.1994.12090737

Evans, J. (2018, 7 September). The mystery of commercial rents: Boom or bust? *Financial Times.* https://www.ft.com/content/6ba2aee8-68d5-11e8-aee1-39f3459514fd

Evers, D. (2002). The rise (and fall) of national retail planning. *Tijdschrift voor Economische en Sociale Geografie, 93*(1), 107–113. https://doi.org/10.1111/1467-9663.00186

Ferguson, N., & Woods, L. (2010). Travel and mobility. In M. Jenks & C. Jones (Eds.), *Dimensions of the sustainable city* (pp. 53–74). Durdecht: Springer.

Guy, C. M. (2000). From crinkly sheds to fashion parks: The role of financial investment in the transformation of retail parks. *International Review of Retail, Distribution and Consumer Research, 10*(4), 389–400. https://doi.org/10.1080/09593960050138949

Guy, C. M. (2010). Development pressure and retail planning: A study of 20-year change in Cardiff, UK. *International Review of Retail, Distribution and Consumer Research, 20*(1), 119–133. https://doi.org/10.1080/09593960903498250

Hardwick, M. J. (2010). *Mall maker: Victor Gruen, architect of an American dream.* University of Pennsylvania Press.

Hipwell, D. (2019, 22 March). High Street chief has eye on next big thing. *The Times.* https://www.thetimes.co.uk/edition/business/high-street-chief-has-eye-on-next-big-thing-2tgr32mlg

Howard, E. (2007). New shopping centres: Is leisure the answer?. *International Journal of Retail & Distribution Management, 35*(8), 661–672.

Investment Property Databank (IPD). (2014). *IPD UK annual digest.* IPD.

Jahshan, E. (2019, 7 March). John Lewis Partnership staff bonus slashed as profit nosedives 45%. *Retail Gazette.* https://www.retailgazette.co.uk/blog/2019/03 /john-lewis-partnership-staff-bonus-slashed-profit-nosedives-45/

Jones, C. (2014). Land use planning policies and market forces: Utopian aspirations thwarted? *Land Use Policy, 38*(May), 573–579. https://doi .org/10.1016/j.landusepol.2014.01.002

Jones, C. (2017). Spatial economy and the geography of functional economic areas. *Environment and Planning B, 44*(3), 486–503. https://doi.org/10.1177 /0265813516642226

Jones, C., & Livingstone, N. (2015). Emerging implications of online retailing for real estate: Twenty-first century clicks and bricks. *Journal of Corporate Real Estate, 17*(3), 226–239. https://doi.org/10.1108/jcre-12-2014-0033

Jones, C., & Livingstone, N. (2018). The "online High Street" or the High Street online? The implications for the urban retail hierarchy. *The International Review of Retail, Distribution and Consumer Research, 28*(1), 47–63. https://doi.org/10.1080/09593969.2017.1393441

Katros, V. (2000). A note on internet technologies and retail industry trends. *Technology in Society, 22*(1), 75–81. https://doi.org/10.1016/s0160-791x(99)00035-4

Local Data Company (LDC). (2019). *Retail and leisure market analysis 2018 full year.* LDC.

McArthur, E., Weaven, S., & Dant, R. (2016). The evolution of retailing: A meta review of the literature. *Journal of Macromarketing, 36*(3), 272–286. https:// doi.org/10.1177/0276146715602529

Odell, M., & Pickford, J. (2013, 21 November). Amazon eye's gap as tube's 24 hour move signals the end for ticket offices. *Financial Times.*

Office for National Statistics (ONS). (2014). Retail sales, April 2014.

Office for National Statistics (ONS). (2019a). Real net households adjusted disposable income per capita CVM SA. https://www.ons.gov.uk/economy /grossdomesticproductgdp/timeseries/crsf/ukea

Office for National Statistics (ONS). (2019b). Retail sales index internet sales.

Portas, M. (2011). *The Portas review: An independent review into the future of our High Streets.* Department for Communities and Local Government.

Savills. (2019). *UK Retail Warehousing, Spring 2019.* Savills.

Schiller, R. (1986, July). Retail decentralisation: The coming of the third wave. *The Planner,* 13–15.

Teale, M. 2013. *The internet leviathan slows.* CBRE.

Thomas, C., Bromley, R., & Tallon, A. (2006). New "High Streets" in the suburbs? The growing competitive impact of evolving retail parks. *International Review of Retail, Distribution and Consumer Research, 16*(1), 43–68. https://doi.org/10.1080 /09593960500453484

Thompson, C. (2012, 28 October). Shops will be "little more than showrooms." *Financial Times.* http://www.ft.com/cms/s/0/e6ef4454-2122-11e2-babb -00144feabdc0.html

PART THREE

The Survival of Mom-and-Pops

7 Small Business Survival: How and Why?

VIKAS MEHTA

The Small-Business Paradox

A wide-ranging body of knowledge in economics, retailing, political science, sociology, American studies, and urban planning argues for the benefits of small independent businesses. It is well established that small businesses are a key component of any economy.[1] In the United States, small businesses have played a vital role in the creation of the American culture of entrepreneurship, self-reliance, and individualism, along with economic development. It is estimated that nearly half of every $100 spent at local businesses stays in the local economy, while a national chain leaves only 13 per cent locally (Cunningham, 2008). In the United States, a significant share of all jobs is still in the small-business sector, which continues to provide opportunities for women and minorities (Blackford, 2003). Small businesses contribute tremendously to innovation in product design and development as well as commercial processes.

Small businesses also sustain local economies by buying from local suppliers and hiring local residents (Carmona, 2015; Fitzgerald & Muske, 2016). There are other social benefits to the neighbourhoods that support small businesses. By way of their small scale, small businesses deliver more interactive facades through increased variety and difference, which provide visual interest and opportunities for social interaction (Mehta, 2011; Heffernan et al., 2014; Mehta & Bosson, 2018). Compared to chain stores, small businesses are also more likely to be considered "third places" in neighbourhoods that provide space for social exchange (Mehta, 2011). Small businesses are often integral to the local community in a manner that is mutually beneficial. For example, Besser (1999), found that a supportive relationship and commitment to local community is associated with perceived business success of small businesses.

The proclamation of the death of small business has almost outlasted the lifetime of independent and mom-and-pop stores. In the United States, the demise of independent retailers started over a century ago by department stores, followed by big-box retailers and now by e-commerce. The contemporary general sentiment in the small-business-owner community aligns with this demise, particularly as they see a clear threat from e-commerce giants. According to a national survey by Advocates for Independent Business and the Institute for Local Self-Reliance (2017), independent business owners say that Amazon is the number one threat to their survival. This situation is not likely to change soon, as the global volume of e-commerce sales has been on a steady increase. Stories of store closures, large and small, abound. Yet some attribute store closings to more of a rightsizing rather than a result of loss of business or clientele, noting that the United States has a disproportionately high per capita amount of retail space. At the same time, 90 per cent of retail sales in the United States still occur at a physical location, only declining to 75 per cent by the end of the 2020s according to industry experts (Townsend, 2018). In fact, many urban researchers point to and predict a further resurgence in urban retail, responding to demographic shifts and the changing nature of retail consumption (Lipsman et al., 2018). Not only are younger, highly educated Americans moving back to urban cores (Florida, 2012), they value physical retail locations as part of their urban lifestyle. As several chapters in this book have noted, the urban storefront is shifting from a transactional to an experiential space, as offline and online brands cater to consumer needs for identity, social relationships, and belonging – both online and in physical space (Gilovich & Kumar, 2015; Goldman et al., 2017). Stores now offer goods and services as a part of a larger experience, be it cultural, recreational, social, or educational.

While these cultural and economic forces, trends, and changes are well known, the response of small-business owners to these factors is largely unknown. This chapter surveys the challenges and opportunities for small retail businesses, as well as the influencing factors and strategies used by small businesses to survive in the age of big-box retailers, e-commerce, and the current experience economy. What do the small-business owners recognize as their main advantage and offerings to their customers? And what mechanisms do they employ to achieve and sustain these advantages? How do they choose where to locate? What social, political, marketing, digital, social media, physical design, and other strategies do they use to attract and retain their clientele? What are the challenges they face? Which policies support such stores and which ones seem detrimental to their survival? And finally, what trends do the business owners see in the future? Broadly speaking, our study aims to unveil *how* small businesses survive in the current retail landscape.

First-Hand Report

Although numerous disciplines – many of which feature in this volume – address many of these questions from the perspective of academics, developers, or designers, this chapter focuses on answering them from the first-hand perspective of the business owners. Using interviews with two dozen independent businesses, this chapter illustrates how and why small businesses are surviving. The interview format also provides insights into the real concerns of and strategies employed by the people who own and operate small businesses. The interviews focus on small businesses that sell goods and services that are in direct competition with e-commerce retailers such as Amazon and big-box stores. Many of these businesses have been eliminated, and the surviving ones remain under constant threat. Yet others have completely reinvented themselves and created new market niches. We studied and interviewed owners of businesses selling a variety of items including apparel and footwear, activewear, jewellery, homewares, antiques, furniture, wine and beer, toys and hobby supplies, books, specialized stationery, hardware, and paint. We studied their websites and other social media presence such as Facebook and Instagram, and we researched the organizations that the businesses and business owners belong to. Many of our conclusions come from the analysis of their verbal and written discourse using established qualitative research methods in the social sciences, such as thematic analysis (Miles & Huberman, 1994; Braun & Clarke, 2006). As this study focuses on small businesses that compete with chains and online retailers, it purposely does not include shops that are primarily in the business of serving prepared food and drinks – with the exception of a wine shop and a coffee shop that both sold products that were also available at online retailers.

Study Areas

This study was conducted in Cincinnati, Ohio, one of the cities in North America that has many distinct neighbourhoods with a strong identity. These neighbourhoods are anchored by commercial corridors and business districts – also known as neighbourhood business districts (NBDs) – that provide a wide range of services and amenities. Of Cincinnati's 52 neighbourhoods, 32 have designated NBDs. Decades of suburbanization and urban decay typical of rust-belt cities has fuelled deterioration, vacancy, and blight in a majority of Cincinnati's NBDs. The City of Cincinnati has created several initiatives to counter this trend, such as the Neighborhood Business District Improvement Program and the Neighborhood Business District Improvement Support Fund, which provide financial support for maintenance and improvements of the NBDs. Regardless of these initiatives, the conditions of Cincinnati's NBDs range from very urban, walkable, and vibrant to automobile dependent and distinctly suburban, or even mostly vacant. To focus on small businesses that are surviving the changing retail landscape, we interviewed business owners in neighbourhoods that are mostly walkable and vibrant.

Characteristics and Qualities

We consider most of the business owners we interviewed as entrepreneurs, as they aim to operate a profitable business – in contrast to professional derision of some small-business owners as hobbyists who obstruct main street success (Gibbs, 2012). Besides being entrepreneurs, however, the majority of these business owners have other identities, traits, and qualities that they project and weave into their business. We found that most small-business owners are civic-minded, optimistic, and benevolent. Most are proactive citizens who, in some way or another, believe in doing their part in positively affecting their communities, neighbourhoods, and cities. Business owners demonstrate this belief in their practices, products, and services, and the image of their business. For many, their civic beliefs are paramount to owning their business. Through our interviews, we have distinguished five main types of small-business owners, all with their own identities, characteristics, and qualities. We classify them as *urbanists and place-makers, communitarians and community builders, passionate hobbyists, storytellers and people-persons, and wholesome sustainables.*

Urbanists and Place-Makers

Since all interview participants in this study were located on mixed-use commercial streets in urban neighbourhoods, we found that many displayed an attachment and affection to their physical environment. We call them *urbanists and place-makers.* These business owners were deeply attached to their place of business, had immense knowledge of it, and cared for the quality of place, from simply routinely maintaining or cleaning their street to helping to plan for making the neighbourhood more walkable and people-friendly.

Many chose to locate their business equally on the basis of their association and attachment to place, and on foot traffic or rent affordability.

Often these sentiments have nostalgic or personal undertones. One business owner writes on the website,

> I located the shop on Ludlow Avenue because of the connection I feel to the neighborhood ... When I walk the beautiful tree-lined streets, admire the old homes or even enjoy a meal, I like to think I'm having the same experiences [my parents] did. They're definitely in my heart as my vintage shop and new life adventure takes root.

Some recognized the role that businesses play in the neighbourhood, as another store owner explained in an interview:

> There's a long history of retail in Walnut Hills [a neighbourhood business district that was known as Cincinnati's second downtown in the early twentieth century]. My business partner grew up in Walnut Hills and we want to contribute to the renaissance.

A game store owner brought up a similar reason to locate: "I wanted to be in an urban redeveloping area. I grew up in the neighbourhood and went to high school nearby. I have friends that own the hair salon nearby." This business owner clearly articulated a preference for the urban: "We wanted to be in an urban area between the University of Cincinnati and Xavier [University] ... Other game stores are in the suburbs but we wanted to be in the city."

Urbanists and place-makers understood the importance of neighbourhood business districts and the unique ambience of such places (Figure 7.1). One business owner said, "People want lots of character in neighbourhood shopping districts. The character and uniqueness of the area must be maintained." In a recent story about ideas for making the neighbourhood commercial street more pedestrian-friendly, this store owner said to the reporter, "In a district where we tout the walkability of the business district, and we talk about how it's part of our lifestyle here, it's really important for us to maintain that."

Communitarians and Community-Builders

Numerous business owners were actively interested in using their business as a vehicle to further the well-being of their community. We call them *communitarians and community-builders*. While we did not interview restaurant, coffee shop, or bar owners – "third place" community spaces, as coined by Oldenburg (1991) – we were surprised to find some non-food businesses that were perceived and used as neighbourhood gathering places.

Figure 7.1. Business owners recognized the unique character of the local streets. (Credit: Photo courtesy the author.)

Records of neighbourhood organizations reveal that *communitarians and community-builders* not only play an active role in business associations, they also participate and lead subcommittees in neighbourhood community groups. These business owners were active participants in organizing small and large neighbourhood events that brought the community together. One shop owner regularly organized fundraisers for the community and had a place for voter registration forms in front of the show window. *Communitarians and community-builders* recognized the role that their business played as a space for the community and the reciprocal benefit. "People have a need for community space. So the more businesses can provide that [community space], it will contribute to their success," noted one. The businesses providing goods and services that were inherently social created easy mechanisms for coming together, such as a gaming store that organizes multiple events and offers a "Play Pass" where patrons can pay a daily or monthly membership fee to use the space. Those fees go directly towards a store credit to purchase games with. "The main advantage is having a community space," explained the owner. In another case, a storeowner opens her store and serves champagne to the public for gathering and socializing during a weekly business block party on Fridays. The business owners of another store explain on their website, "We know a business can't grow without the love and support of its community, so we make sure to return the love you've shown to us." Similarly, one of the "intrinsic ideas" listed for a bakery states, "Engagement with our community is mandatory."

Passionate Hobbyists

Although most business owners we interviewed aimed to generate revenue from their stores, some operated their business as their hobby. Some had enough money to operate the store and even absorb small losses, while others had a separate part- or full-time job to supplement their income. One merchant we spoke to used to manage retail for college bookstores and saw that others shared his interest in gaming. Burnt out on his job and looking to connect with other gamers, he opened his own gaming shop. In another case, a bookshop owner in his mid-70s and in business for decades remained as enthusiastic as on his first day of business, exclaiming, "I love my job. I work 6 days a week." Different hobbies have generated a wide range of neighbourhood stores, selling items as varied as antiques, games, books, fashion, jewellery, and wine.

Storytellers and People-Persons

Aside from being passionate about what goods and services they offer, a large number of business owners operate a business to tell their story. They also highly value human contact to tell this story. We call them *storytellers and people-persons*. Conversations occur not only in stores, as social media and other digital means provide additional platforms for storytelling. Of the businesses that have websites, many have spent time in crafting messages that tell their story to the customer. The website for a store that sells African heritage goods, arts, and crafts, apparel, and jewellery tells the story of the business owner and his family.

> The owner of the store is a wood carver who was born and raised in a small woodcarving village of Wamunyu in eastern Kenya. His [father and grandfather] were woodcarvers. [He] spent his early years working with his father at a family gallery in Nairobi and later worked in various galleries in The Netherlands and Switzerland before coming to the US. [He] travels to Africa every year to buy and collect new products for the store.

Even as they use digital platforms to tell stories, most business owners we interviewed are more interested in conversations at their physical location (Figure 7.2). One owner specifically prefers in-person business over

Figure 7.2. Using branding, signage, and art, many businesses convey stories through their stores, websites, and social media. These business owners were also keenly interested in telling stories of their business and the goods and services they provided. (Credit: Photos courtesy of the author.)

online sales, as she likes to "talk to people and hear their stories – that's the best part." *Storytellers and people-persons* understand storytelling as a part of their customer service. One of them sources his jewellery and artefacts from makers in Mexico whom he knows very well. He regards his stories of the craftspeople and independent artists as well as their jewellery and artefacts as a service that brings joy to his customers – and one of his main advantages. For him it is a priority "to treat everyone in the store specially and give them enjoyment by telling these stories."

Wholesome Sustainables

The experience economy has co-opted wide-ranging issues such as health and well-being, everyday sustainability and environmentalism, ethical business practices, education, and localism. Many small-business owners have aligned their values with these issues and created an image, products, and services that reflect their identities. We call them *wholesome sustainables.*

For example, working directly with farmers to source products has become a common practice in the business community, but these business owners go beyond. After a local coffee shop started importing coffee from Guatemala, they established relationships with a group of farmers and built their business relationship into development work "to improve their products and quality of life" of the farmers. The coffee-shop owners went further and created a non-profit to help the farmers "in a more direct and tangible way" by "enabling our farmers all the way to owning their own processing facility" and "to strategize how we can repeat this model with other countries." In presenting their approach, the coffee-shop owners state, "Our commitment to this excellence starts by partnering with communities of small coffee farmers and trying to improve their quality of life through the purchase of their coffee and collaboration on quality improvement."

Similarly, a local children's bookshop extended beyond reading and took on the issues of the value of real-world experiences in aiding learning and development. The bookshop states,

> We seek to provide the best in children's books and related activities, handpicked for their potential to engage young minds and stand the test of time. We resist the commercialization and digitization of children's "content," [and] we reject the notion of omnipotent celebrity … Most of all, we embrace the glorious analog brain, and the importance of reading, conversation, and real-world experience for children to reach their full potential.

Further highlighting their role in the community and the environment, they say,

> We will strive to provide engaging events tailored to our local audience. We will develop affordable, innovative programs hosted by passionate people, and continually improve them. We will seek partnerships with area schools and non-profit groups to benefit those in need. We will source sustainable products, recycle, and operate in the most energy-efficient way possible. And most of all, we will seek to be a destination for families, an active participant in our community, and a fun place to work; a unique, inspiring place where kids will take their own children someday.

Another *wholesome sustainable,* the seller of heritage furniture, explained, "Our mission is to share our heritage with the rest of the world. Providing a reasonable outlet for the artists to market their products without the common exploitation that is associated with sweatshops." As another example, a lady who grew up in the neighbourhood combined her passion and expertise in health and wellness to open an athletic-wear store. She sees her strength and advantage over e-retailers as her clients come to her not only for the high-quality clothing but also because she advises them on healthy diet, exercise, and wellness. These examples capture the range and types of issues *wholesome sustainables* care for and bring to the business. The website of the aforementioned coffee shop summarizes their overall attitude: "We believe that people return to what is good, and we strive to be good for the people."

Strategies: Key to Survival

One of the biggest advantages of small businesses is their ability to provide unique, customized products and an exceptional and personalized service. In a recent survey of over 850 American independent retailers in 49 states, a majority report that they have distinct attributes that appeal to consumers and give them an edge over chains as the retail landscape shifts. These include their deep expertise, highly personalized service, ability to create community, and rewarding in-store experience – insights that corroborate other research findings (AIB & ILSR, 2017; McGee & Peterson, 2000). In our interviews, we found the business owners reporting similar attributes as their advantage over e-retailers and chain stores. However, we also found other strategies that give small businesses a distinct edge. In order to comprehensively understand the strategies, we classify these as *image, location, appearance and design, partnership and collaboration, combination,* and *engagement and involvement.*

Image

Image may be understood as an all-encompassing message that a business delivers through its location, size, appearance and design, goods and services it provides, customers it caters to, employees it hires and how they are treated, values it projects and adheres to, and practices it employs (Figure 7.3). However, we use image specifically to showcase how a business projects itself and its overarching values through messages and communication. The majority of our studied businesses focused intently on their image, which covered many of the aspects of their identities discussed earlier in this chapter. Excerpts from their websites and social media show the wide range of images projected by businesses:

"undeniable focal point of the community"

"ecosystem of color and creativity"

"we resist commercialization and digitization of children's content"

"help you create a better life at a price you can afford"

"our mission is to share our heritage without exploitation of labor"

"importance of reading, conversation and real-world experience"

"place for families to get together"

"create unique experiences that grow communities"

"simple philosophy of satisfying customers with quality and value"

"striving for excellence in ethical sourcing and crafting"

Location

While the right location has always been key to retail success, the heightened pressure on independent urban retailers due to increased competition and dwindling demand for urban stores has propelled location to a precondition for survival. More than ever, business owners are acutely aware of the need for locating in places that are visible, physically central, and in the path of flows – pedestrian and vehicular, local and regional – and where applicable, culturally significant. Several business descriptions capture how businesses value their location, including proximity to compatible business neighbours. "We are located in the center of the Gaslight Clifton neighborhood of Cincinnati, Ohio, known for its sense of community and pride in diversity … Ludlow Wines is located … one block from The Esquire Theater next to The Proud Rooster Restaurant and across from the history-making rock music venue The Ludlow Garage."

Figure 7.3. Wine shop conveys an image of a welcoming, inclusive place. (Credit: Photo courtesy of the author.)

In our interviews, we found almost all business owners comment on the virtues of their location or relocation, with respect to clustering, visibility, and flows – even at the higher rents some had to pay.

Appearance and Design

The appearance and design of retailers' physical stores – both toward the street and on the sales floor – are crucial to their success, anchoring their online presence as well as any stationery or giveaways. The ability to provide a sensory experience through window displays, signage, sidewalk signs, and other artefacts lends storeowners a means for attracting customers (Figure 7.4). These are carefully crafted to stand out from the rest, project their values, align with or create new trends, and appeal to

Figure 7.4. Vintage shop storefronts convey a unique but cohesive appearance and design. (Credit: Photos courtesy of the author.)

their customer base. Most of the businesses we studied put great effort into their appearance and design in whatever mediums they were comfortable. Some hired designers to decorate their show windows, while others had participated in programs to redesign their signage. One storeowner talked about keeping an attractive window display and changing it every two weeks as his main advantage in attracting people who walked by. During the neighbourhood festival, he hires artists to paint thematic imagery on the sidewalk outside the store to extend his reach.

Partnership and Collaboration

Building on the strategy of location, business owners perceived a clear advantage in collaboration with other businesses. They were very aware that varied businesses complement each other in ways that benefit most – corroborating nearly century-old theories explained in Conrad Kickert's chapter in this volume (Christaller, 1933/1966; Nelson, 1958; Reilly, 1929). Somewhat unexpected, we found that almost all the discussion about competition was related to e-commerce and no one mentioned other nearby stores as a threat. The overall spirit was to collaborate, not compete, and that more businesses were better for all stores in the neighbourhood – even if they were competitive. In the words of one, "More is more. Add more restaurants and reasons for people to drive by and see the shops. Business synergies are nice too." Citing examples of the complementarity and synergies, a custom stationery business owner explained, "Parlour does hair for weddings, O Pie O caters weddings, Urbana Café does coffee carts for weddings."

We found partnerships and collaborations such as a custom furniture store with an interior decoration store, a bookshop with local area schools and non-profits, a fruit and vegetables seller at a wine shop, and even some curated events based on partnerships, such as Saturday breakfast at the furniture store with food from the local bakery and art show by local artists.

Combination

While partnership and cooperation leverage the synergies between compatible business owners in a district, many owners combine compatible goods and services *within* their stores as well.

Owners can combine goods and services that are either complementary to the main business or completely outside the range of the main business – a strategy that also features in Heather Arnold's chapter in this volume. Some examples of complementary combinations include offering first-hand health and wellness expertise within an outerwear store, and a printing service in a specialized stationery store. More surprising combinations include a notary service at a wine shop, art for sale at an exclusive apparel, jewellery, accessories store, and baked goods at a hardware store. The owner of the hardware store explained how this combination helps another small business but also his own – offering a baker new shelf space after a local grocery store closed, but also bringing hungry patrons to his hardware store.

Engagement and Involvement

Personal relationships and friendliness are common traits of most small-business owners and employees. As expected, most business owners valued providing good customer service and understood that "it is key to creating repeat customers." A quarter of the business owners stated confidently, "I know all of my customers" by name, and they greeted everyone. We witnessed one talking to a customer about another: "Oh she just got back from vacation in Ireland. Her daughter just started at Xavier [University]." Most business owners were also very engaged in the community, for example by organizing events for all customers or by accommodating selected groups. The hardware store organizes contractor parties, and the furniture store lets the showroom be used by disabled artists to sell without any commissions. Several other business owners were very involved with local artists, inviting artists to sell in their shops on a rotating basis. One storeowner stated, "As a small business and entrepreneurial endeavour, our focus has always been on helping other small businesses. This means we support small-production wineries and distributors, local breweries and distilleries, and local farmers and producers." This spirit resembles the bookstore's deep engagement with local schools and neighbourhood families as an integral part of their mission to aid children's education and growth.

Future: Bricks or Clicks?

The business owners we interviewed had a range of opinions on the future of small business, retail in general, and brick and mortar stores. Some were oblivious to "the retail apocalypse" while a few, mostly elderly storeowners, were lamenting that "Amazon will wipe out small businesses" and "small-business brick and mortar is a dinosaur." The majority, however, were bullish on the future of small businesses, on the basis of changing trends. One stated, "Quality is increasingly in demand. Consumers will buy less but buy higher quality." Clairvoyant about the future role of the small business in the experience economy, another storeowner proclaimed, "People still want to shop and to interact with people and ask for advice. When people are shopping outside of their home, they want pizazz. The boring shopping, they will do online. Big-box retailers will all close eventually and it will only be online retail for commodities and small business for fun." Most explained that they had adapted to the combined future of "bricks and clicks" even though they clearly articulated the value of their physical store. The bookstore owner cited a key characteristic of the experience economy to demonstrate his optimism about the future of books and his physical store: "People always love to hold a book, love to sit in a favourite chair and read a book." Yet he acknowledged doing a sizeable business selling through third parties on Amazon and eBay to keep up. "You can't have that same experience online," explained another owner, while acknowledging that she is opening an Etsy store beyond her presence on Facebook and Instagram, as online sales are crucial to expanding her audience. Explaining the value of bricks and clicks, she said, "As soon as I post a picture of something, it sells." Corroborating the findings by Elizabeth Mack in this volume, other business owners used their websites and online presence to drive traffic to their physical location rather than to sell items directly. For example, an owner uses Facebook and Next Door, a local neighbourhood social feed, as outlets to tell a teaser story to get people into the store.

Favourable Conditions

In many ways, our research projects a bright future for local small businesses. Most of the business owners we interviewed are entrepreneurs who spend vast economic and emotional resources to do what they love. However, most of these businesses are thriving in a relatively beneficial milieu, as they are located in relatively affluent neighbourhoods of Cincinnati that are walkable and conducive to local businesses. The nearby populations are inclined to support the local shops in their neighbourhoods since they view them as crucial to their quality of life and property values – and they have the disposable income to match their perspective. Furthermore, some shops benefit from affordable space. While some storeowners acknowledged that rising rents forced them to relocate or downsize, a handful explained that they could stay in business only because of community-oriented landlords who are interested in supporting local businesses. One landlord has linked a business owner's profits to rent, another is a local non-profit institution with no interest in profiting from rent. Another storeowner explained that the historic fabric of many urban neighbourhoods comprises small, sometimes quirky retail spaces that are unattractive to chain stores, thus reducing competition and rent even in very desirable neighbourhoods.

Nevertheless, we must recognize that our conclusions derive from a relatively small sample of businesses in one North American city. Although perspectives and inclinations from these few businesses inform us about broader motivations of small-business owners, it is prudent to expand similar research to other urban retail conditions that are markedly different from the ones examined in this chapter. Aside from the contextual differences, we will also have to examine the resilience of small businesses in the face of other significant unpredictable events such as the global pandemic that we face.

Conclusions: *Why,* Not *How!*

When we began this research, we were interested in learning from small-business owners *how* they are able to survive in the age of big-box retailers and e-commerce. As we talked to the business owners, it became clear that their responses were leading toward a question more important than methods of survival. We were learning *why* these individuals were in business. We learned that these businesses were contributing to the quality of life of neighbourhoods, communities, and people – by providing more than just amenities within easy access. Business owners and business associations advocate for neighbourhood improvements and support events throughout the year that enable people of the neighbourhood to interact. Many businesses act as a public canvas for the neighbourhood messages and voices, whether social, recreational, or political. By collectively sustaining a neighbourhood commercial space, these businesses create a centredness in the neighbourhood, producing a public place for daily socializing, thus helping build a sense of community. By providing a vibrant, ever-changing physical setting that also serves as a social space, local small businesses collectively contribute to the identity and social capital of the neighbourhood.

NOTE

1 The US Small Business Administration considers retail firms with 100 or fewer employees small businesses (SBA, n.d.). Although this definition, based on number of employees, is useful, many Americans associate small businesses with their own experience of mom-and-pops and independent stores. In this definition, the relationship between the business and its local community as well as the internal organization and hierarchies are the defining features of the small business in contrast to large businesses. This functional definition of a small business is used for this chapter.

REFERENCES

Advocates for Independent Business and the Institute for Local Self-Reliance. (2017). *Independent retailers and the changing retail landscape: Findings from a national survey.* https://ilsr.org/wp-content/uploads/2017/11/2017 _SurveyFindings_ChangingRetailLandscape.pdf

Besser, T. L. (1999). Community involvement and the perception of success among small business operators in small towns. *Journal of Small Business Management, 37*(4), 16–29.

Blackford, M. (2003). *A history of small business in America.* University of North Carolina Press.

Braun, V., & Clarke, V. (2006). Using thematic analysis in psychology. *Qualitative Research in Psychology, 3*(2), 77–101. https://doi.org/10.1191/1478088706qp063oa

Carmona, M. (2015). London's local high streets: The problems, potential and complexities of mixed street corridors. *Progress in Planning, 100*(August), 1–84. https://doi.org/10.1016/j.progress.2014.03.001

Christaller, W. (1966). *Central places in southern Germany* (C. W. Baskin, Trans.). Prentice Hall. (Original work published 1933)

Cunningham, M. (2008). Economic benefits of locally owned stores. In D. Farr (Ed.), *Sustainable urbanism: Urban design with nature* (pp.144–145). John Wiley and Sons.

Fitzgerald, M. A., & Muske, G. (2016). Family businesses and community development: The role of small business owners and entrepreneurs. *Community Development, 47*(4), 412–430. https://doi.org/10.1080/15575330 .2015.1133683

Florida, R. (2012). Death and life of downtown shopping districts. City Lab. https://www.citylab.com/life/2012/06/death-and-life-downtown-shopping -districts/1925/

Gibbs, R. J. (2012). *Principles of urban retail planning and development.* John Wiley & Sons.

Gilovich, T., & Kumar, A. (2015). We'll always have Paris: The hedonic payoff from experiential and material investments. *Advances in Experimental Social Psychology, 51*, 147–187. https://doi.org/10.1016/bs.aesp.2014.10.002

Goldman, D., Marchessou, S., & Teichner, W. (2017). Cashing in on the US experience economy. McKinsey & Company. https://www.mckinsey.com /industries/private-equity-and-principal-investors/our-insights/cashing-in -on-the-us-experience-economy

Heffernan, E., Heffernan, T, & Pan, W. (2014). The relationship between the quality of active frontages and public perceptions of public spaces. *Urban Design International, 19*(1), 92–102. https://doi.org/10.1057/udi.2013.16

Lipsman, A., Cakebread, C., Cheung, M.-C., Rotondo, A., & Wurmser, Y. (2018). *The future of retail 2019.* New York, NY: eMarketer.

McGee, J. E., & Peterson, M. (2000). Toward the development of measures of distinctive competencies among small independent retailers. *Journal of Small Business Management, 38*(2), 19–33.

Mehta, V. (2011). Small businesses and the vitality of Main Street. *Journal of Architectural and Planning Research, 28*(4), 271–291.

Mehta, V., & Bosson, J. K. (2018). Revisiting lively streets: Social interactions in public space. *Journal of Planning Education and Research, 41*(2). https://doi.org/10.1177/0739456x18781453

Miles, M., & Huberman, A. (1994). *Qualitative data analysis: An expanded sourcebook* (2nd ed.). Sage.

Nelson, R. L. (1958). *The selection of retail locations.* FW Dodge Corporation.

Oldenburg, R. (1991). *The great good place: Cafes, coffee shops, bookstores, bars, hair salons, and other hangouts at the heart of a community.* Marlowe.

Reilly, W. (1929). Methods for the study of retail relationships. *University of Texas Bulletin 2944,* 1–9.

Townsend, M. (2018). Warby Parker and Casper are coming to a strip mall near you. Bloomberg. https://www.bloomberg.com/news/articles/2018-10-22/warby-parker-and-casper-are-coming-to-a-strip-mall-near-you#xj4y7vzkg

US Small Business Administration (SBA). (n.d.). What is a small business. https://www.sba.gov/blog/does-your-small-business-qualify

8 Can Mom-and-Pop Stores Survive? A Survey of Small Retailers in Chicago

EMILY TALEN

In recent years, the retail sector has been described as being in a state of crisis. One in ten American workers is employed in retail, and thus the store and mall closures, bankruptcies, and job losses are being compared to the upheavals that followed deindustrialization in the latter half of the twentieth century (Corkery, 2017; Peterson, 2017). It is possible that this contraction was inevitable, as the United States has 24 square feet more retail space per capita than any other country (Melaniphy, 2018), but the social, economic, and even political impacts of these dramatic changes in the consumption economy cannot be ignored.

The economic effects (on jobs and tax revenue) of retail transformation are one dimension of change; for urban planning, the effect on urban quality – street activation, neighbourhood servicing, social and economic connectedness – is an additional outcome to address. Small-scale, independent, and neighbourhood-oriented retail – what is often called the "mom-and-pop" store – has long been heralded as an important dimension of urban quality. Mom-and-pop stores are favoured for their neighbourhood-serving orientation and their perceived ability to promote sense of community, social connection, and safety and security ("eyes on the street"; Jacobs, 1961). These views persist despite critiques that Jane Jacobs' vision of activated streetlife via small-business ownership is a nostalgic illusion (Gopnik, 2016; Teaford, 2013).

This chapter reports on a survey of 75 small, independent retailers – mom-and-pop stores – in Chicago. We wanted to know, from the perspective of mom-and-pop store owners and employees directly, their perception of the changing landscape of retailing. Two dominant narratives about retailing are that, first, small, independent shops exert a positive impact on streetlife, economic resilience, and social connection, and second, that mom-and-pop retail is threatened by e-commerce, commercial gentrification, and shifting consumer preferences and behaviours. The survey was structured around these two themes: context and connection, and small-business challenges.

We found that most small businesses feel they are a part of their neighbourhoods even as their clientele is not neighbourhood-based, most do not belong to an organization, and owners and employees tend to live outside the neighbourhood. We also found that most small retailers do not feel threatened by e-commerce. Complaints were mostly confined to specific issues like parking and licensing fees, as opposed to larger issues like gentrification or government regulation more generally. And, despite the common narrative that retail is imploding and mom-and-pop stores are dying, our respondents were overwhelmingly optimistic about the future.

Research on Small Retailers

Several surveys of small retailers have emerged in the past decade. Manhattan Borough President Gale Brewer surveyed local retailers in New York and cited several factors that most stressful to small businesses: commercial rent, lack of availability of small retail spaces, more affluent clientele, zoning rules, code inspectors with a "guilty until proven innocent" attitude, and Business Improvement Districts (BIDs) that cater to property owners rather than business owners (Brewer, 2015). Surveys of small businesses might also focus on their social networks, the "entrepreneurial ecosystems" of cities and small towns (Roundy, 2018), and whether factors such as social capital, entrepreneurialism, and the "strategic management of community" can improve small-business success (Campbell & Park, 2016, p. 302).

Scholars working in the domain of "entrepreneurship research" have been interested in the stressors and community relations of family-owned businesses (e.g., Nguyen & Sawang, 2015), particularly how family owned businesses view community social responsibility in relation to corporate owners (Razalan et al., 2017). One survey of small businesses in Iowa found that "perceived business success is significantly associated with business operators' support for and commitment to the community" (Besser, 1999, p. 16). A survey of the "motivations and meanings" of local shopping from the perspective of both retailers and shoppers found that local shopping was "an act of civic engagement," context and place quality mattered, and "camaraderie" amongst small retailers in a revitalizing downtown was significant (Wilson, 2018).

Valuing Small Retailers

Varying theories have supported small-scale retail, especially when positioned along a pedestrian-oriented main street. First, small-scale urban retail is thought to contribute to street activation. Jane Jacobs (1961) famously argued that the exchange fostered by small retailers played an essential role in street activation, as residents interact with a variety of actors. Kickert (2016) explored how small-scale retail promotes facade articulation and transparency, revealing active uses within. The "active frontage" of small businesses is believed to promote social exchange, as "sidewalk contacts are the small change from which a city's wealth of public life may grow" (Jacobs, 1961, p. 72; see also Heffernan et al., 2014).

Second, small retailers are valued for the localized economic connections they produce. Independently owned small businesses recycle local resources by purchasing supplies from local distributers and employing local residents, thus providing stability and resilience and a "basic core" that stimulates local economic interdependence (Fitzgerald & Muske, 2016; Carmona, 2015). Corporate retailing, in contrast, tends to work against local connectedness, "funnelling previously diffuse demand to a central location" (Ellickson & Grieco, 2013, p. 1).

Retail studies conducted in 10 American cities showed that the local economic benefit of independent stores, taken together, was 3.7 times more than the benefit of shopping at chains (American Independent Business Alliance, n.d.). Smick (2017) argued that "main street capitalism" is an economic equalizer, offering wider opportunity for entrepreneurship and economic growth. Local ownership has demonstrated economic value in that it creates jobs (Mack, 2015; Gomez et al., 2015) and counteracts the shift from labour-intensive to capital-intensive retailing that chain store growth fosters. The calculation of a local economic premium by groups like the American Independent Business Alliance and Civic Economics have amassed multiple supporting statistics, such as that 68 out of 100 dollars spent at a local business remain in the community, while just 43 dollars spent at a chain store do so.

A third main category of small-retail valuation concerns social connection. Small retailers provide a social function that helps connect neighbourhood residents (Hester, 2016) and helps sustain local business (Meltzer, 2016). Oldenburg (1999) tapped the social and cultural significance of these "third places" and their ability to promote social connection and neighbourhood stability (Yancy & Ericksen, 1979) above chain stores, which tend to be more footloose (Meltzer & Capperis, 2017). This is where "lowly, unpurposeful and random" face-to-face contact happens (Jacobs, 1961), creating a mutually reinforcing context beneficial to both residents and businesses.

The importance of social contact for casual or spontaneous interaction has been associated with social ties that are "weak" but essential (Granovetter, 1983; Skjaeveland & Garling, 1997; Kaplan & Kaplan, 1989). Some studies have explored the specific role played by Black and ethnic businesses in the redevelopment of marginalized neighbourhoods, such as how barbershops contribute to community building (Mills, 2013), or how small ethnic retailers have successfully merged identity, business acumen, and politics to contribute to neighbourhood regeneration (Krohn-Hansen, 2013; Taylor, 2018).

In sum, small businesses are believed to be the essence of what it means to live local, exerting a significant influence "on the social, cultural, and economic life of the neighborhood" (Meltzer, 2016, p. 59; see also Hyra, 2008; Cachinho, 2014). One study of neighbourhoods adjacent to commercial streets with small retailers found "significantly higher sense of community" than those that lacked this adjacency (Pendola & Gen, 2008). Research in the Netherlands found that the strongest predictor of "community" was the sharing of resources and common activities, in turn fostering mutual interdependence (Völker et al., 2007). Small retail taps "identity values" that extend beyond mere service provision (Corinna, 2011, p. 227; Carrigan et al., 2011). These connections have been shown to encourage certain positive behaviours, including sustainability practices and improved health outcomes (Mitchell, 2009; Samujh, 2011; Bader et al., 2013).

The Decline of Small Retailers

This valuation has not stopped the decline of small-scale retailing. Since the early twentieth century, when retail consisted of small independent shops lining commercial corridors, a series of changes in the way shopping occurs affected retailing – and urbanism – dramatically. Shopping malls, large-format retailers, and more recently, commerce conducted online (e-commerce) have challenged the viability of small-scale retailers. One estimate is that if e-commerce grows to 25 per cent by 2026 (from its current share of 16 per cent), 75,000 more stores in the United States will need to close (Thomas, 2019). Where independent retailers do remain, they often tend to be "art galleries, boutiques, and cafés" – what Zukin et al. (2015) refer to as the "ABCs" of gentrification. Such shops cater exclusively to destination or experiential retailing rather than satisfying daily life needs or stimulating social and economic interdependence.

The urban experience of streets lined with small-scale retailers was in decline well before e-commerce, starting in the middle decades of the twentieth century as a result of broad structural transformations occurring in the economic, technological, and social life of cities. The rise of corporate retailing signalled "creative destruction," where one large retailer could undermine smaller mom-and-pop stores (as chronicled by Levinson, 2012). This impact is still felt, and researchers continue to document the effect of corporate chains on small retail closures (Jia, 2008; Basker, 2005), although the negative effects might be stronger in outlying areas than in walkable downtowns (Sciara et al., 2018).

In tandem with corporate retailing was the technological change that access to goods and services for daily life no longer depended on walking and proximity. Car-based expansion at the urban periphery changed the nature of consumption such that the local corner store was no longer the basis of neighbourhood identity and social connection (Griffiths et al., 2008; Dawson, 1988). By the 1960s, retailers in many once vibrant neighbourhoods no longer played the role of social and economic connector or provider of goods and services needed for daily life (Kunstler, 1994; Fogelson, 2003; Rae, 2005). Loss of neighbourhood-based retail in poor, "inner-city" neighbourhoods has been especially problematic, as retail there has shifted from grocers and hardware stores to liquor stores and moneylenders (Schuetz et al., 2012; Shimotsu et al., 2013; Diao, 2015). The loss of neighbourhood-based retailing has been found to be racially driven rather than the result of demand level or negative externalities (Koebel, 2014).

Government policies played a role in stimulating the changes that were ultimately detrimental to small businesses. Economic development policies were designed to increase employment and had little concern for the social and neighbourhood-serving functions of small, independent retailing (Beauregard, 1989; Sutton, 2010; Florida, 2017). City governments backed subsidies and tax incentives ("public-private partnerships") for corporate-owned chain stores, giving larger stores a significant advantage. This emphasis can be contrasted with the advantages that small-scale retail enjoys in other parts of the world. Throughout Western Europe, for example, small retail is supported not only by a denser, more walkable urban form, but also by government policies that support small retail: subsidies for health care, housing, and education, as well as controls on pricing and hours of operation, which limit market competition in a way that is advantageous for small retailers.

High rents are often cited as a significant stress on small retailers, and a main reason for mom-and-pop store decline. In part, property owners increase rents in response to the changing demographics of walkable urban neighbourhoods, which are in high demand (Myers, 2017; Smart Growth America & George Washington University School of Business, 2019). The irony is that the influx of high-income residents in neighbourhoods whose commercial areas were once lined with mom-and-pop stores is now displacing the very stores incoming residents seek access to. These population trends have brought some degree of "place branding," "class-based consumption," and an emphasis on aesthetics – all of which work to increase rents and undermine smaller retailers (Keatinge & Martin, 2015, p. 867). Commercial areas that offer only the look of traditional retailing without providing what's needed for daily life put a strain on existing residents, who might experience a sense of cultural exclusion (Sullivan & Shaw, 2011).

In response to higher rents, one outcome is that independent retailers get pushed to more affordable locations in car-oriented strip malls (Linovski, 2012). But these locations are less desirable because they tend to have lower surrounding densities and larger interior spaces than what small retailers need (Kickert, 2016). In addition, anti-pedestrian qualities like surface parking lots undermine the physical connectedness between small business and the surrounding neighbourhood.

A few US cities have enacted policies to combat these trends. For example, to maintain "neighbourhood individuality," Section 303.1 of San Francisco's planning code requires chain stores to obtain conditional use permits in certain zones; formula retail is completely prohibited in three neighbourhoods, including Chinatown. Formula retail is defined as a chain store or business with 11 or more locations and "a recognizable 'look' or appearance," and "standardized features" that, although providing "clear branding" for consumers, run counter to the City's General Plan policies that "value unique community character" (San Francisco Planning, 2019). San Francisco has also established a "Landmark Fund" to preserve iconic businesses in addition to buildings (ILSR, 2016).

In New York City, Manhattan Borough President Gale Brewer launched "The Storefronters Campaign" (Brewer, 2015, 2019) with specific policy goals, yet to be implemented: helping small businesses negotiate better lease terms, allowing small retail in neighbourhoods as a way to create more competition for building owners (and thus potentially lowering rents), encouraging "condo-ization" of storefront space, and creating a "ladder of entrepreneurship" for street vendors.

Support for small retailers in Chicago consists of two funding sources: the Small Business Improvement Fund (SBIF), which reimburses building owners for building improvements, and the Neighborhood Opportunity Fund program, which uses revenue from downtown development to finance commercial projects in the under-invested West and South Sides of the city. For the latter, as of April 2019, 191 projects totalling almost $47 million have been dispersed. Grants are to be used for building renovation as well as adjacent property acquisition.[1]

A Survey of Chicago's Small Retailers

Two overall themes emerge from the literature cited above. First, that small, independent retailers are believed to have a positive impact on streetlife, economic resilience, and social connection. Second, that small, independent retailers are threatened by online shopping, commercial gentrification, and shifting consumer preferences and behaviours. The goal of our survey was to better understand these trends from the perspective of small retailers themselves.

We structured the survey around two over-arching themes: context and connection, and small business challenges. The former concerns the degree to which small retailers feel they are a part of their surrounding neighbourhoods, while the second concerns the degree to which small retailers feel challenged by what are thought to be their primary stressors: e-commerce, government regulations, and the cost of sustaining a small business.

We conducted the survey in two phases. In the first phase, we started with a complete list of geocoded business licences from the City of Chicago's Department of Business Affairs and Consumer Protection. These data do not identify chain stores, so we combined several sources to derive a list of chain stores that we then eliminated from the list.[2] In order to facilitate surveying, we included only stores that were within half a mile of a transit stop. This yielded 2,104 small, independent businesses to survey. We then performed a random sample selection of retailers from this list, which resulted in a selection of 379 businesses.

Two research assistants were hired during the summer of 2018 to visit every business on the list. Surveyors approached each business with a standardized introductory script, and then asked respondents (an employee, manager, or owner) if they would be willing to complete a survey on site, or else mail the survey back using a pre-labelled, stamped envelope. They were also asked if they would participate in a recorded interview, and 7 respondents agreed to do so. The survey was also made available in Spanish; one of the surveyors was Spanish-speaking, and another spoke Chinese.

The response rate for this initial approach was very low – only 23 respondents (< 1 per cent). The willingness of retailers to complete a survey varied by type of business. Boutiques and other high-end retailers with low customer volumes were the most likely survey respondents. Those who vended food (e.g., convenience stores) were less accessible during peak hours (such as lunchtime). Convenience store operators, especially those located in the downtown area ("the Loop"), were also reluctant to participate and often expressed limited knowledge of the surrounding neighbourhood or streetscape characteristics (many were located in the lobbies of office towers). Restaurants and bars were especially difficult to access. Even during non-peak hours, employees were usually busy with cleaning or prepping tasks. Interviews with grocery store personnel was also challenging, as employees and managers were engaged in continuous tasks.

In the fall of 2018, we initiated a second survey that included several changes, in an effort to increase the response rate: we omitted bars and restaurants; we included an option for completing the survey online; we included a raffle for the chance to win a $100 Visa gift card; and we eliminated several questions (pertaining to respondent background

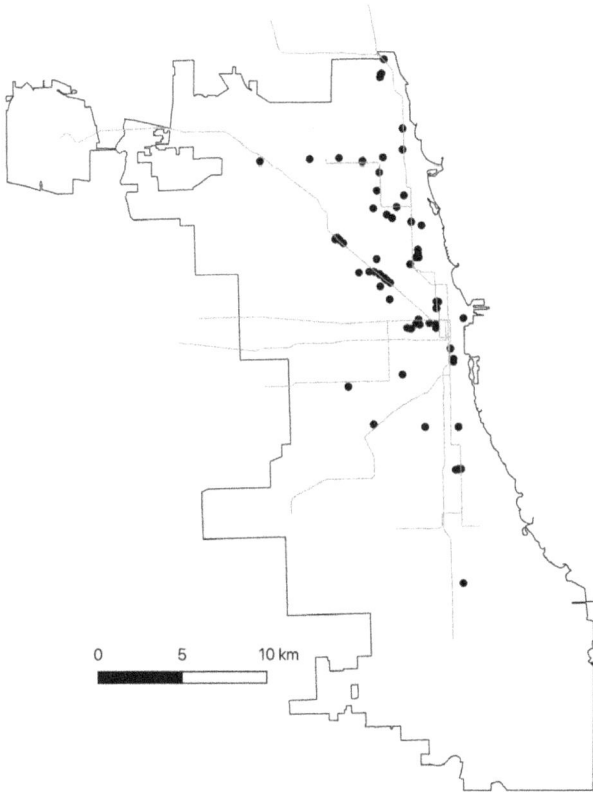

Figure 8.1. Businesses and their associated transit lines. (Source: Created by the author.)

characteristics) in order to make the survey shorter. This time, we did not use a random sample, but targeted all small businesses (except bars and restaurants) that were located within half a mile of a transit stop, which totalled approximately 389.[3] Surveys were distributed throughout the fall (2018), and by early 2019, an additional 52 surveys were returned. The total number of completed surveys was thus 75, which is close to 20 per cent of small businesses near transit stops (excluding bars and restaurants). A map showing the locations of the 75 businesses surveyed (and associated transit lines) is shown in Figure 8.1. Table 8.1 compares demographic information for the 22 community areas in which the surveyed businesses are located (there are 77 community areas in Chicago). The table shows that the contexts of the businesses we surveyed are widely varying, ranging from close to $100K median household income to less

Table 8.1. Demographic characteristics of communities with surveyed businesses

Community area	# Businesses surveyed	Median household income ($)	% Single-family housing	% White	% Hispanic	% Black	% Asian
West Town	13	88,761	8	61	25	7	4
Near West Side	8	80,727	4	42	10	29	17
Lake View	7	86,119	11	79	8	3	6
Lincoln Park	6	99,685	15	79	7	5	7
Logan Square	6	70,339	12	46	44	5	3
Grand Boulevard	4	31,970	6	3	2	92	0
Near North Side	4	93,707	5	71	6	9	11
Rogers Park	4	39,106	12	43	22	26	5
Loop	4	103,336	1	62	7	11	17
North Center	3	104,351	18	77	12	2	5
Albany Park	2	58,128	25	30	48	5	14
Lincoln Square	2	71,737	14	64	17	4	11
Near South Side	2	94,152	5	47	5	23	22
Uptown	2	49,681	6	55	14	18	10
Bridgeport	1	45,671	18	34	24	3	39
Chatham	1	32,597	28	2	1	96	0
Douglas	1	31,907	7	11	3	71	13
Edgewater	1	49,287	13	55	17	13	12
Lower West Side	1	42,458	3	16	76	4	2
McKinley Park	1	44,904	13	17	61	1	20
Portage Park	1	61,393	37	50	41	1	5
South Lawndale	1	32,896	8	4	84	12	0

Data source: American Community Survey (2013–2017), US Census Bureau.

than $40K median household income. Most community areas are racially mixed; 10 out of the 14 areas where at least 2 businesses were surveyed have less than 65 per cent white population.

The survey asked questions about business background, neighbourhood context, and perceived challenges.

Findings

To provide some context regarding the retail landscape of Chicago, Tables 8.2 and 8.3 present summary data on retail establishments in Chicago based on the National Establishment Time-Series (NETS) database,

Table 8.2. Retail establishments in Chicago, 1990–2014

	1990	1995	2000	2005	2010	2014	% Change 1990–2014
Chicago population	2,783,726	2,802,000	2,896,016	2,829,000	2,695,598	2,727,000	–0.02
Stores per 1,000 pop.	10.18	11.06	10.04	11.22	12.24	12.25	17
Basic needs	5,853	6,186	5,563	6,431	7,026	7,144	18
Bars, restaurants, bakeries	9,600	10,250	10,264	11,129	11,956	13,503	29
General merchandise	7,420	8,290	7,411	8,341	8,239	7,435	0
Specialty stores/ boutiques	5,462	6,267	5,848	5,839	5,763	5,333	–2
Total	28,335	30,993	29,086	31,740	32,984	33,415	15
% of total							
Basic needs	20.7	20.0	19.1	20.3	21.3	21.4	–
Bars, restaurants, bakeries	33.9	33.1	35.3	35.1	36.2	40.4	–
General merchandise	26.2	26.7	25.5	26.3	25.0	22.3	–
Specialty stores/ boutiques	19.3	20.2	20.1	18.4	17.5	16.0	–
Total	100.0	100.0	100.0	100.0	100.0	100.0	–

Data source: National Establishment Time-Series (NETS) database, 1990–2014, summarized by Anselin & Farah (2018).

Note: Retail categories are based on the following NAICS codes: Basic needs – grocery stores, health and personal care stores, hardware stores, paint and wallpaper stores, other building material stores; Bars, restaurants, bakeries – food services and drinking places, retail bakeries; General merchandise – electronic and appliance stores, building material and equipment and supplies dealers, clothing and clothing accessories stores, sporting goods, hobby, musical instrument, bookstores, general merchandise stores, home centres, lawn and garden equipment and supplies, office supplies and stationery; Specialty stores/boutiques – florists, pet and pet supplies stores, tobacco stores, art dealers, gift, novelty, and souvenir stores, beer, wine, and liquor stores, furniture stores, specialty food stores, jewellery stores, luggage and leather goods stores

Table 8.3. Independent vs. chain establishments in Chicago, 1990–2014

Size	1990	1995	2000	2005	2010	2014	% Change 1990–2014
Independent (2 or fewer)	20,003	20,985	19,719	21,949	22,864	23,129	14
Regional chain (3 to < 40 stores)	4,098	5,076	4,504	4,838	4,803	4,537	10
Large chain (40 or more)	2,372	3,281	3,587	3,945	4,479	5,038	53
Corporate chain (Forbes 500)	101	221	318	462	452	586	83

Note: 2 per cent of businesses could not be classified.

which is a longitudinal database (1990–2014) of business and non-profit establishments. NETS includes data on type of establishment, employment, years active, and industry classification, among other variables.

The tables reveal several interesting patterns and trends. First, the number of stores per capita has increased significantly since 1990 (17 per cent), with 15 per cent more retail establishments, even as population has declined. It is likely that this growth has continued, as retail sales in Chicago rose 3.5 per cent in 2018, higher than any year since 2014. These trends belie the dominant narrative that the retail sector is in crisis as it tries to compete with online shopping (Gallun, 2019).

Second, most of the increase has been in the categories of bars, restaurants, and bakeries – "experiential" food offerings, fuelled by "millennial-inspired and tech-infused dining trends" that have been growing in recent years even as fast food establishments have declined (CB Insights, 2018). Third, specialty stores and boutiques, which form the smallest retail category, were the only retail to decline between 1990 and 2014. In terms of independent vs. chain establishments (Table 8.3), there was growth in all categories, although independent stores and regional chains experienced much less growth than chains. Large chains and corporate chains saw the most significant growth – 53 per cent and 83 per cent respectively – between 1990 and 2014.

Our 75 respondents constitute a small subset of these establishments, since we limited our sample to small retailers near transit stops and excluded most bars and restaurants. Tables 8.4 and 8.5 summarize some background characteristics of our sample. Most of our sample (64 per cent) came from apparel or boutique stores. The remaining respondents were service establishments (e.g., shoe repair) or basic consumption (e.g., groceries), in addition to a small number of restaurants. Respondents

Table 8.4. Respondent characteristics

	Number	%
Respondent is		
Business owner	37	49.3
Manager or employee	38	50.7
Total	75	100.0
Respondent lives in neighbourhood		
No	48	64.0
Yes	24	32.0
No response	3	4.0
Total	75	100.0

Table 8.5. Business characteristics

	Number	%
Business has more than one store		
No	50	66.7
Yes	25	33.3
Total	75	100.0
Number of employees in the store		
0–4	37	49.3
5–9	26	34.7
10–14	7	9.3
15+	4	5.3
No response	1	1.3
Total	75	100.0
Type of business*		
Apparel	24	32.0
Boutiques	24	32.0
Basic consumption	6	8.0
Services	11	14.7
Restaurants	10	13.3
Total	75	100.0

* Apparel = clothing and shoe stores; Boutiques = gifts, jewellery, flowers, books, furniture, rugs; Basic consumption = grocery, hardware, bakery, wine shop, liquor; Services = eyecare, shoe repair, hair or nail salon, frame store

Table 8.6. Mobility

	Number	%
Years in current location		
Under 2 years	11	14.7
2–4.5 years	20	26.7
5–9.5 years	17	22.7
10+ years	25	33.3
No response	2	2.7
Total	75	100.0
Business moved from another location		
No	44	58.7
Yes*	31	41.3
Total	75	100.0
Reason(s) for moving:		
Affordable rent	23	74.2
Closer to customers	27	87.1
Business expansion	13	41.9
More visible location	31	100.0
Previous location:		
Close by, within neighbourhood	14	45.2
2+ miles farther out	12	38.7
Out of state	2	6.5
No response	3	9.7
Total	31	100.0

were evenly split between owners and employees, and only about a third reported living in the same neighbourhood as the business. Most were single, small stores, although about a third of our sample had more than one store. Most of our sample had 4 or fewer employees, and 84 per cent had fewer than 10 employees.

Our sample represented a fairly stable group, with a third having been in the same location for 10 or more years, and about 50 per cent in business between 2 and 10 years (Table 8.6). Only about 15 per cent were new to the location (under 2 years in business at the current location). Most businesses originated at their current locations, but for the 41 per cent that had moved, all cited "more visible location" as a reason for moving. Most were also seeking more affordable rent as well as proximity to customers – two objectives that can sometimes be at odds. Respondents were also asked to characterize the location they had moved from. Businesses were evenly split between those who moved from within the neighbourhood, and those that moved from farther away. Half of those in the latter category reported moving in order to be in a "hot" neighbourhood.

Context and Connection

A dominant theme in the literature is that small retail is more successful if it is socially connected to the surrounding neighbourhood. Only about a third of our respondents said that their customers come primarily from within the neighbourhood. Most businesses depended on customers from "all over the city." Two factors might be at work. First, many respondents claimed that the neighbourhood had changed significantly in the past 5 years, and this is likely to have had an impact on the ability to sustain a neighbourhood customer base. Second, our survey was limited to businesses near transit stops, which, although likely to sustain a higher adjacent population, also means the customers can be drawn from places outside of the neighbourhood.

Despite this non-neighbourhood clientele, most businesses (79 per cent) reported feeling that they were an important part of the neighbourhood (the 7 businesses responding "no" straddled all retail categories). Several businesses went further and related that their business was involved in proactive community building. One owner stated that she put "people before profit" and viewed her store as an important community-building venue for the neighbourhood. Some businesses were cognizant of their role as a "third place," and one tattoo merchant in a particularly diverse neighbourhood stated that his business was connected because "every walk of life feels safe here, from gangbangers to gays to the police."

Respondents were attuned to the surrounding street context of their business in two ways: the mix of stores on the block, and the quality of upkeep of other businesses (see table 8.7). Most businesses seemed highly cognizant of the surrounding mix, but there seemed to be a tension between perceived competition and the need for greater activity and business clustering. On the one hand, small businesses are susceptible to the impact of other competing businesses opening up and taking away customers – a threat that can apply to large chain stores (a new Target was "obviously worrisome" to one merchant) as well as independent retailers with similar products or services. But on the other hand, business agglomeration was clearly recognized as desirable. Thus while one respondent expressed positive feelings about the fact that "we do not have many boutiques in this side of the neighbourhood," and another felt that his location in a "retail desert" in combination with "excellent foot traffic" was highly beneficial, it was more common for respondents to wish for "more businesses" on the street. Often this applied to "big name businesses" because of the view

Table 8.7. Context and connection

	Number	%
Customers primarily come from		
Within the neighbourhood	23	30.7
All over the city	43	57.3
Both	9	12.0
Total	75	100.0
Is this business an important part of this neighbourhood?		
Yes	59	78.7
No	7	9.3
Unsure	5	6.7
No response	4	5.3
Total	75	100.0
Do you interact with other business owners/managers in this neighbourhood?		
Once a week or more	31	41.3
Once a month	24	32.0
Hardly ever	19	25.3
No response	1	1.3
Total	75	100.0
Does your business belong to a neighbourhood or business association?		
No	45	60.0
Yes	30	40.0
Total	75	100.0

that "people bring people" and success depends on other businesses. One merchant was happy that "new small business boutiques are opening up," noting that "clustering is important to us."

Another aspect of this balancing act involves understanding the trade-off between higher rent and higher revenue potential. Businesses were aware of significant disparities between neighbourhoods, in neighbourhood upkeep and the resources available to the local business association to stimulate commerce. One merchant was waiting for the neighbourhood to "get there" but noted that there was a "tricky trade-off between paying higher rent and getting more customers and paying lower rent but having fewer customers." One owner observed that she was paying "higher rent in my current location – but it's worth it."

Most businesses (73 per cent) reported interacting with other businesses, either once a week or once a month. Many businesses mentioned the importance of connecting to other businesses to share experiences and network. One business owner said that businesses would text each other if they saw the police giving out fines for sidewalk signs. Sometimes these connections were made via a Chamber of Commerce ("the Chamber has been good for events like festivals, and for connecting to other businesses"); more often, social connection was tied to personality type ("My owner knows everyone").

Yet these interactions were, for the most part, not formalized. Most businesses (60 per cent) did not report belonging to a Chamber or a business or neighbourhood association. Many admitted being unaware of any neighbourhood business associations; one respondent relayed that "it would have been nice to have a more welcoming thing when we came here, like a welcome wagon." Another manager was hoping for a "street festival among ourselves and the community," but assumed there was no organization to accomplish it.

Challenges

Our survey did not find strong evidence that e-commerce was perceived as detrimental to small businesses (Table 8.8). Only about 17 per cent reported that online shopping was harming their business "a lot." Another 23 per cent of businesses reported "a little" negative effect, but most businesses did not report any negative impact. In-person conversations revealed that business were looking for ways to take advantage of e-commerce. For example, one business reported developing a website that functioned as a "local Etsy" that could increase access to the handcraft market and encourage more "makers" to pursue their crafts. Further, many businesses related that social media was having a positive impact on their business. The importance of "internet marketing" and tools like Groupon, Grubhub, and Ubereats were regularly cited as having a positive influence. For some businesses, e-commerce was increasing awareness that "brick and mortar businesses" were able to "sell an experience" that customers would not find elsewhere.

Table 8.9 summarizes responses to other challenges that are believed to negatively impact mom-and-pop stores. Somewhat surprisingly, government regulation was not a common concern. One explanation for the low incidence of complaint about government regulation is that the businesses we surveyed had presumably already successfully navigated the most burdensome regulatory constraints, particularly zoning. In addition, we focused on businesses near transit stops, where, in Chicago, parking requirements have been waived; complaints about onerous parking requirements would not come into play.

Table 8.8. Impact of e-commerce

	Number	%
Business has a website		
Yes	54	72.0
No	15	20.0
Yes, but not used much	6	8.0
Total	75	100.0
% of business conducted online		
None	38	50.7
< 25%	23	30.7
25–50%	10	13.3
> 50%	4	5.3
Total	75	100.0
Does online shopping harm your business?		
Yes, a lot	13	17.3
Yes, a little	17	22.7
No negative impact	29	38.7
Unsure	6	8.0
No response	10	13.3
Total	75	100.0

For those 23 per cent of businesses who did cite "regulation" as a concern, most cited was the high cost of licensing fees and the burden of licensing requirements ("Regulations are annoying because you have to update your licence every year, and permits expire at different times." Another remarked, "Laws are changing every year and it's hard to keep up." Another complained that it took "90 days to get a liquor licence"). There were references to "fines for stupid things like smoking near a building." Also commonly cited was a critique of the "bag tax," which many businesses found annoying and "silly" for small businesses and thought it should apply only to larger businesses.

Businesses remarked on population changes in the surrounding area. Some areas experienced loss of population that was blamed on Chicago's high taxes ("High taxes make lots of our customers move out of this area, and that is not good for business") and for a small number of respondents, safety and crime. When gentrification was cited, it was generally not expressed as something negative ("We are quickly gentrifying but that is a positive for my business"). In one response, an employee noted business expansion from gentrification: "We have become more popular and expanded to include wines to suit neighbourhood budgets,"

Table 8.9. Other challenges

	Number	%
Do government regulations harm your business?		
Yes	17	23.0
No	45	60.0
Unsure	13	17.0
Total	75	100.0
Has this neighbourhood changed in the last 5 years?		
Yes, a lot	43	57.3
Yes, a little	17	22.7
No	8	10.7
Unsure	4	5.3
No response	3	4.0
Total	75	100.0
Are any of the following a concern to you? (# reporting "yes")		
Street/sidewalk quality	35	47
The other businesses on this street	36	48
Parking	42	56
Gentrification	9	12
Are you optimistic about the future?		
Yes	51	68
Yes, somewhat	11	15
No	5	7
Unsure	3	4
No response	5	7
Total	75	100

meaning that customers had more buying power. Another embraced neighbourhood social changes, remarking that the number of customers is "still the same, just the people have changed."

Parking was the most cited concern (56 per cent), predominantly in terms of being too expensive, with "too many pay parking spots" (one business claimed to have "lost customers because of the expensive parking") and that parking was too limited as the result of the expansion of apartments (Chicago, like many cities, has witnessed a surge of apartment building in recent years). New construction was often viewed as an annoyance because it was believed to block parking spaces. There were mixed perceptions about the city's reputation for being "ruthless about ticketing" – in one instance thought to be a "good thing" but in another that "you shouldn't have to pay for parking" and that "parking enforcers are way too strict."

Asked whether retailers were optimistic about the future, businesses were overwhelmingly positive, despite the challenges. As one business summarized, "It's hard for small business, but once you are in you are in." Another noted, "We were established in 2002. We have weathered all storms."

Policy Responses

The most important takeaways of this survey can be summarized as follows. On the importance of context and connection, respondents indicated that, first, small businesses feel they are a part of the neighbourhood even as their clientele is drawn from throughout the city and most business owners and employees do not live in the same neighbourhood. Second, small businesses interact with other businesses, but for the majority this interaction lacks formal organization.

On the topic of small business challenges, three points seem most significant. First, most small retailers did not seem threatened by e-commerce, but were instead looking for ways to take advantage of it. Second, small businesses struggled with the need to balance rent and revenue potential on the one hand, and business singularity versus clustering on the other hand. There was tension between the idea that more people moving into the neighbourhood is something positive and the view that more businesses and more competition are a strain on business. Finally, respondent complaints were mostly confined to very specific issues like parking and licensing fees as opposed to larger issues like gentrification or government regulation in general. And, despite the common narrative that retail is imploding and mom-and-pop stores are dying, our respondents expressed optimism about the future. A limitation of our survey, it should be noted, is that it was confined to retailers currently operating in transit-served areas; mom-and-pop stores who have been forced to close or are struggling to open might have a less optimistic view.

How might planners and local governments respond to these viewpoints? One response would be to do nothing: let the market run its course and let urban living adapt along with it. This reasoning is in line with those who believe that Jane Jacobs's love of small retailers is at odds with market realities (Gopnik, 2016; Teaford, 2013), and that government policies that try to counter market forces (such as anti-chain legislation) are politically risky and unsustainable over the long term. Further, empirical studies have shown that planners' attempts to shape market forces have been short-term at best (Jones, 2014); worse, they have produced unintended outcomes (Adams & Tiesdell, 2010). Others argue that corporate ownership is not something to be squelched: it helps pedestrian-oriented, mixed-use commercial areas thrive, and it is not at odds with the cultural values of middle- and low-income neighbourhoods (Chapple & Jacobus, 2009). Additionally, small and independent stores may not be neighbourhood-serving but instead likely to stimulate gentrification and put commerce at odds with the needs of local residents (Deener, 2007; Zukin et al., 2015). Finally, chain grocery stores do a better job than mom-and-pop stores at providing access to fresh fruits and vegetables (Zenk et al., 2009).

But a "do nothing" response ignores the challenges facing mom-and-pop stores – which could be supported through policy. The question is whether cities value the dividends of mom-and-pop retail – streetlife activation, social and economic connectedness – enough for governments to respond proactively to these concerns. If this response is confined to policy suggestions related to the survey results presented here, support would fall into two categories: individual business assistance, and collectivizing small businesses (more aggressive policies would include programs like helping business owners become shop owners and providing financial and legal backing for business owners to "condo-ize" and purchase storefronts).

First, local governments could provide business support that is specifically geared to mom-and-pop retail (as opposed to support for small businesses in general). This assistance could include helping small businesses negotiate better lease terms or assisting with landlord-tenant negotiations. Local governments could also make it easier for mom-and-pop stores to operate by streamlining licensing and permitting requirements. This requires recognizing that mom-and-pop retail should not be treated the same as large corporate chain stores, who have more resources available to navigate city permits and licensing requirements.

In addition, local governments might help small retailers programmatically by offering management and skills development, help with signage design and storefront displays, or assistance with website construction and e-commerce strategies. Cities could provide resources for small businesses to take better advantage of internet marketing; larger businesses have the resources to employ commercial firms who specialize in internet marketing – small establishments need better access to the same marketing tools.

Second, local governments could help collectivize small retailers, ensuring access to business organizational power and improving their social and economic connectedness. Small retailers do not operate in isolation and seem to recognize the importance of neighbourhood connection – yet their ability to organize in a more formal way is underserved. Business associations could also provide support for small retailers who are constantly trying to find the right balance between rent, revenue, and location. The Main Street America management program developed by the National Main Street Center (a subsidiary of the National Trust for Historic Preservation) is an essential model that mom-and-pop stores should have wider access to.

Where business organizations already exist, the City could work to ensure that Business Improvement Districts (or in Chicago's case, Special Service Areas) direct resources to support mom-and-pop retailers specifically. Where such organizations do not exist, local governments could support the creation of business organizations that function like a Chamber of Commerce. This was recently recommended following a City-initiated study of a commercial corridor in one of Chicago's under-invested neighbourhoods (City of Chicago, 2019). One recommendation out of that study was the need to "help organize the landlords and business owners to provide marketing and services to business within the Study Area more specifically."

These support strategies are fairly modest and within reach of local governments who lack large discretionary budgets. However, enacting even modest policies and programs will require determining the level of commitment to mom-and-pop retail, and the degree to which their value – streetlife activation and social and economic connectedness – is enough to justify proactive support. The decline of mom-and-pop retail is not a fait accompli.

NOTES

1 More information can be found at Neighborhood Opportunity Fund (2019). In 2019, one of the largest grants – $2.5 million – was for acquisition and rehabilitation of a site for a "new training campus" to house an urban farm, event rental space, and a cafe.
2 Our sources for chain stores were Technomic (2016), Faran (2019), and Wikipedia (2019).
3 We found this "opportunity" based approach most efficient; in the field, surveyors discovered that many of the businesses identified in the business licence dataset obtained earlier had either closed or changed.

REFERENCES

Adams, D., & Tiesdell, S. (2010). Planners as market actors: Rethinking state–market relations in land and property. *Planning Theory and Practice, 11*(2), 187–207. https://doi.org/10.1080/14649351003759631

American Independent Business Alliance. (n.d.). The local multiplier effect: How independent local businesses help your community thrive. https:// amiba.net/wp-content/uploads/2021/02/The-Local-Multiplier-Effect.pdf

Anselin, L., & Farah, I. 2018. *Methodological issues in the analysis of local patterns in the retail sector* (Center for Spatial Data Analysis Working Paper). University of Chicago.

Bader, M. D. M., Schwartz-Soicher, O., Jack, D., Weiss, C. C., Richards, C. A., Quinn, J. W., Lovasi, G. S., Neckerman, K. M., & Rundle A. G. (2013). More neighborhood retail associated with lower obesity among New York City public high school students. *Health and Place, 23*, 104–110. https:// doi.org/10.1016/j.healthplace.2013.05.005

Basker, E. (2005). Job creation or destruction? Labor market effects of Wal-Mart expansion. *Review of Economics and Statistics, 87*(1), 174–183. https:// doi.org/10.1162/0034653053327568

Beauregard, R. (1989). Between modernity and postmodernity: The ambiguous position of US planning. *Environment and Planning D: Society and Space, 7*, 381–395. https://doi.org/10.1068/d070381

Besser, T. L. (1999). Community involvement and the perception of success among small business operators in small towns. *Journal of Small Business Management, 37*(4), 16–29.

Brewer, G. (2015). Small business big impact: Expanding opportunity for Manhattan's storefronters. New York City. http://manhattanbp.nyc.gov /downloads/pdf/SmallBusinessBigImpactFINAL.pdf

Brewer, G. (2019). The storefronters campaign. https://www.manhattanbp.nyc .gov/issues/storefronters/

Cachinho, H. (2014). Consumerscapes and the resilience assessment of urban retail systems. *Cities, 36*(February), 131–144. https://doi.org/10.1016 /j.cities.2012.10.005

Campbell, J. M., & Park, J. (2016). Extending the resource-based view: Effects of strategic orientation toward community on small business performance. *Journal of Retailing and Consumer Services, 34*, 302–308. https://doi.org /10.1016/j.jretconser.2016.01.013

Carmona, M. 2015. London's local high streets: The problems, potential and complexities of mixed street corridors. *Progress in Planning, 100*(August), 1–84. https://doi.org/10.1016/j.progress.2014.03.001

Carrigan, M., Moraes, C., & Leek, S. (2011). Fostering responsible communities: A community social marketing approach to sustainable living. *Journal of Business Ethics, 100*(3), 515–534. https://doi.org/10.1007/s10551 -010-0694-8

CB Insights (2018, 17 July). 11 new restaurant concepts reimagining fast food & casual dining. *Research Briefs.* https://www.cbinsights.com/research/new -restaurant-concepts-fast-food-casual-dining/

Chapple, K., & Jacobus, R. (2009). Retail trade as a route to neighborhood revitalization. In N. Pindus, H. Wial, & H. Wolman (Eds.), *Urban and regional policy and its effects* (pp. 19–68). Brookings.

City of Chicago. (2019). North Park commercial corridor study. Chicago Department of Planning and Development. https://www.chicago.gov /city/en/depts/dcd/supp_info/north-park-study.html

Corinna, M. (2011). Retail and public policies supporting the attractiveness of Italian town centres: The case of the Milan central districts. In Attractive places to live [Special issue]. *Urban Design International, 16*(3), 227–237. https://doi.org/10.1057/udi.2010.27

Corkery, M. (2017, 15 April). Is American retail at a historic tipping point? *The New York Times.* https://www.nytimes.com/2017/04/15/business/retail -industry.html

Dawson, J. (1988). Futures for the high street. *The Geographical Journal, 154*(1), 1–22. https://doi.org/10.2307/633470

Deener, A. (2007). Commerce as the structure and symbol of neighborhood life: Reshaping the meaning of community in Venice, California. *City & Community, 6*(4), 291–314. https://doi.org/10.1111/j.1540-6040.2007.00229.x

Diao, M. (2015). Are inner-city neighborhoods underserved? An empirical analysis of food markets in a US metropolitan area. *Journal of Planning Education and Research, 35*(1), 19–34. https://doi.org/10.1177/0739456x14562283

Ellickson, P. B., & Grieco, P. L. E. (2013). Wal-Mart and the geography of grocery retailing. *Journal of Urban Economics, 75*, 1–14. https://doi.org /10.1016/j.jue.2012.09.005

Farfan, B. (2016). Top 100 largest US based retail companies. https://www .thebalance.com/largest-us-retailers-4045123

Fitzgerald, M. A., & Muske, G. (2016). Family businesses and community development: The role of small business owners and entrepreneurs. *Community Development, 47*(4), 412–430. https://doi.org/10.1080/15575330 .2015.1133683

Florida, R. (2017). *The new urban crisis: How our cities are increasing inequality, deepening segregation, and failing the middle class – and what we can do about it.* Basic Books.

Fogelson, R. M. (2003). *Downtown: Its rise and fall, 1880–1950.* Yale University Press.

Gallun, A. (2019). Struggling retail landlords will be happy to hear this. Crain's Chicago Business. https://www.chicagobusiness.com/commercial-real -estate/struggling-retail-landlords-will-be-happy-hear

Gomez, R., Isakov, A., & Semansky, M. (2015). *Small business and the city: The transformative potential of small scale entrepreneurship.* University of Toronto Press.

Gopnik, A. (2016, 26 September). Jane Jacobs's street smarts. *The New Yorker.* https://www.newyorker.com/magazine/2016/09/26/jane-jacobs-street-smarts

Granovetter, M. S. 1983. The strength of weak ties: A network theory revisited. *Sociological Theory, 1*, 201–233. https://doi.org/10.2307/202051

Griffiths, S., Vaughan, L., Haklay, M., & Jones, C. (2008). The sustainable suburban high street: A review of themes and approaches. *Geography Compass, 2*(4), 1155–1188. https://doi.org/10.1111/j.1749-8198.2008.00117.x

Heffernan, E., Heffernan, T., & Pan, W. (2014). The relationship between the quality of active frontages and public perceptions of public spaces. *Urban Design International, 19*(1), 92–102. https://doi.org/10.1057/udi.2013.16

Hester, J. L. (2016, 9 August). The storefronts that survive. CityLab. http://www.citylab.com/navigator/2016/08/storefront-survivors-project-new-york-city-small-businesses/494576/

Hyra, D. S. (2008). *The new urban renewal: The economic transformation of Harlem and Bronzeville.* University Of Chicago Press.

Institute for Local Self-Reliance (ILSR). (2016). Affordable space: How rising commercial rents are threatening independent businesses, and what cities are doing about it. https://ilsr.org/wp-content/uploads/downloads/2016/04/ILSR-AffordableSpace-ExecutiveSummary.pdf

Jacobs, J. 1961. *The death and life of great American cities.* Random House.

Jia, P. (2008). What happens when Wal-Mart comes to town: An empirical analysis of the discount retail industry. *Econometrica, 76*, 1263–1316 https://doi.org/10.3982/ecta6649

Jones, C. (2014). Land use planning policies and market forces: Utopian aspirations thwarted? *Land Use Policy, 38*, 573–79. https://doi.org/10.1016/j.landusepol.2014.01.002

Kaplan, R., & Kaplan, S. (1989). *Experience of nature: A psychological perspective.* Cambridge University Press.

Keatinge, B., & Martin, D. G. (2015). A "Bedford Falls" kind of place: Neighbourhood branding and commercial revitalisation in processes of gentrification in Toronto, Ontario. *Urban Studies, 53*(5), 867–883. https://doi.org/10.1177/0042098015569681

Kickert, C. (2016). Active centers – interactive edges: The rise and fall of ground floor frontages. *Urban Design International, 21*(1), 55–77. https://doi.org/10.1057/udi.2015.27

Koebel, C. T. (2014). Analyzing neighborhood retail and service change in six cities [Technical report]. Virginia Center for Housing Research. https://vtechworks.lib.vt.edu/handle/10919/48643

Krohn-Hansen, C. (2013). *Making New York dominican: Small business, politics, and everyday life.* University of Pennsylvania Press.

Kunstler, J. H. (1994). *The geography of nowhere: The rise and decline of America's man-made landscape.* Free Press.

Levinson, M. (2012). *The great A&P and the struggle for small business in America.* Hill and Wang.

Linovski, O. (2012). Beyond aesthetics: Assessing the value of strip mall retail in Toronto. *Journal of Urban Design, 17*(1), 81–99. https://doi.org/10.1080/13574809.2011.646247

Mack, E. A. (2015). Small business and the city: The transformative potential of small scale entrepreneurship. *Journal of Regional Science, 55*(5), 874–875. https://doi.org/10.1111/jors.12230

Melaniphy, J. (2018). *Melaniphy 2018 retail sales report: Chicago Metropolitan Area & City of Chicago.* https://www.melaniphy.com/node/140

Meltzer, R. (2016). Gentrification and small business: Threat or opportunity? *Cityscape: A Journal of Policy Development and Research, 18*(3), 57–85.

Meltzer, R., & Capperis, S. (2017). Neighbourhood differences in retail turnover: Evidence from New York City. *Urban Studies, 54*(13), 3022–3057. https://doi.org/10.1177/0042098016661268

Mills, Q. T. (2013). *Cutting along the color line: Black barbers and barber shops in America.* University of Pennsylvania Press.

Mitchell, S. (2009). Neighborhood stores: An overlooked strategy for fighting global warming. Grist. http://grist.org/article/2009-08-19-neighborhood-stores-strategy-for-fighting-global-warming/

Myers, D. (2017). Peak millennials: Three reinforcing cycles that amplify the rise and fall of urban concentration by millennials. *Housing Policy Debate, 26*(6), 928–947. https://doi.org/10.1080/10511482.2016.1165722

Neighborhood Opportunity Fund. (2019). https://neighborhoodopportunityfund.com

Nguyen, H., & Sawang, S. (2015). Juggling or struggling? Work and family interface and its buffers among small business owners. *Entrepreneurship Research Journal, 62*, 207–246. https://doi.org/10.1515/erj-2014-0041

Oldenburg, R. 1999. *The great good place 2 Ed: Cafes, coffee shops, community centers, beauty parlors, general stores, bars, hangouts.* De Capo Press.

Pendola, R., & Gen, S. (2008). Does "Main Street" promote sense of community? A comparison of San Francisco neighborhoods. *Environment and Behavior, 40*(4), 545–574. https://doi.org/10.1177/0013916507301399

Peterson, H. (2017). The retail apocalypse has officially descended on America. Insider. http://www.businessinsider.com/the-retail-apocalypse-has-officially-descended-on-america-2017-3

Rae, D. W. (2005). *City: Urbanism and its end.* Yale University Press.

Razalan, D. M., Bickle, M. C., Park, J., & Brosdahl, D. (2017). Local retailers' perspectives on social responsibility. *International Journal of Retail & Distribution Management, 452*, 211–226. https://doi.org/10.1108/ijrdm-01-2016-0006

Roundy, P. T. (2018). "It takes a village" to support entrepreneurship: Intersecting economic and community dynamics in small town entrepreneurial ecosystems. *International Entrepreneurship and Management Journal, 15*, 1443–1475. https://doi.org/10.1007/s11365-018-0537-0

Samujh, H. (2011). Micro-businesses need support: Survival precedes sustainability. *Corporate Governance, 11*(1) 15–28. https://doi.org/10.1108 /14720701111108817

San Francisco Planning (2019). Chain stores. https://sfplanning.org/permit /chain-stores-businesses

Schuetz, J., Kolko, J., & Meltzer, R. (2012). Are poor neighborhoods "retail deserts"? *Regional Science and Urban Economics, 42*(1–2), 269. https:// doi.org/10.1016/j.regsciurbeco.2011.09.005

Sciara, G.-C., Lovejoy, K., & Handy, S. (2018). The impacts of big box retail on downtown: A case study of Target in Davis (CA). *Journal of the American Planning Association, 841*, 45–60. https://doi.org/10.1080/01944363.2017 .1404926

Shimotsu, S. T., Jones-Webb, R. J., MacLehose, R. F., Nelson, T. F., Forster, J. L., & Lytle L. A. (2013). Neighborhood socioeconomic characteristics, the retail environment, and alcohol consumption: A multilevel analysis. *Drug and Alcohol Dependence, 132*, 449–456. https://doi.org/10.1016/j.drugalcdep .2013.03.010

Skjaeveland, O., & Garling, T. (1997). Effects of interactional space on neighbouring. *Journal of Environmental Psychology, 17*, 181–198. https:// doi.org/10.1006/jevp.1997.0054

Smart Growth America and George Washington University School of Business. (2019). Foot traffic ahead: 2019. https://smartgrowthamerica.org/resources /foot-traffic-ahead-2019/

Smick, D. M. (2017). *The great equalizer: How Main Street capitalism can create an economy for everyone.* Public Affairs.

Sullivan, D. M., & Shaw, S. C. (2011). Retail gentrification and race: The case of Alberta Street in Portland, Oregon. *Urban Affairs Review, 47*(3), 413–432. https://doi.org/10.1177/1078087410393472

Sutton, S. A. (2010). Rethinking commercial revitalization: A neighborhood small business perspective. *Economic Development Quarterly, 24*(4), 352–371. https://doi.org/10.1177/0891242410370679

Taylor, H. L. Jr. (2018). Stores in the hood: Neighborhood-based business and community development in underdeveloped neighborhoods of color. *Journal of Urban History, 44*(5), 1025–1031. https://doi.org/10.1177 /0096144218772286

Teaford, J. C. (2013). Jane Jacobs and the cosmopolitan metropolis: 2012 UHA presidential address. *Journal of Urban History, 395*, 881–889. https://doi .org/10.1177/0096144213479311

Technomic. (2016). Top 500 chain restaurant report. https://www.technomic .com/reports/industry-reports/top-500.

Thomas, L. (2019). 75,000 more stores need to close, UBS estimates, as online sales and Amazon grow. https://www.cnbc.com/2019/04/09/75000-more -stores-need-to-close-ubs-estimates-as-online-sales-grow.html

Völker, B., Flap, H., & Lindenberg, S. (2007). When are neighbourhoods communities? Community in Dutch neighbourhoods. *European Sociological Review, 23*(1), 99–114. https://doi.org/10.1093/esr/jcl022

Wikipedia. (2019). List of restaurant chains in the United States. https://en.wikipedia.org/wiki/List_of_restaurant_chains_in_the_United_States

Wilson, J. L. (2018). *Shopping locally: An exploration of motivations and meanings in the context of a revitalized downtown* [Unpublished doctoral dissertation]. University of North Carolina.

Yancey, W. L., & Ericksen, E. P. (1979). The antecedents of community: The economic and institutional structure of urban neighborhoods. *American Sociological Review, 44*(2), 253–262. https://doi.org/10.2307/2094508

Zenk, S. N., Lachance, L. L., Schulz, A. J., Mentz, G., Kannan, S., & Ridella, W. (2009). Neighborhood retail food environment and fruit and vegetable intake in a multiethnic urban population. *American Journal of Health Promotion, 23*(4), 255–264. https://doi.org/10.4278/ajhp.071204127

Zukin, S., Kasinitz, P., & Chen, X. (2015). *Global cities, local streets : Everyday diversity from New York to Shanghai.* Taylor & Francis.

Business Profile: Armitage Hardware

Chicago, Illinois

Armitage Hardware is a business owned and operated by father and son in the heart of the Lincoln Park neighbourhood on the north side of Chicago. The business opened in 1969. It is a veritable institution and exemplifies a "third place," where neighbours stop by, and Dan O'Donnell (father) and Brian O'Donnell (son) seem to know everyone.

Business during the pandemic is thriving at Armitage Hardware. In fact, the O'Donnells did more business in the first 6 months of 2020 than they did in all of 2019. They cite several reasons. First, they were already well versed in online sales (especially for niche products like the "Big Green Egg" Weber grill), and with online shopping going up, that part of their market has grown substantially. Second, a lot of the "corporates" – the large hardware chain stores like Lowe's and Home Depot – have been frequently running out of basic supplies, whereas this small, family-run hardware store has been more adept and agile at securing supplies (the "corporates" have less flexibility and more bureaucracy). Third, Armitage Hardware is able to "switch things around" quickly, as needed. One of those needs is to service local restaurants that need to adapt to outdoor dining; they need products that this hardware store can secure quickly, like heat lamps for outdoor dining.

Figure B8.1. Brian O'Donnell. (Credit: Image courtesy of Emily Talen.)

Figure B8.2. Armitage Hardware.
(Credit: Image courtesy of
Emily Talen.)

9 What's in a Chain? On Hipness, Corporate Stores, and False Dichotomies in Urban Life

JEFFREY NATHANIEL PARKER

The figure of the hipster and the concept of hipness[1] are conspicuously under-theorized in social scientific writing about cities and neighbour-hoods. To the extent that they are discussed, hipsters and hipness are typically conceived in relation to capitalism, with the story told of either resistance or co-optation, depending on whether or not capital emerges victorious in the form of gentrification. Debates about the role of hipness in gentrifying neighbourhoods often takes the form of discussion about commerce (for further discussion of commercial gentrification specifi-cally, see Meltzer in this volume). Specifically, the emergence of chain stores is often characterized as intrinsically oppositional to the hipness of neighbourhoods.[2] Chain stores are not only harbingers of gentrifica-tion, the story goes, but are symbolically set against hipsters and hipness. For example, Aubrey, who manages a clothing resale shop in the repu-tationally hip Chicago neighbourhood of Wicker Park (Schippers, 2002; Lloyd, 2006; Parker, 2018), claims that "an operational definition of hip-ster would be too cool for school," and that she "would love to see, like, you know, an Anthropologie, like a J. Crew, something that sort of fights the like hipster Wicker Park-ness, something just to kind of break up the, it's so impossibly cool here." When discussing what would break up the "hipster Wicker Park-ness," she names chains. This assessment of the role of chains as a break from hipness aligns with a more general understand-ing of neighbourhood hipness being existentially allergic to chains in popular culture, as in a note from the satirical lifestyle guide *The Hipster Handbook* that notes that hipsters "are very selective about where they drink" and that "franchises and chains are routinely avoided" (Lanham, 2002, p. 35). The hipster – and by extension, the hipster neighbour-hood – can be associated with what Veblen (1899) calls "invidious distinc-tion," in which individuals consume products in order to demonstrate social superiority, and it is notable that Veblen speaks of the idea of an "honorific crudeness" and the idea that "hand labour is a more wasteful method of production; hence the goods turned out by this method are more serviceable for the purpose of pecuniary reputability" (p. 97). On the basis of this standard, non-chains are honorific because they do not have the mark of being "machine-like" or related to mass production.

Ocejo (2017) has written about how traditionally low-status manual labour jobs like bartenders and butchers have acquired a veneer of cool in recent years, but I would argue it is not even necessary for hipsters to actually consume products at an establishment for it to be culturally valuable to them. Brown-Saracino (2007), for example, identifies a phenomenon she terms "virtuous marginality" in which "people associate authenticity with and highly value traits they do not share, and consequently, out of a desire to preserve the authentic, come to regard their distance from it – their marginality – as virtuous" (p. 439). Brown-Saracino is discussing people – whom she refers to as old-timers – but we might reasonably apply the same logic to retail establishments. As such, mom-and-pop stores can lend a veneer of authenticity to a neighbourhood that is crucial for its reputation for hipness, even if people we might call hipsters do not actually frequent them.

In this chapter I argue that this narrative that sets up mom-and-pops and chain stores as diametrically opposed is overly simplistic in its conception of gentrification and naive about how change happens. Chain stores are not *necessarily* antithetical to hipster neighbourhoods, which are typically associated in the popular imagination with independent, "mom-and-pop" stores, bars, and cafes. Without disputing the idea that resistance and co-optation are two possible outcomes of the prospect of gentrification in hip neighbourhoods, I suggest a third social process that can characterize such cases: coordination. Specifically, chain stores and mom-and-pop businesses – typically set up in opposition to each other – can coordinate toward a specific goal of maintaining hipness as a social and economic commodity in neighbourhoods where such a reputation is existent and financially remunerative. Stores typologized by Zukin as "'new entrepreneurial' retail capital (boutiques)," "'corporate' retail capital (chain stores)," and "old, 'local' retail stores" (Zukin et al., 2009, p. 58) can and sometimes do work together toward common goals.

Using the case of Chicago's Wicker Park neighbourhood, I show how chain stores can play a crucial role in the persistence of hipness in neighbourhoods. This is not because chains do not provide a sort of existential threat to what made those neighbourhoods hip in the first place – they often do – but rather because they can also provide material and strategic resources for the perpetuation of reputational hipness as a commodity, and in doing so provide a prophylactic against complete takeover by chains in the neighbourhood.

I conclude by discussing how rhetoric in both the academic and popular imagination often assign mom-and-pop stores value in terms of sentiment and chain stores value in terms of naked economics. As a result, we come to understand independent stores in terms of their symbolic value to the neighbourhood and corporate stores as purely instrumental interlopers. This heuristic obscures the actual social processes going on in neighbourhoods where actors, regardless of type, operate in economies of both money and sentiment.

Ignoring the Hipster

The main thing to know about academic treatment of the hipster is that, until recently (see especially le Grand 2020, Murray 2020, Steinhoff 2021) there has not been much of one. As Schiermer (2014) amusingly observes in an article addressed to this lack, "[T]he entry 'hipster' yields 75 million hits on Google – and thus exceeds the entry 'sociology' (73 million)" (p. 168). He is making the point that "this lack of interest contrasts glaringly with the enormous amount of attention devoted to the hipster phenomenon outside academia. There exists an immense number of opinions and observations on the hipster phenomenon made by journalists, bloggers, and layman experts of all categories" (p. 168). Indeed, sociology – my own discipline – has been so anaemic in its treatment of the hipster that a book of essays on the phenomenon of the hipster (Greif et al., 2010) – cited often by the few academic articles on the subject – there are zero articles written by sociologists.[3]

To the extent that the figure of the hipster has been theorized by academics, it has been analysed mostly in terms of its relationship to authenticity, whether through a lack of it or a valorization of it. They have been situated in terms of both Simmelian dialectics of individuality and imitation (Schiermer, 2014) and Bourdieuian (1984) fields of capital (Michael, 2013; Pederson et al., 2018), as well as incorporated into understandings of consumer behavior (Zeynep & Thompson, 2011). Monson (1995) offers an interesting history of the emergence of the hipster within the specific context of the racialized consumption of jazz music by whites in the early twentieth century, and Frank (1997) explains the rise of hip consumerism in the 1960s as the confluence of the counterculture with Madison Avenue. Other authors offer their own history of hipsters (for one example, see Schiermer, 2014, pp. 168–169), although perhaps the most extensive treatment of the subject comes from a trade paperback (Leland 2004).

It is not clear why there has been inadequate academic attention to theorizing the figure of the hipster, especially within urban sociology, considering the fact that urbanists invoke the hipster ethnographically fairly often (Lloyd, 2006; Zukin, 2011; Deener, 2012; Ocejo, 2014). Schiermer (2014, p. 170) stumbles upon one possibility when discussing the way

> even the hipsters themselves dismiss their proper belonging to the category, hurt by all categorizations implying the existence of imitative behaviour attributes. No hipster has written proudly about his culture proper. He might take himself seriously – not "hipster culture." The hipster has no *cause*; hipster culture possesses no manifestos (but an enormous number of manifestos against it); it has no instituted leaders; it has no clear borders; it is more inclusive and less uniform than the traditional subculture; it does not promote drug-use; it does not battle the police or the authorities; it does not market itself discursively as a distinct alternative or rebellious lifestyle – thus, finally, it does not try to settle issues with the previous generation. In short, hipster culture is no "real" subculture.

Regardless of the reason social scientists have failed to adequately engage with the topic, the figure of the hipster is a conduit to interesting potential areas of research. While there is definitional dispute over the figure of the hipster, Michael provides us with a good starting point when she tells us that

> the hipster can be seen as the ideal type of a trendy person: he or she is on top of current trends, owning vintage items before their remake appears in mainstream clothing chains, inhabiting the trendiest areas of urban centres and listening to the latest bands before they become popular, then quickly dismissing them when they get widely known. (2013, p. 164)

Their presence in "the trendiest areas of urban centres" ought to make them crucial to our study of neighbourhoods – and of the main streets running through them. What do hipsters and hipness actually *do* in neighbourhoods, particularly in relation to retailing, development, and gentrification?

Two Ways of Thinking about Hipness and Gentrification

Conceptualizations of hipness typically place hipness and capitalism in dialectical opposition, with the former rebelling against the latter and the latter co-opting the former. Specifically, hipness is a quality associated with counterculture that either resists and is overtaken by capitalism, or is co-opted by capitalism and thus becomes complicit in gentrification. Often such accounts are applied simultaneously to the same neighbourhood, as some lament decline while others are implicated in it. While not disputing the usefulness of such accounts, I will present a third way of thinking about the relationship between hipness and gentrification that neither of these paradigms would predict.

Resistance and Loss

One prevailing narrative of the role of hipness and hipsters in gentrifying neighbourhoods focuses on resistance to gentrification as represented by corporate retailers, usually followed by a sort of noble loss. This is part of what Wellman and Leighton (1979) refer to as a "community lost," in which social bonds are weakened in an increasingly bureaucratized society – in this case, characterized by the rise of chains over mom-and-pop stores. Lloyd (2006) documents the notion among some in hip neighbourhoods that everything is "always already over" (p. 253), suggesting that a sense of nostalgia – and an assertion that things used to be better before everything got ruined – is somewhat baked into hip neighbourhoods. Perhaps the most famous example of this general conflict between community and development – and the anxiety it provokes – is found in Jacobs (1961/1992), in which cohesive mixed-use neighbourhoods are under threat from top-down development. This conception of city life has cast a long shadow. For example, Zukin (2010) laments "the way [New York City] has morphed from a lumbering modern giant to a smooth, sleek, more expensive replica of its former self" (p. x). In a different piece, she and a group of co-authors (Zukin et al., 2009, p. 48) note that

> the arrival of chain stores in areas that had previously depended on small, individually owned shops disrupts social bonds, as long-term residents must decide between shopping at the corner bodega – whose owner may offer credit – or switching to a well-stocked but impersonal supermarket. Residents may not even face a choice if old stores are unable to pay rising rents and disappear when their lease ends, or if the stores are replaced by new "luxury" condos.

They also note "Williamsburg's perceived resistance, as a neighbourhood of hipsters, to most chain stores" (p. 60). Sure enough, though, a year later the *New York Times* ran an article on the neighbourhood about the arrival of chain stores (Bagli, 2010).

Co-optation and Complicity

In terms of co-optation, Lloyd (2006) provides an illustrative example in his treatment of 1990s Wicker Park, which he describes as a synthesis of bohemia and capitalism. At work are two seemingly contradictory forces, coming together in the 1990s, resulting in "slackers" who do not want straight jobs (that increasingly do not exist in a post-Fordist economy in any case), providing a reserve army of hipster labour. Mele (2000) tells the story of the "Selling [of] the Lower East Side" in which the diversity of the East Village in New York City is what was used to sell it by real estate developers. Hubbard (2016) goes even further and implicates hipsters and hipness in a systematic process of valuer extraction from neighbourhoods traditionally inhabited by the working class and people of colour, noting,

> [W]hat sells is a form of consumption in which knowledgeable white entrepreneurs turn "sketchy" neighbourhoods into ones characterised by a more navigable and "safer" version of cosmopolitanism. Incoming hipster businesses are complicit in this process given they trade on the cache [7] of being in an edgy multicultural neighbourhood, but offer forms of consumption easily intelligible to the white middle class readers of the Sunday supplements and style magazines. This suggests an almost complete enrolment of hipster cultures within an infrastructure of gentrification that involves international lifestyle commentators, restaurant reviewers, and real estate agents discoursing particular neighbourhoods as cosmopolitan and cool. (Hubbard, 2016)

In this vision of urban hipness and hipsters, capitalism co-opts a subcultural value and turns it into something that one can use to make money. Put succinctly by Greif (2010a), "the hipster is that person, overlapping with declassing or disaffiliating groups ... who in fact aligns himself *both* with rebel subculture *and* with the dominant class, and opens up a poisonous conduit between the two" (p. 9). Such a structural position – perhaps not the main antagonist, but a sort of class traitor or stooge of capital – has provoked protests of certain businesses associated with gentrification (Wilkinson, 2016) by people opposed to cultural appropriation, gentrification, displacement, or some combination of the three.

Wicker Park as an Illustrative Case

Like Frank (1997), who observed that the counterculture was not so much co-opted by Madison Avenue as co-constituted with it, I would like to draw attention to the way capitalism and neighbourhood hipness do not always work in opposition to each other, and indeed sometimes work in tandem. Using the case study of the same Chicago neighbourhood where Lloyd documented bohemia being co-opted by capitalism in the 1990s, I suggest that the maintenance of Wicker Park's reputation for hipness has depended partially on the capitalist agents of neighbourhood change that are thought to represent its very destruction, and particularly on the cooperation of chain stores with mom-and-pops in propagating the neighbourhood's reputation.

The insights in this chapter are drawn from a larger project about place reputation. While that project spanned three neighbourhoods and drew from interviews with 100 merchants and other stakeholders from 2012 to 2019, I draw from one of those neighbourhoods – Wicker Park – and 29 interviews, all conducted in 2012 and 2013. The interviews were supplemented with ethnographic fieldwork, historical research, and analysis of media coverage of the neighbourhood, but most of what I will be discussing in this chapter comes from the interviews themselves.

Being interested in social processes surrounding place reputation, I targeted merchants as what I call "frontline reputational actors," figures both particularly attuned to and active in the construction of place reputation. They are particularly attuned to place reputation because they have to be – their livelihoods depend on their knowledge of what potential customers think of their neighbourhoods and what draws them to it – and they are active in its construction to the extent that they strategically manipulate it to funnel more people towards their stores. Storefronts themselves are both economic drivers of urban neighbourhoods and recognizable symbols of community culture (see Cortwright & Mahmoudi, 2016).

Wicker Park is a neighbourhood northwest of Chicago's downtown Loop, lying along the Blue Line elevated tracks that shuttle people to and from O'Hare International Airport (Figure 9.1). As documented elsewhere (Lloyd, 2006; Parker. 2018), Wicker Park is historically a site of first settlement for immigrant groups in Chicago, transitioning from a neighbourhood of Germans and Poles at the end of the nineteenth century to a neighbourhood of Mexicans and Puerto Ricans in the middle of the twentieth century. A period of steep decline caused by patterns common to American cities in the late twentieth century – deindustrialization, white flight, and disinvestment – created a neighbourhood attractive to artists and musicians on the lookout for cheap rent and large spaces to create, and Lloyd (2006) documents the resultant emergence of Wicker Park as a "Neo-Bohemia" in the 1980s and 1990s. A neighbourhood once known primarily for light manufacturing and Nelson Algren's writing about hustlers and junkies was now known nationally as

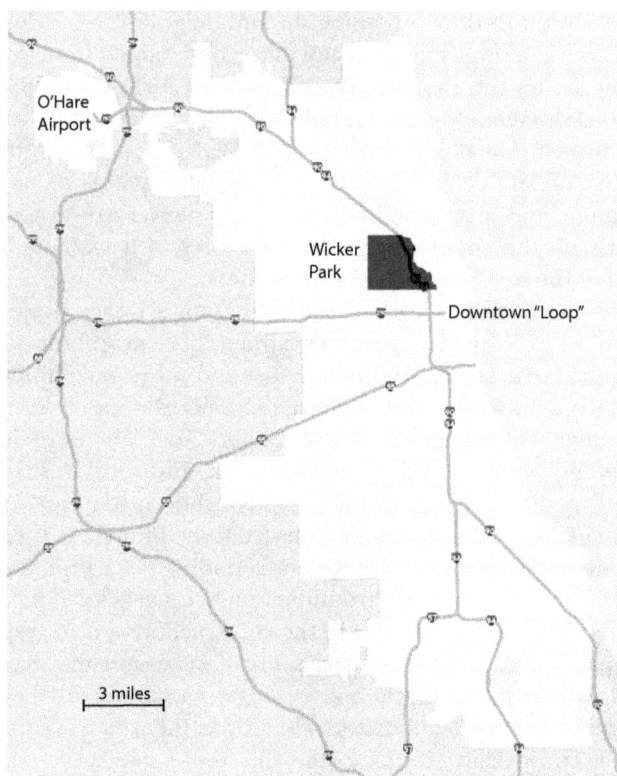

Figure 9.1. Wicker Park.

the subject of Liz Phair's record *Exile in Guyville* and the site of an early season of MTV's ur-reality show *Real World*. In the years since, Wicker Park has only become richer and more inaccessible to the artists Lloyd wrote about. Chains have moved into the neighbourhood en masse, and homes in the area sell for millions of dollars (Rodkin, 2019). At this point, the neighbourhood is a mix of chains and mom-and-pops. As I have documented elsewhere (Parker, 2018), merchants in Wicker Park often bemoan the decline of the neighbourhood's hipness, and yet the reputation persists: in 2012, the neighbourhood was named a top five "hippest hipster neighbourhood" by Forbes (Brennan, 2012), and in 2017, the city of Chicago's marketing branch identified the neighbourhood as one of three "hipster havens" in the city (Ward 2017).

I argue, however, that this maintenance of reputation has not been in spite of the chains, but partially because of them, as actors whom social scientists have normally drawn strict lines between – specifically what Zukin calls "'new entrepreneurial' retail capital (boutiques),'' "'corporate' retail capital (chain stores),'' and "old, 'local' retail stores" (Zukin et al. 2009, p. 58) actually work in tandem in the pursuit of joint reputational goals. While I use data from one neighbourhood in Chicago, it is illustrative of social processes that can occur across local contexts.

Coordinating Hipness

In Wicker Park, chain stores have been maintaining the neighbourhood as a place open to traditionally hip, that is, *not* corporate, places. This is not dependent on any sort of altruism of chains, and in fact one can find instrumental calculation behind the actions and attitudes without looking too hard. What this findings reveals is that (1) people involved with chain stores recognize that it is in their interest to maintain a neighbourhood's reputation for hipness, given how financially remunerative it is, (2) they recognize that such maintenance is contingent on protecting non-chains in the neighbourhood, and (3) as a result, it is possible for stores that differ across multiple dimensions to coordinate in order to maintain a reputation that makes them all money.

While I discuss actions and attitudes common to chain and independent stores in Wicker Park elsewhere (Parker, 2018), the most important thing to note about Wicker Park chains for the purposes of this chapter is that they have been maintaining a mix of different types of retail in the neighbourhood. Much of this has been done under the auspices of the local Special Service Area (SSA). As the city website puts it, "[K]nown as Business Improvement Districts or BIDs in other cities, [SSAs] are local tax districts that fund expanded services and programs through a localized property tax levy within contiguous areas" (City of Chicago, 2022). In the case of the Wicker Park–Bucktown SSA, that money has gone toward things like bright orange bicycle racks, street festivals with indie rock bands in the summer, and a master plan put together by urban planners to identify the neighbourhood's resources and plan its trajectory. The people involved with the SSA are overwhelmingly local merchants – at one meeting I attended, when someone asked who in the room owned a business, everyone but one other person and me raised their hand, despite the fact that these meetings are open to the public and the SSA is transparent about what it does. When I interviewed Jamie, an urban planner who worked on the master plan, she told me that there were a variety of people on the planning committee. She identified one group of people who complained:

> Oh, the neighbourhood is so different from what it used to be. There used to be all this great stuff going on and now it's not anymore ... People would say this used to be such a great art scene here 20 years ago when I first moved in and there were all these great cafes and people would talk for hours in these cafes and the music scene – blah, blah, blah.

Another group she identified as "very business-minded, developer-minded."

While this seems like a recipe for intractable conflict, Jamie says it actually came together because the second group recognized the value of what the first group wanted, saying, "They would try to care about the art scene and the music scene because they did recognize that it did bring a unique flavour to the neighbourhood and it did draw people in." Notice that this recognition does not necessarily mean that pro-development people began to value art and music intrinsically. Instead, they recognized its financial value – tied up in Wicker Park's reputation for hipness – and acted to preserve that reputation in their own interest. This also manifested itself in managers of chains talking about the value of independents in the neighbourhood and even disparaging chains, despite the fact that they themselves were associated with them (Parker, 2018).

Discussion and Conclusion

What do these attitudes of chain store operators tell us about how we conceptualize hipness and its connection to neighbourhood development? Primarily, they tell us that the two major ways neighbourhood hipness and counterculture has been portrayed – as a resistant force likely to lose out to gentrification or a dupe/handmaiden of capitalism – do not exhaust the possibilities of this relationship. Contemporary Wicker Park is partially a result of recognition *by pro-development merchants* that things like art and music contribute to a reputation that is financially remunerative and thus worthy of protecting from unfettered economic development. As a result, a mix of mom-and-pop and corporate retail has been maintained in Wicker Park, a neighbourhood whose affluent residential profile would suggest a complete takeover by those businesses with the most capital, for example, chain stores.[4] In this instance, chains' recognition that it was in their interest to preserve space for mom-and-pop stores served as a prophylactic against a more extreme version of gentrification.

Of course, this understanding is open to its own critique. Those who hold to the model of co-optation – that independence is used as a resource for larger forces of capital – can reasonably say that this is an instance where capital still wins, and the security of remaining mom-and-pop stores exists only because of the symbolic value they have to those with more money. Even if this is true, it still represents a different vision of the interplay of capital and hipness in the urban political economy landscape. Instead of capital seeking to control hipness and using it to sell a neighbourhood, representatives of hipness have forced capital to scale back from an otherwise total transformation of a neighbourhood, not out of the goodness of their hearts, but because a case can be made that capital still needs hipness in Wicker Park. Instead of just setting the groundwork for more capital-intensive actors to move into the neighbourhood as part of a stage version of gentrification, we see that hipness has made itself irreplaceable in neighbourhoods primarily defined by it. Tourists will not come to the neighbourhood if *all* the coffee shops are replaced with banks, and while this does not take the sting out of the very real turnover and displacement of mom-and-pop stores, it does suggest a strategy to save at least some of them. Corporations have found a way to monetize mom-and-pops, but at least the importance of mom-and-pops to the reputation of the neighbourhood is being recognized. This may not reassume people whose main concern is the larger forces of capitalism and globalization and its effect on cities, but it does offer a strategy for maintaining otherwise vulnerable mom-and-pop retail, albeit in a decidedly neoliberal context.

A final point to be made is that the rhetoric surrounding gentrification, both in academic writing and in the popular imagination, is missing some of this complexity. Essentially, in narratives of neighbourhood change and commercial gentrification, the characters in the play are set as mom-and-pop stores as holders of sentiment and chain stores as holders of rational choice economics. We read independent stores in terms of their symbolic value to the neighbourhood and corporate stores as purely instrumental interlopers. While this understanding indeed describes much of what we see in cities – after all, resistance and co-optation have become tropes because they are often true – we do ourselves a disservice if we fail to see how mom-and-pops can be strategic and chain stores can be guided by sentiment. While the moral and economic calculuses might differ from case to case, every business – no matter the type – balances financial incentives with cultural commitments. Moreover, merchants of all stripes – so often analytically divided to better explicate their divergent interests – can and do coordinate when their interests align. While there are certainly cases where chains *do* pose a threat to mom-and-pop retail in hipster neighbourhoods, and merchants react negatively to them (Parker, 2018), it is not always the case, and their mutual interests in reputation maintenance make for strange bedfellows. They operate in an ecosystem in which, in some cases, they can benefit from the same outcomes, even if their incentive structures are different. In order to better understand the crucial role of merchants in the landscape of changing neighbourhoods, we must recognize this fact, and the resultant truth that hipness does not have to lose to or be co-opted by capital in the struggle over Main Street. It can also set its own terms and demand some concessions.

NOTES

1 Note that "hipster," while usually a noun, can also be used as an adjective. For example, Irvin (2017), in her typology of food trucks in New Orleans, discusses hipster food trucks. For the purposes of this chapter, I will be discussing a larger bundle of associated phrases – "hipster" as both noun and verb, as well as "hip" and "hipness."

2 "Hip" and "hipster" is notoriously difficult to define (Greif, 2010c), and often in academic sociology, urban ethnographers invoke the figure without carefully defining it. Systematically defining this figure, therefore, is an important task for those of us who talk about hipsters, but for the purposes of this paper, I rely on accounts of hipsters and hipness from my respondents and popular culture.

3 Greif (2010b), one of the conveners of the colloquium and editors of the volume, notes that "the authorities missing in this account, I regret to say, are professional social scientists."

4 Given that chain stores typically have more financial backing to help weather economic shocks than mom-and-pops, we should keep an eye on the potentially changing mix in neighbourhoods like Wicker Park in the aftermath of COVID-19. As more mom-and-pops become unable to afford to stay open, it will be interesting to see whether coordination between mom-and-pops and corporate chains to give them a lifeline actually happens, or if a new cultural milieu, more dominated by chains, emerges.

REFERENCES

Arsel, Z., & Thompson, C. J. (2011). Demythologizing consumption practices: How consumers protect their field-dependent identity investments from devaluing marketplace myths. *The Journal of Consumer Research, 37*(5), 791–806. https://doi.org/10.1086/656389

Bagli, C. V. (2010, 12 November). As a neighborhood shifts, the chain stores arrive. *The New York Times.* https://www.nytimes.com/2010/11/13 /nyregion/13metjournal.html

Bourdieu, P. (1984). *Distinction: A social critique of the judgment of taste.* Routledge.

Brennan, M. (2012, 20 September). America's hippest hipster neighborhoods. *Forbes.* http://www.forbes.com/sites/morganbrennan/2012/09/20 /americas-hippest-hipster-neighborhoods/

Brown-Saracino, J. (2007). Virtuous marginality: Social preservationists and the selection of the old-timer. *Theory and Society, 36*(5), 437–468. https:// doi.org/10.1007/s11186-007-9041-1

City of Chicago. (2022). Economic development. https://www.chicago.gov /city/en/depts/dcd/provdrs/ec_dev.html

Cortwright, J., & Mahmoudi, D. (2016). The storefront index. City Report. https://cityobservatory.org/wp-content/uploads/2016/04/Storefront _Index_April_2016.pdf

Deener, A. (2012). *Venice: A contested bohemia in Los Angeles.* University of Chicago Press.

Frank, T. (1997). *The conquest of cool: Business culture, counterculture, and the rise of hip consumerism.* The University of Chicago Press.

Greif, M. (2010a). Positions. In Greif, Ross, & Tortotici, *What was the hipster? A sociological investigation* (pp. 4–13).

Greif, M. (2010b). Preface. In Greif, Ross, & Tortotici, *What was the hipster? A sociological Investigation* (pp. vii–xvii).

Greif, M. (2010c, 22 October). What was the hipster? *New York.* http://nymag .com/news/features/69129/

Greif, M., Ross, K., & Tortotici, D. (Eds.). (2010). *What was the hipster? A sociological investigation.* n+1 Foundation.

Hubbard, P. (2016). Hipsters on our High Streets: Consuming the gentrification frontier. *Sociological Research Online, 21*(3). https://doi.org /10.5153/sro.3962

Irvin, C. (2017). Constructing hybridized authenticities in the gourmet food truck scene. *Symbolic Interaction, 40*(1), 43–62. https://doi.org/10.1002 /symb.267

Jacobs, J. (1992). *The death and life of great American cities.* Vintage. (Original work published 1961)

Lanham, R. (2002). *The hipster handbook.* Anchor Books.

le Grand, E. (2020). Moralization and classification struggles over gentrification and the hipster figure in austerity Britain. *Journal of Urban Affairs.* https://doi .org/10.1080/07352166.2020.1839348

Leland, J. (2004). *Hip: The history.* Ecco.

Lloyd, R. (2006). *Neo-bohemia: Art and commerce in the postindustrial city.* Routledge.

Mele, C. (2000). *Selling the Lower East Side: Culture, real estate, and resistance in New York City*. University of Minnesota Press.

Michael, J. (2013). It's really not hip to be a hipster: Negotiating trends and authenticity in the cultural field. *Journal of Consumer Culture, 15*(2): 163–182. https://doi.org/10.1177/1469540513493206

Monson, I. (1995). The problem with white hipness: Race, gender, and cultural conceptions in jazz historical discourse. *Journal of the American Musicological Society, 48*(3): 396–422. https://doi.org/10.2307/3519833

Murray, M. A. (2020). White, male, and bartending in Detroit: Masculinity work in a hipster scene. *Journal of Contemporary Ethnography, 49*(4), 456–480. https://doi.org/10.1177/0891241620907126

Ocejo, R. (2014). *Upscaling downtown: From Bowery saloons to cocktail bars in New York City*. Princeton University Press.

Ocejo, R. (2017). *Masters of craft: Old jobs in the new urban economy*. Princeton University Press.

Parker, J. N. (2018). Negotiating the space between avant-garde and "hip enough": Businesses and commercial gentrification in Wicker Park. *City & Community, 17*, 438–460. https://doi.org/10.1111/cico.12294

Pederson, W., Jarness, V., & Flemmen, M. (2018). Revenge of the nerds: Cultural capital and the politics of lifestyle among adolescent elites. *Poetics, 70*, 54–66. https://doi.org/10.1016/j.poetic.2018.05.002

Rodkin, D. (2019, 17 June). Peek inside a 19th-century Wicker Park house with a 21st-century addition for sale at almost $3 million. *Crain's Chicago Business*. https://www.chicagobusiness.com/residential-real-estate/peek-inside-19th -century-wicker-park-house-21st-century-addition-sale

Roschen, M. (n.d.). Special Service Area (SSA) Program. City of Chicago. https://www.chicago.gov/content/city/en/depts/dcd/supp_info/special _service_areassaprogram.html

Schiermer, B. (2014). Late-modern hipsters: New tendencies in popular culture. *Acta Sociologica, 57*(2), 167–181. https://doi.org/10.1177/0001699313498263

Schippers, M. (2002). *Rockin' out of the box: Gender maneuvering in alternative hard rock*. Rutgers University Press.

Steinhoff, H. (Ed.). (2021). *Hipster culture: Transnational and intersectional perspectives*. Bloomsbury.

Veblen, T. (1899). *The theory of the leisure class*. Dover Publications.

Ward, J. (2017, 8 August). Do you Llive in a "hipster haven"? Tourism Bureau names 3 hippest 'hoods. *DNAInfo*. https://www.dnainfo.com/chicago/20170808 /bridgeport/chicago-hipster-havens-bridgeport-choose-chicago-wicker-park -bucktown-logan-square-janet-scanlon-hardscrabble/

Wellman, B., & Leighton, B. (1979). Networks, neighborhoods, and communities: Approaches to the study of the community question. *Urban Affairs Quarterly, 14*(3), 363–390. https://doi.org/10.1177/107808747901400305

Wilkinson, E. (2016). "Let us devastate the avenues where the wealthy live": Resisting gentrification in the 21st century city. *Sociological Research Online*, *21*(3), 156–162. https://doi.org/10.5153/sro.4026

Zukin, S. (2010). *Naked city: The death and life of authentic urban places*. Oxford University Press.

Zukin, S. (2011). Reconstructing the authenticity of place. *Theory and Society, 40*, 161–165. https://doi.org/10.1007/s11186-010-9133-1

Zukin, S., Trujillo, V., Frase, P., Jackson, D., Recuber, T., & Walker, A. (2009). New retail capital and neighborhood change: Boutiques and gentrification in New York City. *City & Community, 8*(1), 47–64. https://doi.org/10.1111/j.1540-6040.2009.01269.x

PART FOUR

Retail, Place, and Place-Making

10 Retail Scenes

HYESUN JEONG AND TERRY NICHOLS CLARK

Retail is generally thought of as a form of production, driven by investors, entrepreneurs, and profits. Yet recent statistics show dramatic differences on the consumption side of retail, with some types growing and others declining. What drives these consumer preferences? In this chapter we get at this question by exploring clusters of retail activities that combine to create retail "scenes," which are defined as structured types of social consumption measured by specific local characteristics (Silver & Clark, 2016). We start with some simple empirical results to lay the groundwork, and then use case studies to show how clusters of cafés and related amenities create street scenes that define place quality. Our aim is to show that retail should not be treated as an individual transaction measured by sales volume, but rather is best understood as being part of a retail scene.

Personal Consumption Is the Largest Sector of US GDP

The main components of the US gross domestic product (GDP) demonstrate the importance of consumption. In 2018, consumer spending accounted for 69 per cent of GDP, while business investment was just 18 per cent. This spending has two parts: consumer goods and services. Goods are the classic icons of the industrial economy, like televisions and refrigerators, but represent only about 25 per cent of GDP. Many of these goods have become commodities that facilitate global market competition and can be produced anywhere and ordered online. This has led to a decline in some types of local retail, such as traditional department stores, bookstores, and music stores, which have all significantly dropped in sales and employment (Hortaçsu & Syverson, 2015).

While consumer services are almost twice as large as consumer goods, comprising 45 per cent of GDP, they are understudied (Table 10.1).

Table 10.1. Components of real US GDP (2018)

Component	Amount ($ trillions)	%
Personal consumption	12.89	69
Goods	4.55	25
Services	8.36	45
Business investment	3.39	18
Government	3.18	17
Federal	1.23	7
State and local	1.95	10
Total GDP	18.57	100

Source: "Table 1.1.6. Real GDP," Bureau of Economic Analysis. "Concepts and Methods of the US National Income and Product Accounts."

Notes: Net imports minus exports is –4 per cent and is omitted here. Adapted by the authors.

Consumer services are more complex than consumer goods, as service firms are inherently more local and personal and largely work in smaller and more specific sub-markets of customers. From barbershops to restaurants, services are more visible to the local consumer who can judge a haircut or taste the food, that is, *experience the service*, and assess the price. However, the reasons behind a customer's choice of a service tend to be relegated to "consumer preferences" by economists. Further, economists typically leave the content of preferences to "other disciplines." Many economists at Chicago such as Gary Becker, James Heckman, and Ali Hortaçsu have pioneered the theorization of amenities and consumer behaviour. However, most economists and sociologists of culture tend to atomistically analyse localities by focusing on just one amenity or only a few, such as restaurants, museums, or bookstores, assuming *ceteris paribus*, that "all else is equal." This atomism neglects the context of other surrounding amenities and combinations of amenities that differently affect consumer behaviour (Silver & Clark, 2016).

Bankers, business consultants, accountants, lawyers, and large corporate participants who buy, sell, and manage their own retail outlets are often insensitive to the preferences of local consumers. Many sales professionals stress advertising and marketing, using language like selling "product" X to "market segment" Y, which plays down the specifics of consumer tastes.

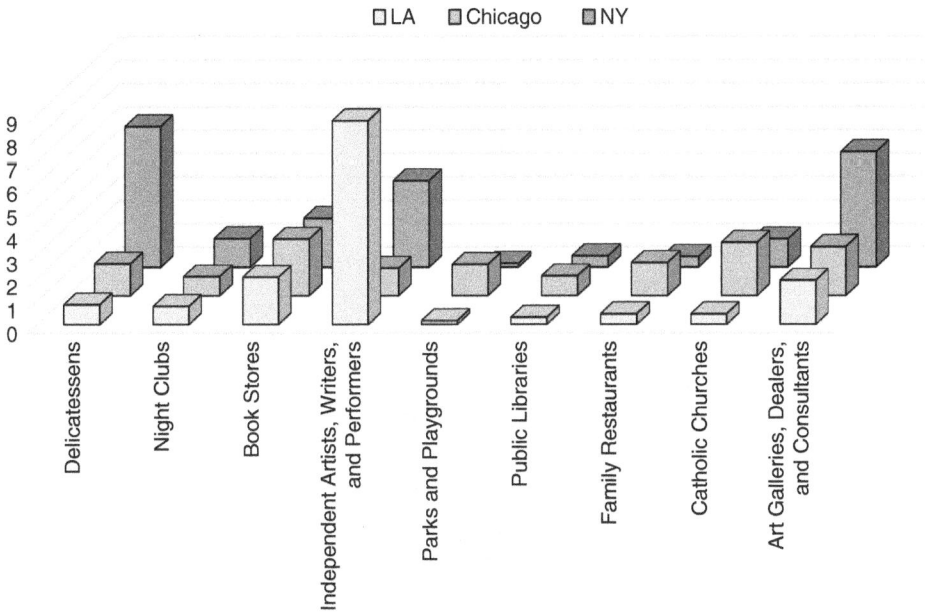

Figure 10.1. Retail types differ across cities. Mean number of firms by NAICS industrial codes per ZIP code in county areas of LA, Cook, and 5 NYC county boroughs, 2008. (Data source: Silver and Clark [2016].)

Local Retail Variations Are Substantial

Besides ignoring taste, some retail research ignores local context, since it is thought to be disappearing in favour of corporate commodification. This perspective has been summarized as Hollywoodization or McDonaldization, illustrated by Ritzer (1993). But even if some commodification is undeniable, ignoring local and personal variations in consumption patterns is a poor analytical strategy for anyone concerned with how retail works.

There is much simpler evidence that local preferences matter. Figure 10.1 illustrates similarities and differences among consumption preferences for the three largest US cities – and the differences are clear. Many banks and multinational corporations have not incorporated these local consumption concerns, instead continuing to use simple standard measures in their lending and location practices.

The growing importance of the consumer is critical, not simply in the United States but globally. Figure 10.2 provides data for retail in Canada.

Live Theatres and Other Performing Arts Presenters with Facilities
Performing Arts Promoters and Presenters without Facilities
Other Performing Arts Companies
Independent Artists, Writers, and Performers
Dance Companies
Musical Groups and Artists
Festivals without Facilities
Other Spectator Sports
Sports Teams and Clubs
Agents and Managers for Artists, Athletes, and Entertainers
Musical Theatre and Opera Companies
Fitness and Recreational Sports Centres
Sports Stadiums and Other Presenters with Facilities
Sports Presenters and Other Presenters without Facilities
Total
Historic and Heritage Sites
Sound Recording Studios
Full-Service Restaurants
Non-commercial Art Museums and Galleries
Museums Except Art Museums and Galleries
Book Stores and News Dealers
Musical Instruments and Supplies Stores
Horse Race Tracks
Amusement and Theme Parks
Motion Picture and Video Exhibition
Drinking Places, Alcoholic Beverages
Bowling Centres
Pre-recorded Tapes, Compact Disc, and Record Stores
Zoos and Botanical Gardens
Amusement Arcades

−50%   0%   50%   100%   150%

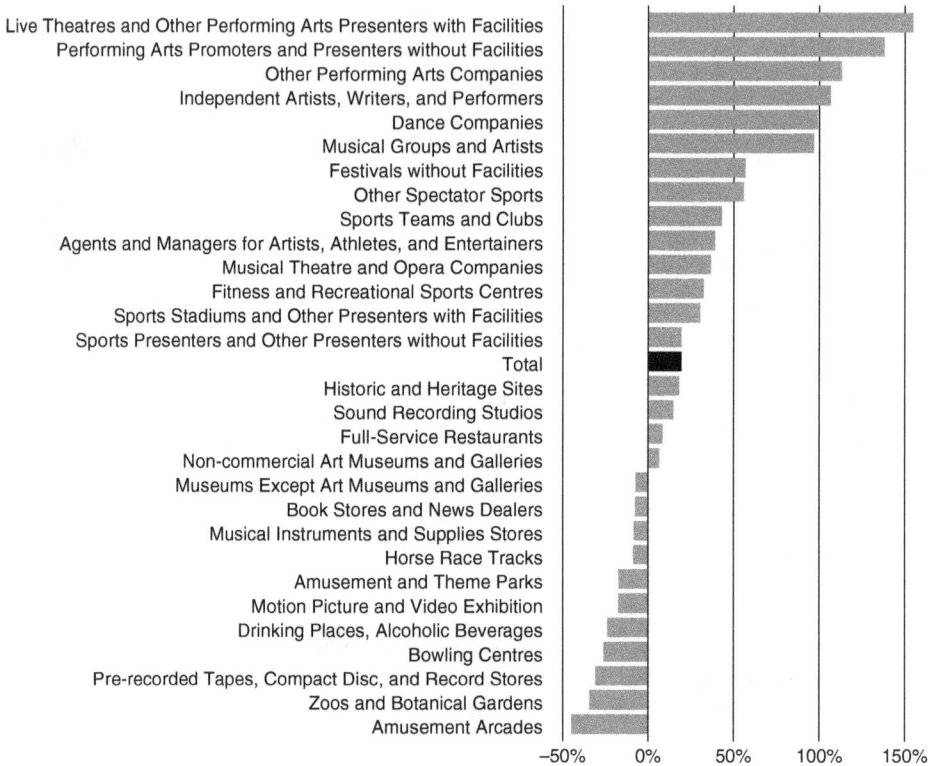

Figure 10.2. Differences in Canada in growth or decline by type of firm, 1999–2008. This figure shows the percentage change in the number of arts, culture, and leisure establishments across Canada. For example, dance companies and musical groups nearly doubled over this period. The black bar shows the average growth in total business as a benchmark: everything above it grew faster and everything below it either grew more slowly or contracted during this period. The figure uses six-digit NAICS categories and can be found at http://www23.statcan.gc.ca/imdb/p3VD.pl?Function=getVD&TVD=118464. (Data source: Statistics Canada, Canadian Business Patterns (1999–2008); chart by Daniel Silver. Canadian NAICS codes are nearly identical to those in the United States. Figure adapted from Silver and Clark [2016].)

The figure shows highly varied patterns in growth and decline for different types of retail consumption. Such variations are not unique to the United States or Canada; even China, which has long had a strong state production-oriented export economy, is rapidly incorporating more domestic consumption-driven concerns.

Street Scenes and Retail Clusters

A café does not stand alone but is part of a larger local street scene. It is a retail form that often includes active street frontage with transparent windows and doors that give patrons a view of daily life. Storefronts that include decorations, themed displays, or community information – not blank walls – are able to engage pedestrians on the street. These physical elements aim to create a constant social exchange on the sidewalk that then leads to conversation inside the café (Gehl et al., 2006). As Jane Jacobs argued, a city's wealth of public life grows from sidewalk contacts, and such street activity fosters incremental urban growth (Jacobs, 1961). Mehta & Bosson (2010) measured the physical characteristics of cafés and found that seating and shelter facilitated the sociability of the community.

Creating public spaces to enhance social interaction has become crucial in urban design. Jan Gehl's transformation of Times Square (2009) to a more car-free space with chairs and tables was devoted to this purpose of creating human-scale and sociable urban space. Public space enhancement – that is, design strategies to promote social interaction – is increasingly seen as important for retail. For example, in the San Francisco Parklet project, temporary platforms extend sidewalks and replace parking spots with public space for programs such as outdoor sitting, bicycle storage, and pop-up stores. The first parklet was built near the Mojo Bicycle Café in 2010 (Figure 10.3). Metropolitan Planning Council (n.d.) conducted a survey in 2014 and found that 80 per cent of the merchants with a parklet near their storefront reported increased pedestrian traffic, and 34 per cent of those pedestrians made spontaneous food or beverage purchases. Besides attracting more business, they quickly became community meeting places. The success of small-scale interventions, also known as Tactical (or Lean) Urbanism, has spread to many other cities including New York, Chicago, and Dallas as a simple way to promote local activities and engage citizens.

Parklets are an application of Jane Jacobs' ideas about the importance of pedestrian quality and street life (Talen, 2005). Richard Florida (2002) similarly identified small amenities as being important for street interaction and "creative" firms. Clark (2004) suggested that amenities of this sort are an increasingly important part of lifestyle choices and cultural activities, especially as the city becomes an "entertainment machine."

Importantly, it is the clustering of consumption amenities – for example, cafés, art galleries, and fusion restaurants – that is critical to the creation of street scenes. Consumption amenities provide entertainment

Figure 10.3. Mojo Bicycle Café with parklet, San Francisco. (Credit: Photo by Jeremy Shaw.)

often continued to street-level activities like festivals, biking and walking events, and flea markets. Aggregated data on cafés, breweries, bars, bookstores, barbershops, religious organizations, museums, art galleries, theatres, and news stands can be used to analyse how these establishments combine to generate distinct entertainment patterns. It is an analysis that extends Oldenburg's (1999) unitary concept of third places by adding civic, religious, and arts institutions (we do not assume that all such places share a cultural style, as Oldenburg did). Data for most of these amenities are available online from the US Census by ZIP Codes Business Patterns (ZBP). We illustrate these patterns with 10 years of changes in all ZIP codes in the United States. The data show that the retail firms providing non-alcoholic beverages and snacks (the category including cafés), barbershops, and breweries substantially increased after 2012 (Figure 10.4). In contrast to the rise of breweries (NAICS 312120), conventional bars selling alcoholic beverages (NAICS 722410) have decreased. While these findings may appear contradictory, they reflect a rise in more refined consumer tastes that prefer local specialty coffee, food, and "craft" beer, suggesting increased awareness regarding health and environmental concerns.

Coffee shops are not reported as a distinct category at the ZIP code level by the US Census, but instead as a subcategory of "snacks and non-alcoholic beverages" (NAICS 722515; NAICS 722213); however, data for coffee

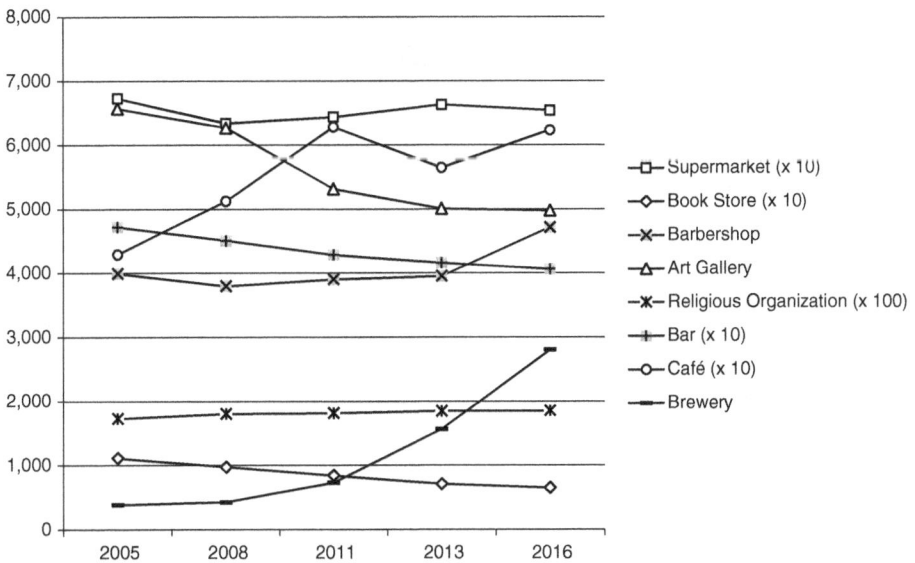

Figure 10.4. Change of selected amenities, 2005–2016, for all US ZIP codes. (Data source: US Census Bureau, County Business Patterns 2005–2016, chart created by the authors.)

shops are publicly reported at the national level. Figure 10.5 shows the national results for coffee shops separated from snacks and non-alcoholic beverages. Both categories greatly increased from 2002 to 2012. Coffee shops almost doubled in the first five years, but grew slower thereafter.

Figure 10.6 shows that some retail types strongly overlap. In particular, cafés cluster with bakeries, bookstores, music stores, health clubs, art galleries, and supermarkets. In contrast to neoclassic economics, which treats transactions atomistically, this clustering implies that combinations of activities are highly relevant for analysis. Marketing analysts treat buying a book as unrelated to buying a cup of coffee, as these transactions occur in two different shops, yet empirically, certain retail activities cluster geographically. Treating retail in terms of individual sales transactions overlooks this effect – that is, that combined activities attract consumers to particular neighbourhoods. The right mix of commercial shops creates a non-market amenity: a neighbourhood street scene. Consumers don't just buy, they also look and discuss their preferences with friends,

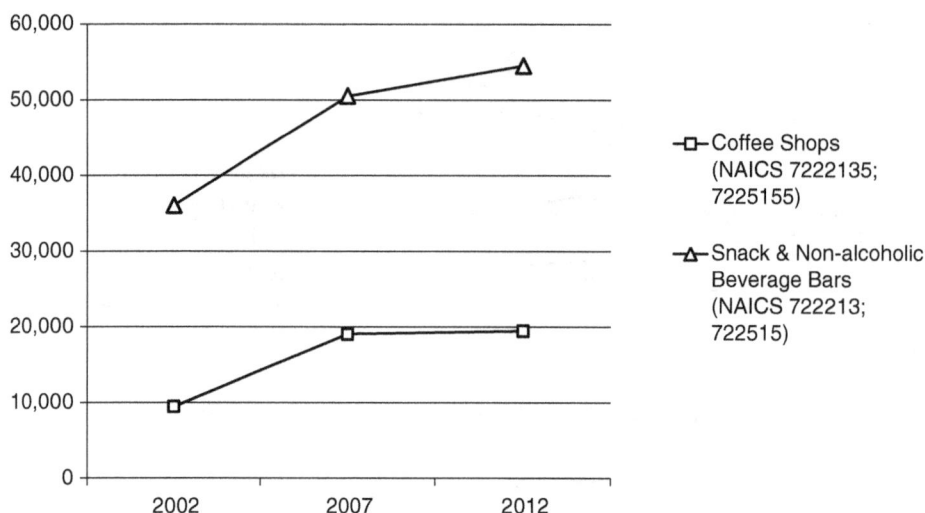

Figure 10.5. Changes in coffee shops and snack and non-alcoholic beverage bars in the United States, 2002–2012. (Data source: US Census Bureau, County Business Patterns, modified by the authors.)

walking by multiple storefronts and sitting in cafés. The retail associations shown in Figure 10.6 become the building blocks of distinct scenes.

Case Studies of Chicago Scenes

We elaborate this idea in three cases of Chicago café scenes. In Chicago, cafés concentrate in Lincoln Park (N = 36), Streeterville and West Loop (N = 28), Lakeview (N = 28), and Wicker Park (N = 26) on the North Side. On the South Side, the University of Chicago in Hyde Park has a fair number of cafés (N = 21) or 0.123 per 1,000 residents (Figure 10.7).

Three Types of Café Scenes

We conceptualize three types of scenes and their cafés: (1) Bohemia, (2) Corporate/Utilitarian, and (3) Traditional/Local/Neighbourly. The three types use distinct combinations of 15 scene dimensions (Silver & Clark, 2016) as foundational elements for contrasting different café environments. These are measured with ZCTA scores for each of the 15 scenes dimensions, using data from the Yellow Pages and ZIP Codes Business Patterns (ZBP).

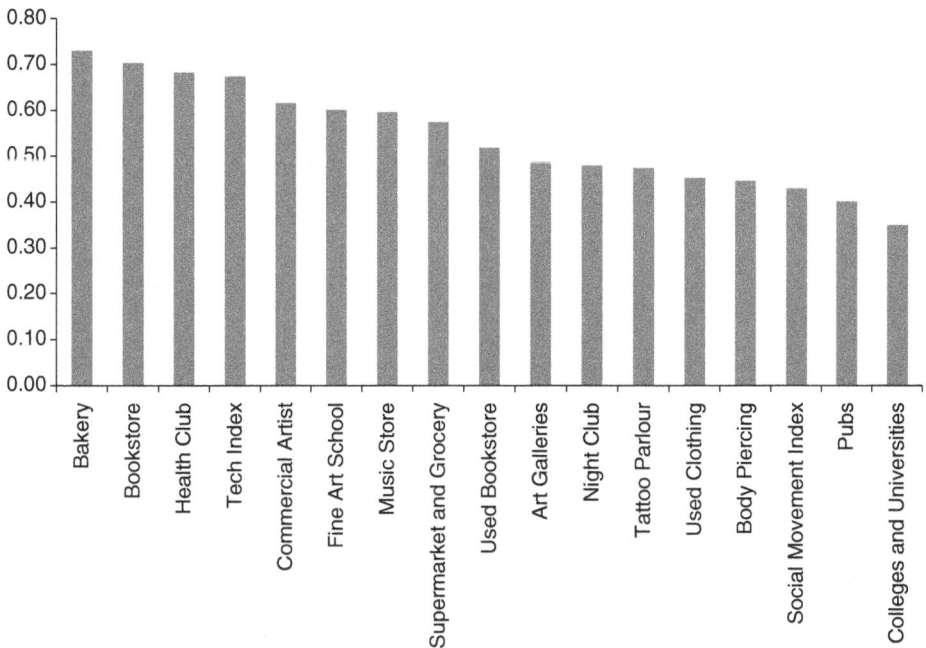

Figure 10.6. Pearson correlation between café and other amenities at ZIP code level, significant above 0.01 level. (Data source: ZIP code sums of individual listings electronic Yellow Pages [2007]. Chart created by the authors.)

1. Bohemian: The bohemian scene is often associated with clusters of small hip cafés, tattoo parlours, night entertainment, and artistic cultural amenities. In *Scenescapes* (2016), Silver and Clark characterize the bohemian neighbourhood as having higher scores for transgression and self-expression, and lower utilitarian and corporate scores. Gravitating to areas with lower rent, neo-bohemians and artists often develop or repurpose older industrial spaces to create artistic innovations that support both a countercultural ethos and aspects of capitalism (Lloyd, 2002). Cafés, bookstores, pubs, and other amenities in Bohemian areas often stress egalitarianism in political messages for anti-establishment or social justice movements, like women's rights, LGBTQ rights, and environmental justice. Bohemian neighbourhoods show more alternative transportation usage (Jeong, 2019), so consumers might stroll or stop by cafés with their bicycles as a countercultural, environmentally sustainable, and cheaper mode of transportation. Also, cafés and other amenities in such neighbourhoods tend to be

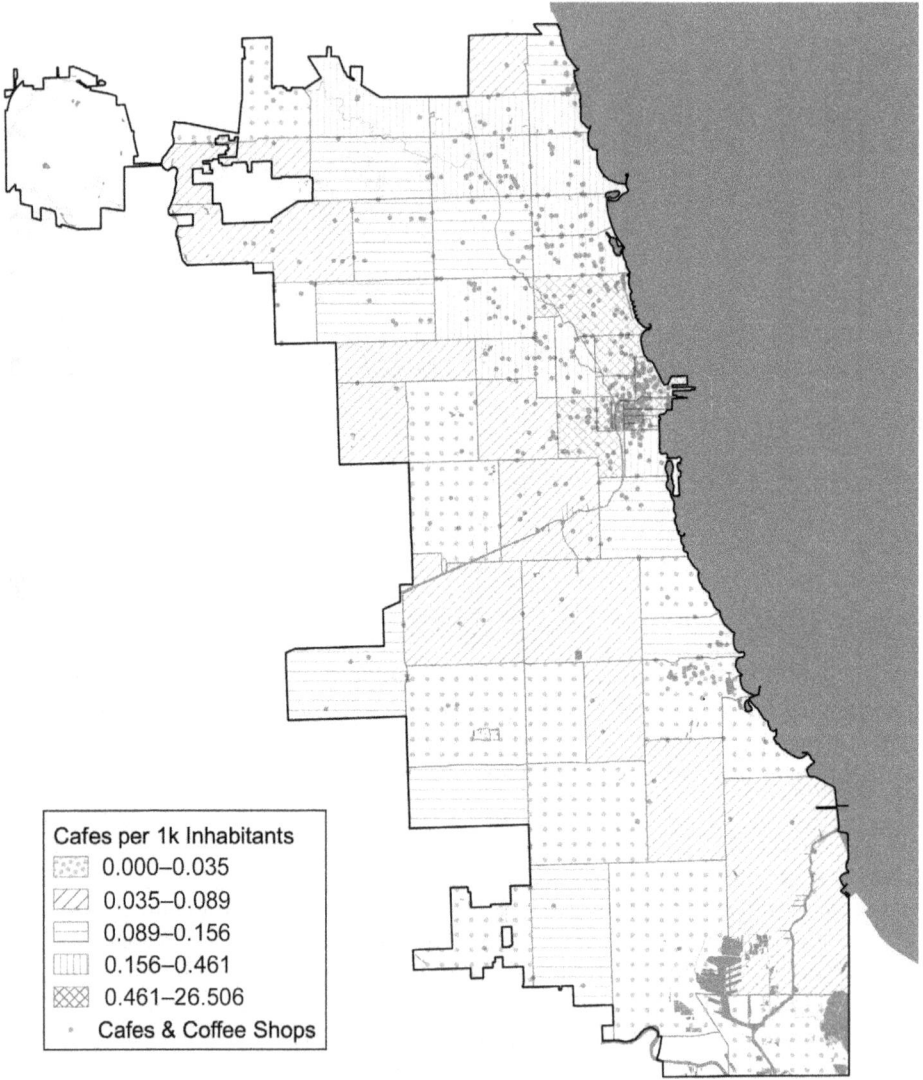

Cafes per 1k Inhabitants
- ▦ 0.000–0.035
- ▨ 0.035–0.089
- ▤ 0.089–0.156
- ▥ 0.156–0.461
- ▧ 0.461–26.506
- • Cafes & Coffee Shops

Figure 10.7. Density of cafés (number of cafés per 1,000 habitants) in ZIP codes of Chicago. (Source: Created by the authors.)

open later, fostering a more active pedestrian scene throughout the day and night. Personal identity, political/social ideology, and retail chosen by retail managers and consumers overlap.

2. Corporate and Utilitarian: Typically located in the downtown core, this second type of scene accommodates tourists and business workers. Following a typical work schedule, customers mainly visit cafés to pick up coffee on their way to work or to meet other colleagues and discuss work-related issues. This scene has a concentration of corporate venues and franchise cafés and stores in a Central Business District (CBD). In larger cities, this type of café is often placed in or surrounded by high-rise buildings and landmarks. Skyscrapers, corporate businesses, and tourists symbolizing capitalism could be seen as the antithesis of the liberal, transgressive bohemian scene. Downtown streets are active during the day, but mostly empty after business hours since most office buildings, and the cafés inside them, close around 5 p.m.

3. Traditional, Local, and Neighbourly: Cafés in this scene have more family activities and religious gatherings. The neighbourhood often has churches, family-oriented restaurants, schools, and picnic places that stress a sense of community. Café owners and patrons know and greet each other personally. Cafés and other amenities serve the neighbourhood residents more than tourists and may have local artwork and host community events. This scene resembles Tönnies' (1957) *Gemeinshaft*, with more intimacy and kinship. In the United States, these traits are often linked with distinct ethnic groups. Italian and Chinese neighbourhoods are classic examples where food consumption combines retail business and local cultural identity.

How does the café function within these three cultural types of scenes? Below we show the distribution of civic, social, and commercial amenities, the transportation networks within a quarter mile of the café, and the socio-economic composition and subculture of each neighbourhood.

A. Bohemia: Volumes Bookcafe (1474 N Milwaukee Ave, Chicago, IL 60622)

Volumes Bookcafe opens onto North Milwaukee Avenue, the main vibrant commercial corridor in Wicker Park (Figure 10.8). Located on the northwest side of the city with a population of around 26,000, Wicker Park is well known for hip culture, community festivals, and nightlife. Richard Lloyd's ethnography labelled Wicker Park *Neo-Bohemia*, where artistic innovation joins economic development (Lloyd, 2002, 2006). The Wicker Park Historic District has landmark Victorian and Art Deco

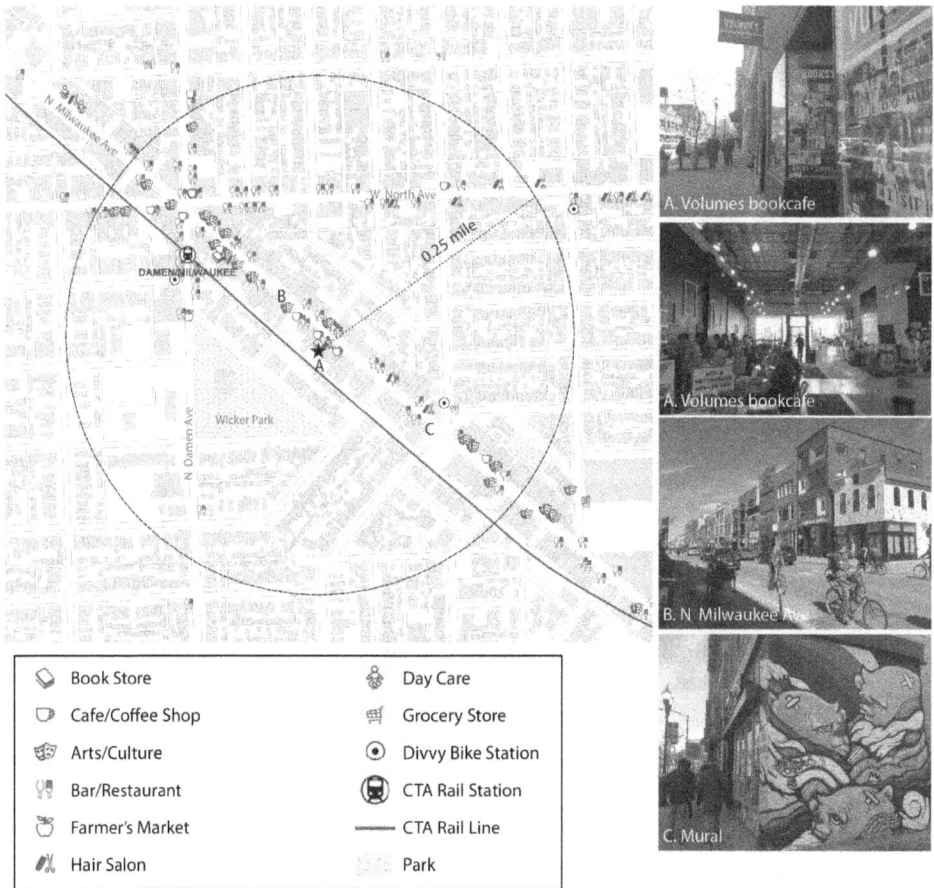

Figure 10.8. Amenities and streetscape around Volumes Bookcafe, Wicker Park. (Source: Created by the authors.)

buildings. A main commercial street, North Milwaukee Avenue has a distinctive artistic vibe created by wall murals and lively interactions among strolling pedestrians. It also has an active night scene: bars, pubs, and performance theatres stay open late, inviting a constant flow of pedestrian traffic. Buildings adjacent to the bookcafé have a wide range of small retail shops on the ground level with residential units above, indicative of the mixed live and work use of the neighbourhood. Transparent storefronts and variation in architectural features allows active visual interaction between shops and pedestrians. Housed in a 3-story brick building, *Volumes* has a large storefront window with display of event posters and new books as well as messages that advocate for social equality. It is open every day until 10 p.m. and hosts regular social events such as book launches, comedy nights, and open mic nights that promote women's, environmental, and LGBTQ rights.

Volumes' proximity to good mass transit options and its hybrid structure of bookstore and café creates easy interactions and the circulation of people and capital throughout the day. Meanwhile, the transformation of the space for multiple social programs continues this circulation well into the night, effectively establishing it as a central place in the Wicker Park community. Consumers' comments on Yelp document its success as a local bookcafé (https://www.yelp.ca/biz/volumes-bookcafe-chicago). Although price increases have driven out many artists as residents, retailers report that they consciously maintain hip style offerings and discuss this policy with other retailers (Parker, 2019). Several dozen art galleries continue monthly evening shows that attract bohemian-dressed crowds.

B. Corporate & Utilitarian: Intelligentsia (53 W Jackson Blvd, Chicago, IL, 60604)

Intelligentsia coffee shop is known as a major representative of the third wave coffee movement that provides an experience of artisanal forms of high-quality coffee and espresso. The one located in Chicago's downtown Loop is a typical example of a corporate and utilitarian café scene. There are many historic skyscrapers and contemporary starchitects' icons that house franchises, retail stores, corporate offices, hotels, and government buildings (Jeong & Patterson, 2020). In contrast to Wicker Park, where retail dominates the streets, the large buildings in the Loop mix commercial and employment spaces with retail shops at street level. Several large cultural institutions are nearby, including the Art Institute of Chicago, Millennium Park, Cultural Center, Chicago Symphony Orchestra, and Federal Center Plaza. The area is highly accessible to workers and consumers by walking and alternative transit; there are 10 rail stations and 8 bike share stations within a quarter mile of the café. While residents are few, half (52 per cent) of them in this ZIP code area commute on foot (US Census, 2008–2012). Also, within this quarter mile there are 27 cafés, 172 bars and restaurants, 10 hair salons, 4 bookstores, 9 art and culture establishments, and 1 farmer's market (Figure 10.9). While there are many more amenities here than in Wicker Park, they are weakly clustered and not open late.

As a Chicago-based coffee chain, *Intelligentsia* has several branches, but this location embodies the corporate and utilitarian downtown scene. A historic landmark, the Monadnock building houses this coffee shop. It is

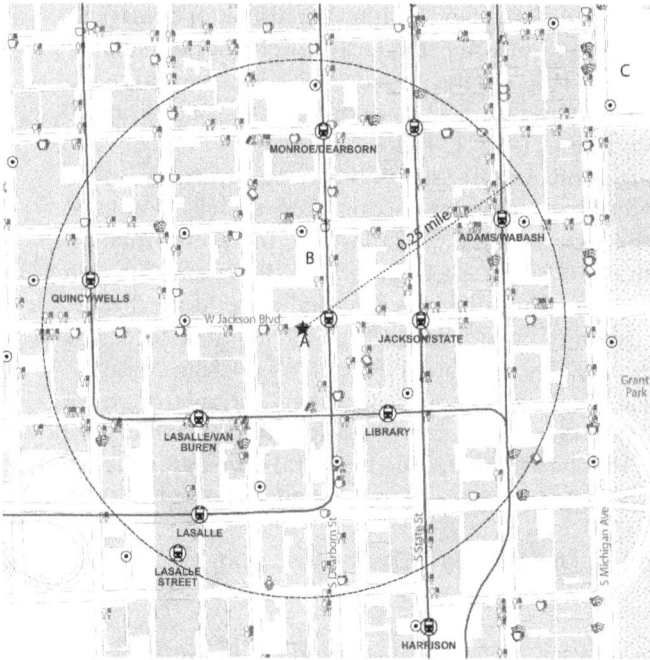

Figure 10.9. Amenities and streetscape around Intelligentsia, the Loop. (Source: Created by the authors.)

the tallest load-bearing brick building, and when built in 1893, was the largest office building in the world. As it was built for offices, the brick facade is simple and ornamentation is more internalized. Likewise, the café matches with minimal signage and an unassuming storefront. The inside of the café has simple and modern decorations such as marble round tables and countertops. During business hours, 6 a.m. to 6 p.m., there are two main types of customers: the professional regulars who often talk about work-related issues, and tourists with maps and shopping bags. While the historic and commercial location of the café and its great accessibility to downtown attracts consumers throughout the day, this café is not a place for "community." There is no obvious way the café promotes particular community values or supports community interaction and engagement beyond providing good coffee and a space to consume it in for business meetings.

Compared to *Volumes Bookcafe,* customers' comments on *Intelligentsia* (https://www.yelp.ca/biz/intelligentsia-coffee-chicago-8) stress the high quality of the coffee and service, and the walkable location. Many also mention that they visit because their office is in the same building or walking distance.

C. Traditional, Local, & Neighbourly: Sip & Savor (528 E 43rd St, Chicago, IL 60653)

Black Metropolis–Bronzeville is a historic district on Chicago's South Side, well known for African-American migration, civil rights, and jazz. It has lost about half of its population in the last three decades (US Census). There are few retail or cultural amenities in this now mainly residential area. The restaurants are mostly fast food franchises.

Sip & Savor is one of the few cafés in Bronzeville. A quarter-mile map of this café contrasts dramatically with the two other examples (Figure 10.10). There are just two other commercial businesses nearby – *Ain't She Sweet,* a café next to *Sip & Savor,* and *Agriculture,* a men's clothing boutique on the same block with *Sip & Savor.* The owner of *Sip & Savor* committed to using the coffee house to support Black entrepreneurship and collaborative business efforts. The owner of *Sip & Savor,* Trez Pugh III, said,

> Whole concept behind this coffee house is where coffee and community meet. I want people to come in and get to know the people in their community. In North Side, you have everything – walk out from your house and go to nice eatery, dog room, and other amenities. South Side did not have that kind of amenity. In this coffee shop, people from all different backgrounds can come together and learn more about architecture, buildings, websites, and marketing in the community, like business incubator. Some of kids in South Side I met have never been to downtown, Sears Tower, and never seen the lake. I do feel that if we show how youth adds value and makes difference, streets will make difference too. (The Jam TV Show, 2017)

The interior displays art by local residents, and the menu offers a senior discount. Many patrons are regulars who discuss community issues and the development of their businesses. Comments on Yelp (https://www

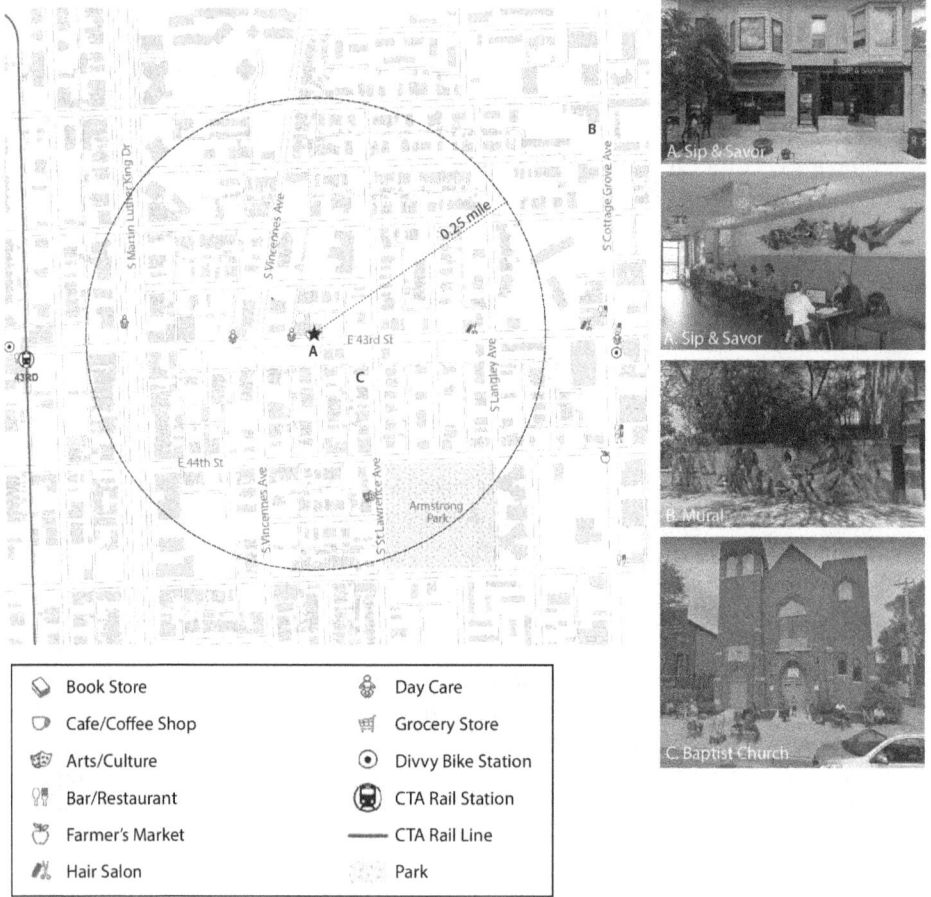

☉	Book Store	♟	Day Care
⌒	Cafe/Coffee Shop	🛒	Grocery Store
🎭	Arts/Culture	⊙	Divvy Bike Station
🍴	Bar/Restaurant	Ⓜ	CTA Rail Station
🍎	Farmer's Market	——	CTA Rail Line
💈	Hair Salon	⬚	Park

Figure 10.10. Amenities and streetscape around the Sip & Savor café, Bronzeville. (Source: Created by the authors.)

.yelp.ca/biz/sip-and-savor-chicago-8) show its community connection, with customers noting that the café has symbolic meaning as a successful Black-owned business that nurtures entrepreneurship in Bronzeville. The café's local art is praised for adding a "homey vibe." Yelp comments highlight the comfortable seating space and daylight that creates a public sphere to sit and talk with neighbours, heightened by baristas and the owner's friendly approach to customers, which spark conservations. Several customers mention that the exhibition of art works and presence of local residents symbolize a collective ownership of the café that enhances a sense of community.

Comments on Yelp feature the eyes of consumers on the three distinct types of cafés. Comments on *Volumes Bookcafe* in Wicker Park highlight the combination of two main programs: bookstore and café, which synergistically combines programmed activities linked to the community's arts and social movements. Comments also noted that night events at *Volumes* enhance walking and exploring shops, bars, and taverns nearby.

Intelligentsia in downtown has comments more about a high quality of espresso and barista services than the surrounding context. Comments stress a main reason to visit the café; people appreciate how the baristas make specialty coffee into a form of sophisticated artwork. This high quality of coffee joined with sustainability and local production bring office workers, professionals, and tourists to experience this "Chicago-based" chain coffee shop in a central district, easily accessible by walking and mass transit.

Compared to two other cafés, *Sip & Savor* in Bronzeville draws more customers who are community residents nearby than tourists or workers residing elsewhere. Comments also show that consumers are in favour of "where community and coffee meet," a main concept of this café. Photos and comments illustrate that regulars use the coffee shop as a space to brainstorm a local business or entrepreneurship, while neighbours discuss other life issues related to Bronzeville. Comments also applaud the comfortable seating that helps people stay longer.

Conclusion

With the global COVID-19 pandemic shutdown, the world's economy has been largely disrupted. Almost across the board, the retail industry was hit hard. With the closing of establishments, retailers faced the loss of both laborers and consumers. Even with the reopening of the economy, news and media report daily surging cases. Many scientists and economists forecast that it will take at least several years to see true recovery, while also advising us to live with this "new" normal. Retailers are heeding this warning. Many cafés and restaurants have implemented new rules such as wearing masks and limiting the number of customers in the store with the government-recommended 6 feet of social distancing. They also have started to have multiple menu options for take-out or delivery to continue their business. Amid this global challenge, a number of organizations and media have issued "calls for proposals" to drive new urban design ideas from architects and planners. Many proposals for post-pandemic urbanism suggest that retailers and business owners consider utilizing the outdoor space in their patio or on the sidewalk when indoor seating is not allowed. When implemented, such options could be a way to keep up with the social and economic cycles while being compliant with health concerns.

During this uncertain crisis, it seems that local retailers need to find flexible and innovative ways to sustain the life of their business. In the post-pandemic era, how should we balance and sustain economic life with health? While there is no definite answer for this question, this reminds us of the value of face-to-face human activities that take place in all social and commercial places outside of home and work. For these reasons, rebuilding the retail establishments will be an important part of recovering urban life in all cities. Stakeholders in public and private sectors need to closely work together in finding new ways of keeping the balance of fundamental values of daily lives in society.

One key strategy for recovering the economy, therefore, is to see the trend of consumer demands in different retail sectors. As documented early in this chapter, consumer services, including retail, is one of the largest categories comprising the US GDP – yet it is under-analysed compared to production and investment. Using data for some 40,000 ZIP Code Tabulation Areas (ZCTAs), we showed how retail amenities cluster across the United States. We then presented case studies of three types of neighbourhoods and cafés to illustrate dramatic differences in their context, pattern, and clustering.

From our case studies, we find all three cafés highly successful. But each would fail if it were moved to a different location or cultural context. Each café succeeds for reasons that are highly place-specific, joining retail and neighbourhood themes to resonate coherently. Consumer comments illustrate how these patterns attract and retain consumers in specific, contrasting ways. Table 10.2 results similarly document the sharp differences among the three, and compare each to Chicago averages to help generalize. National and international data are available for these variables to permit comparing other locations with these three cases. For example, Table 10.2 shows that the ZIP code of the bohemian scene has more cafés and related amenities than the other two scenes. Also, while there is a tiny number of residents in the downtown ZIP code compared to two others, 83 per cent of these downtown residents have professionals with bachelor's degrees. This contrasts with just 27 per cent in traditional Bronzeville and 60 per cent of bachelor's degree holders in Bohemian Wicker Park. In terms of arts amenities, however, Wicker Park is more active than the two other areas. The ZIP code in Wicker Park has 36 art establishments, while the one in downtown has 8 and Bronzeville only 2 art establishments. Besides, about 7 per cent of residents in Wicker Park are artists, whereas downtown has zero and Bronzeville slightly above

Table 10.2. Descriptive statistics of three café streets in Chicago

| | | Types of café scene | | |
| | | Bohemia | Corporate utilitarian | Traditional, local, neighbourly | |
Variables		ZIP 60622 (Wicker Park)	ZIP 60604 (Loop)	ZIP 60653 (Bronzeville)	Chicago average
Demographic variables					
Population		52,548	570	29,908	48,400
% bachelor's degree		59.4	82.5	26.9	39.3
Median HH income ($)		67,493	105,984	25,512	54,953
Art & culture	Art establishments	36	8	2	11
	% artists	7.23	0	1.27	2.77
Social movement	Social movement ratio*	0.118	0.127	0.114	0.100
Street activation	% walk, bike, transit to work	45	63	41	39
Retail amenities					
Café/coffee shop	Yellow Pages (2007)	21	3	1	4.9
	US Census (2011)	21	7	3	8.6
	City of Chicago Business licence data (2017)	26	13	3	10.6
Tavern/bar/ pub	Yellow Pages (2007)	2	0	0	1.2
	US Census (2011)	56	1	2	12.5
	City of Chicago Business licence data (2017) (bars and restaurants)	366	81	43	154
Bookstore	Yellow Pages (2007)	14	5	1	5.7
	US Census (2011)	1	3	0	1.3
	City of Chicago Business licence data (2017)	2	1	0	0.9
Hair salon	US Census (2011)	18	5	1	10.6
	City of Chicago Business licence data (2017)	35	4	16	25.5
Religious organization	Yellow Pages (2007)	28	1	48	27.6
	US Census (2011)	22	5	25	17.9

(*Continued*)

Table 10.2. (Continued)

		Types of café scene		
		Bohemia	Corporate utilitarian	Traditional, local, neighbourly
Variables		ZIP 60622 (Wicker Park)	ZIP 60604 (Loop)	ZIP 60653 (Bronzeville)	Chicago average
Public library	City of Chicago GIS data	1	0	1	1.6
Park	City of Chicago GIS data	11	2	18	10.8

Sources: (1) Demographics: US Census American Community Survey (2008–2012). (2) Amenities: US Census ZCTA Business Patterns (2011), Yellow Pages (2007), and Business licence data from City of Chicago (2017).

* The social movement ratio is the total number of human rights, social advocacy, and environmental organizations divided by all types of organizations, including religious, political, business, and others extracted from ZIP Codes Business Patterns' data for 2011.

1 per cent. Historic Bronzeville's churches generate a stronger religious scene. The downtown ZIP code has just 1 religious organization, while Wicker Park has 28 and Bronzeville 48. These numbers come to life in our quotations from participants about each case.

The case studies of these three cafés illustrate three different consumer scenes. First, the bohemian café is embedded in a powerful cultural landscape of surrounding local amenities. It succeeds by creating a hybrid of bookstore/café/community centre in one central location on North Milwaukee Avenue, the main street of the neighbourhood. Activities change from day to night. Bohemian shoppers can find clothes and paintings by day, while residents and visitors can hear civic activists debate environmental or LGBTQ issues by night. The bookstore features related books, and windows offer posters of multiple nearby events hosted by the neighbourhood residents. Yelp comments stress how the night activities enhance street vitality, attracting more eyes to the street, especially after dark. The second case study features the corporate café scene, with more business-oriented relations among the café owner, baristas, and customers. It resembles other worldwide franchise food retails. In its simple modern architecture, the business-scene café is successful in emphasizing high-quality, hand-crafted coffee for young professionals, business people, and tourists.

Third, the communitarian café scene promotes friendly conversation with the owner, baristas, and among customers. Local artwork as decoration and signs about nearby events provide a visual sense of community. The owner talks proudly about Black entrepreneurship to fill in the gap of low access to food and jobs in Bronzeville. He mentions successfully opening branches in other depopulated neighbourhoods on the South Side of Chicago. This combination of neighbourly, egalitarian, and creative social activities helps small-scale retail to thrive as a third place in the Bronzeville community, with its nationally important legacy of jazz, blues, and African-American cultural leadership. In light of the Black Lives Matter movement that recently exploded across the United States, the post-pandemic discussion regarding the distribution of resources providing education and jobs will be more crucial than ever for minorities. Local coffee shops like *Sip & Savor* or other small-scale amenities could play a critical role in rebuilding legitimacy and supporting the youth as a place for learning or sharing new ideas in the community.

This chapter illustrates how retail can be studied in new ways. To understand retail vitality, we focused on specific consumption patterns and niche-like consumer preferences rather than abstract terms like products, investment, and marketing. Our analysis disaggregates the components of consumer preferences to understand which amenity can work and why in a specific place. In addition to looking at national or regional averages, our study also aims to detail how these change at the neighbourhood and block level. It calls for specific policies like the combination of products such as a joint bookstore/café/meeting place rather than just one of these alone. Combinations of three types may succeed while the same three separated by a few hundred feet might not. Finally, as illustrated above in the graphics on types and dimensions of scenes, this chapter elucidates how specific retail activities connect with surrounding amenities in the street and neighbourhood. An isolated café may have few customers, while the same type of café in a vibrant neighbourhood with lots of foot traffic and other amenities or street arts nearby can offer a completely different experience. What makes one neighbourhood and café vibrant is distinctive to that specific scene. But unique does not mean random or by chance. Many established, general patterns combine to make local retail firms unique and successful.

What the retail sector will look like in a post-pandemic world is an open question. However, our research stresses the importance of how retail establishments fit into the urban landscape within locally important and distinct cultural contexts for rebuilding cities and neighbourhoods.

REFERENCES

Bureau of Economic Analysis. (2018). Concepts and methods of the US National Income and Product Accounts. https://www.bea.gov/system/files/2019-12 /Chapter-1-4.pdf

Clark, T. N. (Ed.). (2004). *The city as an entertainment machine: Research in urban policy*, Vol. 9. Elsevier – JAI Press.

Florida, R. (2002). *The rise of the creative class*. Basic Books.

Gehl, J., Kaefer, L. J., & Reigstad, S. (2006). Close encounters with buildings. *Urban Design International*, *11*, 29–47. https://doi.org/10.1057/palgrave .udi.9000162

Hortaçsu, A., & Syverson, C. (2015). The ongoing evolution of US retail: A format tug-of-war. *Journal of Economic Perspectives*, *29*(4), 89–112. https:// doi.org/10.1257/jep.29.4.89

Jacobs, J. (1961). *The death and life of great American cities*. Random House.

Jam TV Show, The. (2017). *Real Chicagoans: Sip and Savor Coffee House*. Retrieved July 14, 2022. https://www.youtube.com/watch?v=UOXKeyXeNEw&ab _channel=TheJamTVShow.

Jeong, H. (2019). The role of the arts and bohemia in sustainable transportation and commuting choices in Chicago, Paris, and Seoul. *Journal of Urban Affairs*, *41*(6), 795–820. https://doi.org/10.1080/07352166.2018.1516510

Jeong, H., & Patterson, M. (2020). Starchitects in Bohemia: An exploration of cultural cities from the "top-down" and "bottom-up." *Urban Affairs Review*. https://doi.org/10.1177/1078087420934047

Lloyd, R. (2002). Neo-Bohemia: Art and neighborhood redevelopment in Chicago. *Journal of Urban Affairs*, *24*(5), 517–532). https://doi.org/10 .1111/1467-9906.00141

Lloyd, R. (2006). *Neo-Bohemia*. Routledge.

Mehta, V., & Bosson, J. K. (2010). Third places and the social life of streets. *Environment and Behavior*, *42*(6), 779–805. https://doi.org/10.1177/0013916509344677

Metropolitan Planning Council. (n.d.). Chicago's people spots. https://www .metroplanning.org/work/project/12/subpage/4?utm_source=%2fpeoples pots&utm_medium=web&utm_campaign=redirect.

Oldenburg, R. (1999). *The great good place* (2nd. ed.). De Capo Press.

Parker, J. N. (2019). That kind of neighborhood: Creating, contesting, and commodifying place reputation [Doctoral dissertation, University of Chicago]. ProQuest Dissertations and Theses.

Ritzger, R. (1993). *The McDonaldization of society*. Pine Forge Press.

Silver, D., & Clark, T. N. (2016). *Scenescapes*. University of Chicago Press.

Talen, E. (2005). *New urbanism and American planning: The conflict of cultures*. Routledge.

Tönnies, F. (1957). *Community and society*. Michigan State University Press.

US Census Bureau. (2012). "Means of transportation to work by selected characteristics," 2008–2012 American Community Survey 5-Year Estimates. https://data.census.gov/cedsci/table?q=United%20States&t=Commuting&g =0100000US_860XX00US60601&y=2012&tid=ACSST5Y2012.S0802

Business Profile: Inclusion Coffee

Arlington, Texas

Inclusion Coffee is a new social condenser located in Downtown Arlington, just across from the campus of the University of Texas at Arlington. It opened in May 2020 – in the midst of the pandemic – and has been thriving. The store keeps the surrounding streets active and stimulates civic gatherings even with social distancing and mask-wearing rules in place. Other nearby businesses close at 5 p.m.; this coffee shop extended its hours until 8 p.m., allowing community residents to rent the space for small events such as flea markets and music performances. The coffee shop has extra seating space in its mezzanine, and the interior design is industrial-chic – all designed and constructed by its family owners and employees together. The coffee shop has already been a star on Instagram with hundreds of photos and stories featuring its unique spatial character and the quality of its coffee. This shop is rightfully considered a model for how to thrive during a pandemic.

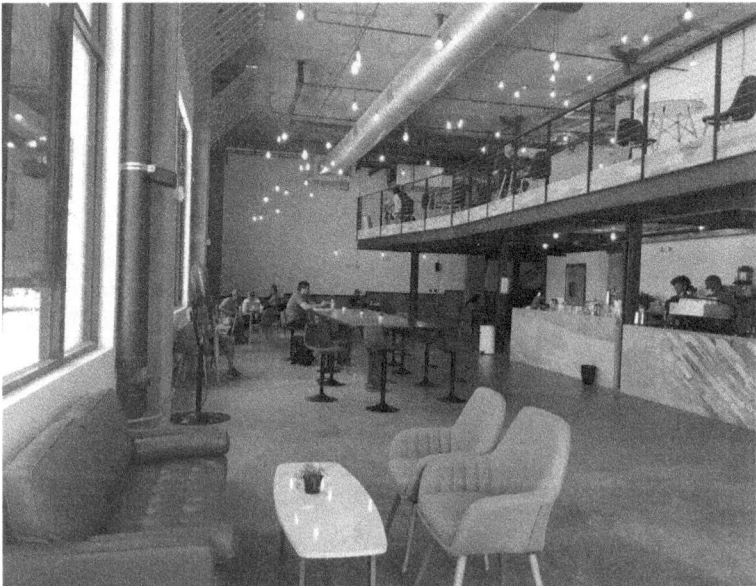

Figure B10.1. Inclusion Coffee in Arlington, Texas (interior). (Credit: Photo courtesy of the authors.)

Figure B10.2. Barista at Inclusion Coffee in Arlington, Texas. (Credit: Photo courtesy of the authors.)

11 Main Street Morphology, Adaptability, and Resilience

ROSA DANENBERG

Introduction

The idea of a traditional commercial main street usually invokes vivid memories of lively street scenes with small-scale economic establishments and bustling interactions of locals and strangers. However, there is more at work with main streets than nostalgia. Neighbourhoods with such streets are recognized by a typically distinct physical design that, as research shows in this book and elsewhere, produces significant positive local and city-wide impacts (Carmona, 2015; Griffiths, 2015). These findings are a testament to the enduring importance of main streets. Although socio-economic conditions surrounding different main streets are rapidly changing, we now see that their specific physical features offer important supportive frameworks for diverse social and economic activities (Orvell, 2017).

This chapter focuses on the urban form of main streets – specifically, the spatial structures that shape and limit social and economic activities, and vice versa. While chapter 16 by Conrad Kickert will study the economic spatial structures that shape the functioning of urban commercial streets, this chapter will study the physical structures that help or hinder main streets – an endeavour also known as urban morphology. After surveying the urban morphology literature and scientific body of knowledge on urban commercial streets, we will synthesize this theory with an interdisciplinary expert panel discussion of urban designers, sociologists, planners, and economists on "the future of main streets" that took place in Stockholm, Sweden, in November 2018. This synthesis will help us understand the current knowledge gap on the relationship between urban morphology and urban retail, and to explore new research directions for filling this gap.

This is not to say, however, that there is a shortage of research about main streets. This book itself is a testament to that. As other authors have mentioned, much of the recent research reflects growing concern for the loss of the traditional role that main streets have played in city life. In-depth case studies focus broadly on the characteristics, qualities, and issues of these streets and their surrounding neighbourhoods (Carmona, 2015; Griffiths et al., 2008; Mahmoudi Farahani et al., 2018; Talen & Jeong, 2019). Other research also focuses on the economics of commercial streets (Kickert, 2019b). However, it is harder to find mature research specifically focused on the role of the physical built form as it affects the social and economic dynamics of urban commercial streets. Here, spatial characteristics include urban elements such as streets, blocks and plots that define pathways of connection, interaction, and exclusion, and thereby shape human interaction and exchange. It follows that the built form alters socio-economic interactions over time. Conversely, changes in non-spatial dynamics such as social and economic activities can prompt adaptations in spatial structures too (Hillier, 1996; Hillier & Hanson, 1984; Carmona et al., 2003). These systems – urban form on the one hand, and socio-economic activity on the other – therefore tend to follow a natural feedback cycle of alteration and adaptation (Porta & Romice, 2010).

Understanding how built form interacts with socio-economic patterns is crucial to plan and design for more resilient and adaptive public spaces in contemporary cities. Urban commercial streets often share common spatial characteristics that have the capacity to constantly evolve and adapt to change, as they have done so for centuries (Carmona, 2015; Talen & Jeong, 2019; Törmä et al., 2017). We need to understand how commercial streets can survive our current roster of socio-economic change. Urban morphologists Porta et al. (2014) stress the need for "further research into the essential logic and principles of what generates a more adaptable, hence resilient, urban fabric. This essential logic has to do with urban morphology: by better understanding the critical relationships between urban streets and plots" (pp. 3398–3399). If we understand the morphological components of more resilient and adaptable commercial main streets, we hold a key to ensuring more resilient urban neighbourhoods around them.

Literature on Urban Morphology

In order to understand how urban form can help build more resilient commercial streets and surrounding neighbourhoods, we first survey the body of knowledge on this form and its study. Urban morphology seeks to understand how variations in urban built form generate different social and economic dynamics, and vice versa. The elements commonly studied within urban morphology are streets, blocks, and plots (Moudon, 1997). In essence, morphologists seek to unveil the "interconnections between the materialities of urban space and the socialities of urban life" (Dovey & Pafka, 2016, p. 10). This underscores the tight and dynamic interrelationship between physical form and social, cultural, political, and economic forces that give shape to cities over time (Moudon, 1997). The ability of urban forms to adapt to change has long been recognized as a key component of the rich morphological character and complexity of cities. As early as the nineteenth century, Viennese urbanist Camillo Sitte (1889) understood that form gives a place its character and resilience through evolutionary adaptations over what he termed *longue durée* ("long duration"). He and others saw the city as a constantly evolving organism that reinvents itself to accumulate quality and value over time. Meaningful places are understood to be generated as urban elements form linked scales at different rates (Kärrholm, 2018). One consequence is that the urban fabric changes more slowly and the buildings and activities taking place in it change more quickly (Conzen, 1960; Hillier, 2007) as the functions taking place within the form of the city are subjected to varying speeds of change (Porta et al., 2014).

The Field of Urban Morphology

The field of urban morphology includes several main schools of thought, reflecting different disciplinary and cultural contexts. Among urban morphologists, the Italian and the British schools tend to focus on the diachronic or evolving dimensions of urban form. The Italian school focuses on the continuity that appraises the persistence of building types and harmonization of the old and new, whereas the British school places more emphasis on the analysis and description of form at the urban scale, often for explanatory and conservation purposes (Sanders & Baker, 2016). The more recent Space Syntax school, also based in the United Kingdom, is concerned with the computation of local and global spatial relationships and their impacts (Hillier, 2007). This school presumes a *social logic of space*, which explains the reciprocal relation between spatial structures and society in terms of spatial configuration: "How we organise space into configuration is the key both to the forms of the city, and how human beings function in cities" (p. 113). In other words, the physical urban elements generate non-physical urban dynamics, such as social and economic activities; simultaneously these dynamics transform the structure of cities. This particular reciprocal relationship helps to understand genius loci, as "[t]he epistemological basis of urban design suggests the need to draw upon the evidence of urban morphology as a methodology of design that evidences the character of a place" (Sanders & Baker, 2016, p. 214). The emerging North American movement seeks an interdisciplinary synthesis of these differing schools, applied to the American context (Scheer, 2016).

Despite the differences in methods, foci, theories, and validity criteria employed across the different schools of spatial morphologists, there are significant overlaps. Urban morphologist Anne Vernez Moudon argues that morphological analysis is based on three principles: "[U]rban form is defined by three fundamental physical elements; urban form can be understood at different levels of resolution; and urban form can only be understood historically" (1997, p. 7). Morphologist Brenda Scheer proposes a common framework that includes definitions for the most studied elements of urban form: buildings, plots, and streets. The framework also includes three general kinds of morphological study: pattern recognition, theories of change, and relations to non-formal conditions (2016, p. 13). Understanding the combination of these types of study helps to identify how specific places are created and transformed. Urban morphologists meet annually to share their methods and insights at the International Seminar on Urban Form.

Morphology of Urban Commercial Streets

Research on the establishment and performance of urban retail has focused upon the key role of streets, which are so central to the overall structure of cities. By the nineteenth century, Barcelona's urban designer Ildefons Cerdà already acknowledged that "streets as connections between people and buildings constitute the basis of urbanization" (Dovey & Pafka, 2016, p. 5). Urban morphologists have studied the relationship between streets and the urban economy for decades, most notably a correlation that has especially been made by the Space Syntax movement (Hillier, 2007; Hillier & Hanson, 1984b). Their models demonstrate a *movement economy*, in which spatial configuration dictates movement of people, which in turn enables economic opportunity such as storefront retail – a key insight for the functioning of urban commercial streets. The spatial configuration creates a network of urban streets with varying levels of connectivity. Street connectivity affects the functions on ground floors and subsequent public life. Urban street connectivity therefore influences the economic performance of retailers (Parker et al., 2014). Beyond retail, well-connected urban streets support a diversity of small-scale activities, which in turn support a diversity of people (Vaughan et al., 2015).

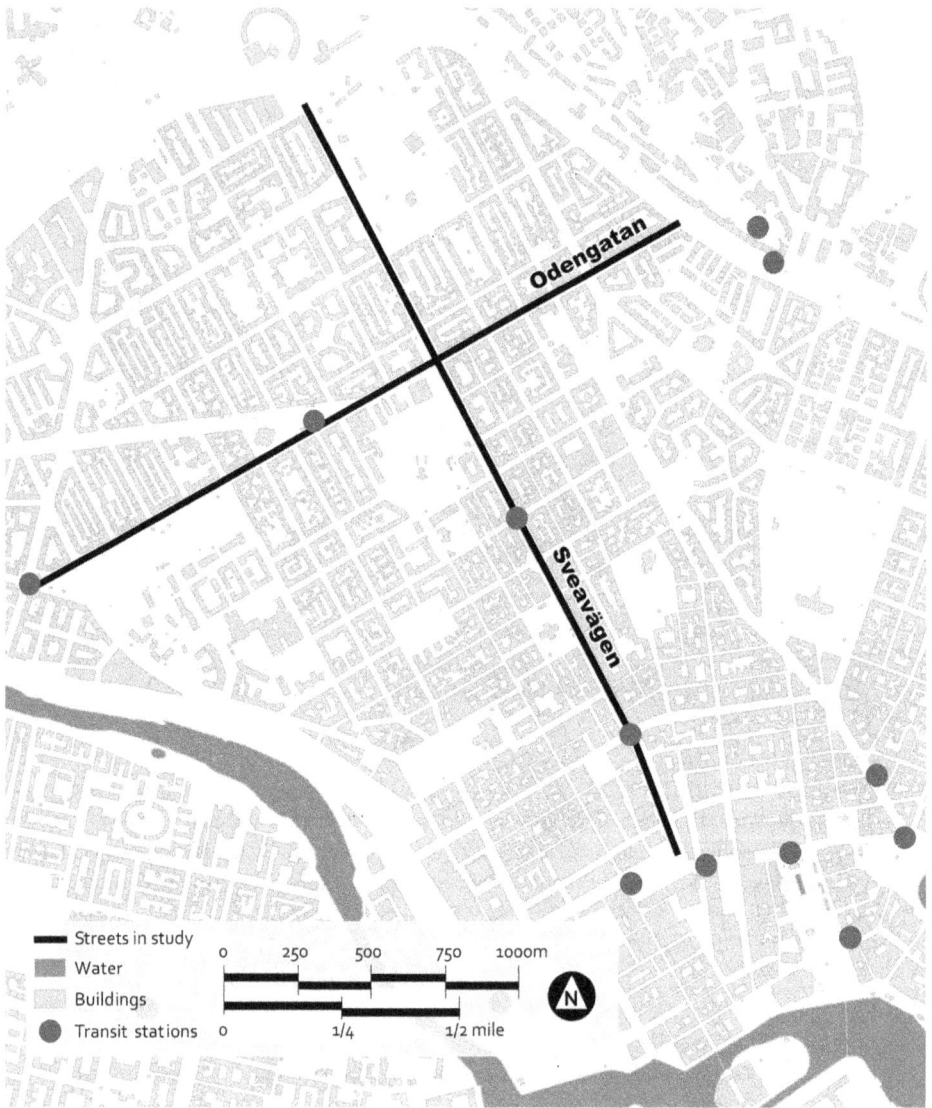

Figure 11.1. Two main streets in Stockholm (Odengatan, east-west connection, and Sveavägen, north-south connection) exemplify fine-grained neighbourhood streets that are well connected to their surrounding neighbourhoods. (Source: Created by the author.)

As cities and their street networks have experienced a significant alteration in scale in the automobile era, the adaptability of retail streets to social and economic changes has diminished. While urban designers have advocated for finer-grained street networks for decades, a morphological study of a hundred urban street networks across the world shows that the distance between main urban streets has doubled between historical and contemporary cities. As these streets tend to contain the urban retail apparatus, most retailers are now located beyond a comfortable five-minute walk, making walkable urbanism significantly less feasible (Porta et al., 2014). Furthermore, street networks in general have become coarser, limiting pedestrian options to move through cities, ultimately lowering street walkability and access – a precondition for successful retail (Carmona et al., 2003; Jones et al., 2007). According to Jacobs, short blocks generate pedestrian flow, which in turn generates more street-level economic opportunities (Jacobs, 1961; Sayyar & Marcus, 2013).

Nevertheless, today's urban commercial street does demonstrate resilience gained from its continuous evolution. Kickert argues that despite the transformation of urban economies, the rise of e-commerce and globalized retail brands, and automobile erosion of urban retail, "retail frontages have mostly remained on downtown's most central and well-connected streets – places they are most likely to be seen by passersby" (2016, p. 73). Griffiths attributes intrinsic social value to the historic built environment that has generated traditional main streets (2015). He argues that main streets are resilient environments and acknowledges their adaptive nature as a possible source for "their long-term durability as places frequently capable of reproducing socio-economic activity of one kind or another over successive generations" (p. 34). Indeed, the variation and evolution of the built form is what generates streetlife (Jacobs, 1961; Sayyar & Marcus, 2013). Our main commercial streets have survived and remained relevant for centuries, so there is reason to believe they will adapt to unknown futures.

Beyond the connectivity and permeability of streets themselves, the form of the blocks and plots they access has a significant impact on the functions situated there, which in turn influences urban resilience. Independently controlled plots that vary in size enable a large diversity of entrepreneurial opportunities and allow for a healthy balance of mixed ground floor functions (Danenberg et al., 2018). The presence of small businesses on the ground floor contributes to the vitality of main urban streets through personalized frontages and extension of territory to outdoors (Mehta, 2011). These frontages bridge the relation between form and function as "[g]round floor facades provide an important link between these scales and between buildings and people" (Gehl et al., 2006, p. 29). Commercial

Figure 11.2. Interactive frontages add to the social life on streets. (Source: Created by the author.)

frontages form the transition and permeable space between the building and the street, significantly raising the level of activity in public space and contributing to its perception of comfort (Gehl, 2010; Heffernan et al., 2014). Furthermore, the number of doors reflects the level of integration of public and private spaces (Gehl et al., 2006; López, 2003). In general, interactive frontages comprise an ecosystem of form, function, and meaning, expressed in the their physical transparency, functional permeability, and perceptual hospitality (Kickert, 2016).

As they can house a wider variety of functions, small plots also improve the resilience of urban commercial streets. In a longitudinal study of a London suburban commercial street, Vaughan (2015) demonstrates that buildings with shorter frontage length, a smaller footprint, smaller plot size, and higher plot efficiency are more susceptible to change, allowing them to adapt to changing external circumstances. Indeed, businesses that are generally small, lean, and able to adapt to change are situated in smaller plot sizes that are often more affordable (Carmona, 2015; Danenberg et al., 2018). They comprise resilient urban fabric, with activities

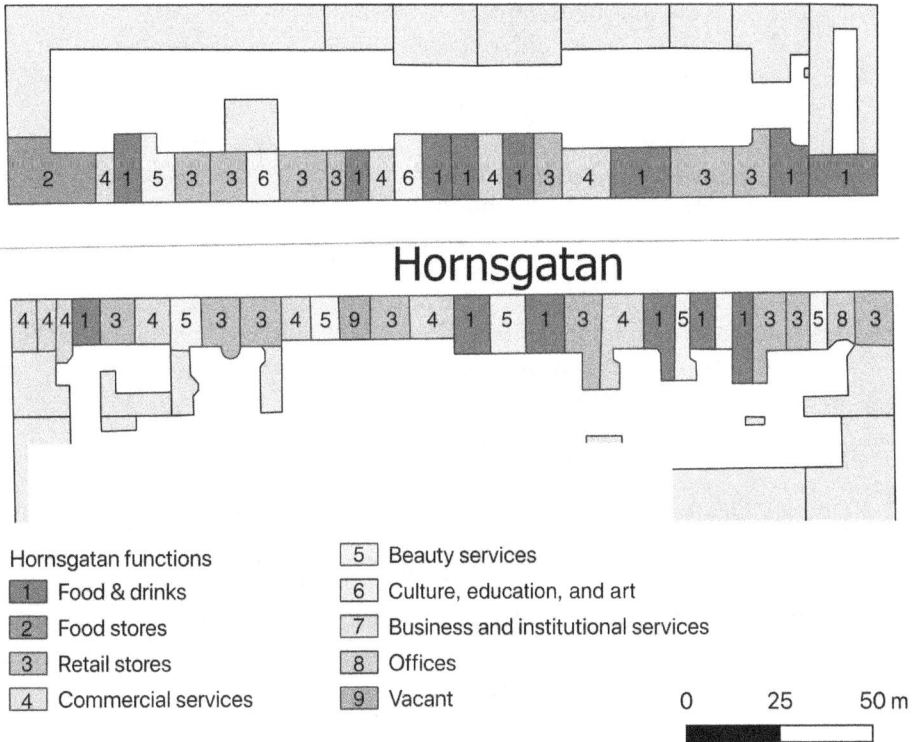

Figure 11.3. Street-level commerce comprises more than shops, as storefronts increasingly house bars, restaurants, and personal services. (Source: Created by the author.)

that can respond and adapt to socio-economic changes (Porta & Romice, 2010; Wrigley & Dolega, 2011). Beyond the benefits of smaller plots, a mix of small and large businesses will ultimately provide the diversity that a resilient urban commercial street requires (Litvin & Rosene, 2017). Unfortunately, the size of blocks and plots is rising in existing and new urban form as the result of the accumulation of land and capital. This has grave potential consequences for the resilience of the urban fabric and its street-level economy (Gehl et al., 2006; Kickert, 2019a).

Urban Morphology from an Interdisciplinary Perspective

To demonstrate the role of urban morphology in this variety of perspectives, an expert panel was held in November 2018 in Stockholm,

Sweden, that focused on the future of urban commercial streets. As this book demonstrates, many different disciplines are involved in the study of urban commercial streets, and the morphological perspective represents only one perspective. However, the morphological perspective relates to more disciplinary perspectives than we might think. The expert panel discussed the future of urban commercial streets and the role of urban morphology in studying this future. It included seven scholars and practitioners from the fields of urban design, urban planning, sociology, entrepreneurship, and economy. Selection of the participants was based on the "information-richness" of the individuals as well as the beneficial difference among the individuals (Krueger & Casey, 2014). Finding participants with diverse and published knowledge of urban commercial streets warranted a "purposive selecting technique" (Cameron, 2005).

Expert Panel as a Method

Expert panels function as focus groups to gather the knowledge of a number of experts (Liamputtong, 2011) and to generate knowledge from the synergy of the group itself (Cameron, 2005; Krueger & Casey, 2014). Initiated by social scientists in the early 1930s as a critical response to positivistic research methods, focus groups were later embraced and developed by market research specialists, and again rediscovered by academics (Krueger & Casey, 2014). More recently, focus group discussions have been used by geographers to better understand complex people-place relations and processes. It has been argued that they are helpful to stimulate the exploration of different viewpoints and the generation of a common language to talk about a phenomenon (Cameron, 2005). This discussion format encourages participants to contribute with varying disciplinary viewpoints that help to explore issues that would not emerge in an conventional single-person interview setting (Liamputtong, 2011). At the same time, focus groups have been criticized as overly situational with limited generalizability (Cameron, 2005), and concerns have been raised regarding the tendency of participants to over-intellectualize (Krueger & Casey, 2014).

Accordingly, the objective of the expert panel was to obtain viewpoints and encourage discussion among the participants regarding the future of urban commercial streets, and a significant part of the discussion focused on the role of urban morphology and urban design. After an introduction by this chapter's author and a presentation on urban commercial street research, the discussion focused on the urban morphology and design features of urban commercial streets, knowledge gaps, and the continued focus of design professionals on improving these streets.

The outcomes of this discussion are framed in a taxonomy of four common morphological themes of research, identified by morphologist Brenda Scheer: data collection, pattern recognition, theories of change, and linkages to non-formal conditions (2016). Distilling the outcomes into these four themes allows us to see a range of disciplinary alignments and misalignments between urban morphologists and other disciplines that study urban commercial streets.

Results of the Expert Panel Discussion

The participants are identified herein by their disciplines: urban designers (UD1 and UD2), urban planner & designer (UPD), sociologist (S), economist (E), and practitioner (P). All the participants are in various degrees engaged and interested in urban form and urban morphology. Urban morphologists are most interested in collecting data and analysing built form elements (Moudon, 1997; Scheer, 2016), closely aligning with the urban analysis of urban designers and urban planners in their practice. Sociologists and economists are arguably the least form oriented. Nevertheless, a consensus on the definition of main commercial streets in this interdisciplinary group did emerge around their form. The urban designers in the group interpreted the main commercial street as "one of the main components of urban form in general including also plots, blocks, buildings etcetera. [These] streets are a part of urban form ... as a contribution to urban life" (UD1). However, the common form of urban commercial streets produces "common structures and endless variations" (UD1). Thus, similarities across form are not limited to producing the same result: "[Y]ou really need to take into account the different contexts and that the city forms each give the conditions for each street to develop and emerge into something" (UD2). For example in Sweden, defining "main street or the high street, it's not quite as clear" (P).

The participants agree that physical components of urban commercial streets should be *humanly scaled*: the streets should be comfortable to walk along and not too wide, the plots should be small enough to house small businesses, and the buildings should be facing the public spaces. This human scale interconnects at all levels, as "the fine structure of ownership of the land" determines that "different scales going up to the fine grain of the plots, fine grain of the blocks, fine grain of this street and all the way up" (UD1). Furthermore, on a neighbourhood scale, the main street is part of a connected street network arranged according to a clear hierarchy of streets, between main and side streets.

Participants also agreed that urban morphology is a crucial perspective for studying urban change. The urban elements with greater

endurance (i.e., paths and plots) give more permanence to the structure of the built form (i.e., objects, buildings and infrastructure)
(Scheer, 2016). The physical shape and form of urban elements can
only be understood historically through their continuous transformation and replacement (Moudon, 1997; Porta & Romice, 2010).
The participants acknowledge the inertia of urban form, as past patterns of blocks, plots and buildings condition the present functioning of commercial streets, since "you inherit specific morphological
topological things, you inherit the history, the culture, you inherit
different layers" (UPD). This layered evolution generates a variety
of conditions for main streets to provide for urban life and attract a
diversity of people so that "a lot of people actually stay and walk the
streets" (UD2). It is clear that when these morphological conditions
are missing, "this area will not be good in time because it doesn't
have the physical prerequisites for a nice main street. There are no
store fronts, there is nothing going on in the ground floors" (P).
Commercial street plots change over time, although "the problem
with morphology is that it doesn't deliver day one. It takes 10 years
or something before you can see the effect" (E). Plot functions move
slowly, "and merging or separating plots happens every 15–20 years
maybe" (UD1).

Lastly, morphologists focus on the reciprocal and complex linkages
between urban form and non-formal conditions in the city (Carmona et
al., 2003; Scheer, 2016). The sociologist in the focus group framed this
linkage as the interchange between hardware (built form) and software
(such as society, economy, culture), which means that "urban form, I
believe, gives certain conditions that is encouraging in certain businesses" (UD2). More precisely, "Main streets form a habitat for small
firms. Small firms on main streets I would say have a value that goes
beyond what they turn over. They have a social value" (E). Therefore,
the mix of software is conditioned by the hardware, which results in
ground floors with different uses that attract visitors and serve regulars,
so that strangers and locals meet (E). This adds up to the layered scales
of "the small footprint of the shops and the diversity of the uses and
the density of businesses for local provisioning" (S). Beyond the right
kind of urban form, a nearby density of residential and working population is important to build a critical mass to support certain functions
and build local socio-economic power (UD2). The strong link between
the urban form and social life became a clear point of consensus in
the focus group, as main streets are "an extremely important element
of urban design and of how you compose and shape cities, but also I
see it as a sort of a very important link in how do you do social life of

urban form" (UPD). The different disciplines view this link from their own perspectives, which can conflict, as the focus group's sociologist questioned the urban design perspective on adaptability: "Why would you think a street would not be adaptable? It's a shell, I mean sorry I'm a sociologist not a designer, it's a shell; it's a physical structure like a box" (S), whereas the urban designer comments, "Certain activities we have to move in a certain directions, … something will be defined by the design actually" (UD1).

Taken together, this focus group with various disciplinary perspectives acknowledge the strong relevance of urban morphology, especially in relating urban form and structures to social and economic life. While participants with a non-design background can take the ability of urban forms to adapt and shape different social and economic conditions for granted, designers strongly believe that urban form creates – or fails to create – conditions for social and economic activities.

Conclusion

Morphological research and our focus group agree that the physical form and structure of cities, and their interactive relationship with the social, economic, and cultural structures of cities, are critical generators of the character, viability and resilience of urban commercial streets. These main streets in turn are important contributors to the life and form of cities. Yet there is clearly a knowledge gap regarding the interrelation between the spatial structure of urban commercial streets and their non-spatial economic and social dynamics. Most morphological work remains normative or descriptive on the relation between urban form and commercial street function, especially with regard to long-term resilience. Further morphological research is essential for understanding urban commercial streets and their role in supporting resilient urban fabric. This literature review and expert panel provide an early assessment of where we stand in this process.

The literature review demonstrates the continuous evolution of physical urban elements and their capacity to adapt to change, which in turn are key contributors to a resilient urban fabric. This is a reciprocal relationship: urban form shapes social and economic activities, and simultaneously these dynamics affect structures and spatial relations in cities. The evolutionary process of urban form, driven by the variations in speed of change of each urban element and scale, adds to the creation of layered and meaningful places. Streets are the fundamental physical elements that enable or hinder urban commerce. In particular, small-scaled urban plots that house commercial activities and dynamically interact with the street are crucial elements of urban life; yet they have come under threat by the expansion of building sizes, decreased pedestrian accessibility, and declining viability of small businesses that provide ground floor functions.

Several lessons can be drawn from our exploratory and interdisciplinary panel discussion between researchers and professionals. First, they agree on defining main commercial streets by common characteristics of urban form, even though variations have evolved through contextual and historical conditions. Second, they agree that urban fabric changes more slowly than the ground floor functions, which means that in most cases buildings must allow for variations of activities, and for the attraction of a diversity of people. Third, for the functioning of a main street as a whole, its physical components must be human-scaled. The fine-grained scale across the various physical elements and ownership structures of the city contribute to the diversity of activities and people. Fourth, main streets are key links between a range of urban scales, and therefore have the unique ability to contribute both locally and city-wide to accessibility and connectivity. Fifth, the adaptive nature of main streets affords a habitat of small-scale economic activities that can generate unique value and benefit, including economic resilience.

A final lesson is the importance of planning and design for adaptability. Considering the key role of urban form and its evolutionary capacities for the survival of urban commercial streets, it seems clear that adaptability must be built into their design.

Planning and design for adaptability and resilience becomes even more prominent in light of the COVID-19 pandemic. The swift and drastic change of public life as a result of (semi) lockdowns and other governmental pandemic responses has created unprecedented consequences for street-level social and economic activities like urban commerce. Even beyond the global virus outbreak, several questions can and should be raised: How can adaptability and resilience be better understood and integrated in contemporary planning and design? How can we ensure that adaptability and resilience is not just taken for granted, but planned and designed, in order to withstand inevitable social and economic changes? As the above lessons and questions demonstrate, the important role of urban commercial streets, and the requirements for developing and maintaining them, should be seen as essential topics in the debate on adaptability and resilience in contemporary cities.

Appendix to Chapter 11

List of Participants in Expert Panel

UD1   Professor in urban design, United Kingdom. Research focuses on urban morphology and spatial analysis.

UD2   Professor in urban design, Sweden. Research focuses on applied urban design with methods such as Space Syntax.

S   Professor in sociology, United States. Research focuses on global culture and gentrification.

UPD   Associate professor in urban planning and design, Sweden. Research focuses on contemporary urbanism.

P   Planner in practice, Sweden. Works in a municipality in Stockholm County.

E   Economist, Sweden. Experience in global networks on small-scale entrepreneurship, public space, and urbanism.

B   Associate professor in business studies, Sweden. Research focuses on entrepreneurship and social innovation.

REFERENCES

Cameron, J. (2005). Focusing on the focus group. In *Qualitative research methods in human geography* (pp. 156–174). Oxford University Press.

Carmona, M. (2015). London's local high streets: The problems, potential and complexities of mixed street corridors. *Progress in Planning, 100*, 1–84. https://doi.org/10.1016/j.progress.2014.03.001

Carmona, M., Heath, T., Oc, T., & Tiesdell, S. (2003). *Public places, urban spaces: The dimensions of urban design.* Architectural Press.

Conzen, M. R. G. (1960). Alnwick, Northumberland: A study in town-plan analysis. *Transactions and Papers (Institute of British Geographers), 27*, iii–122. https://doi.org/10.2307/621094

Danenberg, R., Mehaffy, M., Porta, S., & Elmlund, P. (2018). Main street plot scale in urban design for inclusive economies: Stockholm case studies. *Proceedings of the Institution of Civil Engineers: Urban Design and Planning, 171*(6), 258–267. https://doi.org/10.1680/jurdp.18.00031

Dovey, K., & Pafka, E. (2016). The science of urban design? *Urban Design International, 21*(1), 1–10. https://doi.org/10.1057/udi.2015.28

Gehl, J. (2010). *Cities for people.* Island Press.

Gehl, J., Kaefer, L. J., & Reigstad, S. (2006). Close encounters with buildings. *Urban Design International, 11*(1), 29–47. https://doi.org/10.1057/palgrave.udi.9000162

Griffiths, S. (2015). The high street as a morphological event. In L. Vaughan (Ed.), *Suburban urbanities: Suburbs and the life of high streets* (pp. 32–50). UCL Press.

Griffiths, S., Vaughan, L. S., Haklay, M. M., & Jones, C. E. (2008). The sustainable suburban high street: A review of themes and approaches. *Geography Compass, 2*(10), 1–34. https://doi.org/10.1111/j.1749-8198.2008.00117.x

Heffernan, E., Heffernan, T., & Pan, W. (2014). The relationship between the quality of active frontages and public perceptions of public spaces. *Urban Design International, 19*(1), 92–102. https://doi.org/10.1057/udi.2013.16

Hillier, B. (1996). *Space is the machine: A configurational theory of architecture.* Cambridge University Press.

Hillier, B. (2007). *Space is the machine.* Space Syntax.

Hillier, B., & Hanson, J. (1984). *The social logic of space.* Cambridge University Press.

Jacobs, J. (1961). *The death and life of great American cities.* Random House.

Jones, P., Roberts, M., & Morris, L. (2007). *Rediscovering mixed-use streets.* The Policy Press.

Kärrholm, M. (2019). Scale alignment: On the role of material culture for urban design. *Urban Design International, 24*(1), 7–15.

Kickert, C. (2016). Active centers – interactive edges: The rise and fall of ground floor frontages. *Urban Design International, 21*(1), 55–77. https://doi.org/10.1057/udi.2015.27

Kickert, C. (2019a). Dream city: Creation, destruction, and reinvention in downtown Detroit. MIT Press.

Kickert, C. (2019b). Retail. In E. Talen (Ed.), *A research agenda for new urbanism* (pp. 35–48). Edward Elgar Publishing.

Krueger, R. A., & Casey, M. A. (2014). *Focus groups: A practical guide for applied research* (5th ed.). SAGE.

Liamputtong, P. (2011). *Focus group methodology: Introduction and history; Focus group methodology: Principles and practice.* SAGE Publications.

Litvin, S. W., & Rosene, J. T. (2017). Revisiting Main Street: Balancing chain and local retail in a historic city's downtown. *Journal of Travel Research, 56*(6), 821–831. https://doi.org/10.1177/0047287516652237

López, T. G. (2003). *Influence of the public-private border configuration on pedestrian behaviour: The case of the city of Madrid.* La Escuela Tecnica Superior de Arquitectura de Madrid.

Mahmoudi Farahani, L., Beynon, D., & Garduno Freeman, C. (2018). The need for diversity of uses in suburban neighbourhood centres. *Urban Design International, 23*(2), 86–101. https://doi.org/10.1057/s41289-017-0052-x

Mehta, V. (2011). Small businesses and the vitality of Main Street. *Journal of Architectural and Planning Research, 28*(4), 271–291.

Moudon, A. V. (1997). Urban morphology as an emerging interdisciplinary field. *Urban Morphology, 1*, 3–10.

Orvell, M. (2017). A short history of the idea of "Main Street" in America. What It Means to Be American. https://www.whatitmeanstobeamerican.org/ideas /a-short-history-of-the-idea-of-main-street-in-america/

Parker, C., Ntounis, N., Quin, S., & Grime, I. (2014). High Street research agenda: Identifying High Street research priorities. *Journal of Place Management and Development, 7*(2), 176–184. https://doi.org/10.1108/jpmd -06-2014-0008

Porta, S., & Romice, O. (2010). Plot-based urbanism: Towards time-consciousness in place-making. In C. S. Mäckler & W. Sonne (Eds.), *New civic art* (pp. 82–111). Verlag Niggli.

Porta, S., Romice, O., Maxwell, J. A., Russell, P., & Baird, D. (2014). Alterations in scale: Patterns of change in Main Street networks across time and space. *Urban Studies, 51*(16), 3383–3400. https://doi.org/10.1177/0042098013519833

Sanders, P., & Baker, D. (2016). Applying urban morphology theory to design practice. *Journal of Urban Design, 21*(2), 213–233. https://doi.org/10.1080 /13574809.2015.1133228

Sayyar, S. S., & Marcus, L. (2013). Designing difference : Interpreting and testing Jane Jacobs' criteria for urban diversity in space syntax terms. In Y. O. Kim, H. T. Park, & K. W. Seo (Eds.), *Proceedings of the Ninth International Space Syntax Symposium* (pp. 1–15). Sejong University Press.

Scheer, B. C. (2016). The epistemology of urban morphology. *Urban Morphology, 20*(1), 5–17.

Sitte, C. (1889). *Der Städte-Bau nach Seinen Künstlerischen Grundsätzen: Ein Beitrag zur Lösung Moderner Fragen der Architektur und Monumentalen Plastik Unter Besonderer Beziehung auf Wien.* C. Graeser & Co.

Talen, E., & Jeong, H. (2019). Does the classic American main street still exist? An exploratory look. *Journal of Urban Design, 24*(1), 78–98. https://doi.org /10.1080/13574809.2018.1436962

Törmä, I., Griffiths, S., & Vaughan, L. (2017). High street changeability: The effect of urban form on demolition, modification and use change in two south London suburbs. *Urban Morphology, 21*(1), 5–28.

Vaughan, L. (2015). High Street diversity. In L. Vaughan (Ed.), *Suburban urbanities: Suburbs and the life of High Streets* (pp. 152–174). UCL Press.

Vaughan, L., Törmä, I., Dhanani, A., & Griffiths, S. (2015). An ecology of the suburban hedgerow, or: How high streets foster diversity over time. In *SSS 2015–10th International Space Syntax Symposium* (pp. 1–19).

Wrigley, N., & Dolega, L. (2011). Resilience, fragility, and adaptation: New evidence on the performance of UK high streets during global economic crisis and its policy implications. *Environment and Planning A, 43*(10), 2337–2363. https://doi.org/10.1068/a44270

Business Profile: Anderssons Skomakeri

Stockholm, Sweden

Jack Yacoub Isaac's shoemaker workshop is located in a historic establishment from 1936. It was previously run as a family business until Jack took over approximately 8 years ago. Jack has maintained the original interior of the shop but has lately started to consider modernizing the outlook to attract new customers. The business has undergone a lot of changes. At one point there were up to 13 people working in the workshop, whereas today he works by himself. In his 25 years as cobbler, Jack has witnessed a great decline in the demand for the profession. He accounts this partly to the changing seasons. He notes that today, winters are not as harsh as they used to be. The absence of snow and dirt that usually damage the shoes has meant a decline in the need for shoemaker services. Jack hoped that the repair-and-replace trend promoted by the younger generation will bring new vigour to his business and the industry as a whole. However, he notes that shoes are more often bought new. Hence, it is predominantly the older generation that keeps the profession alive. This has been felt heavily during the pandemic when many elders have decided to isolate. Although e-commerce is on the rise, Jack argues that his approach to the profession of a cobbler relies on tailored repairs that require offline assessment and personal interaction with the customer.

Figure B11.1. Jack Yacoub Isaac at work. (Credit: Photo courtesy of Rosa Danenberg.)

Figure B11.2. Jack Yacoub Isaac. (Credit: Photo courtesy of Rosa Danenberg.)

Figure B11.3. Anderssons Skomakeri. (Credit: Photo courtesy of Rosa Danenberg.)

12 Retail in the Mix

MATTHEW CARMONA

The Mixed Street Corridor

High streets are the backbone of the British retail scene. Often taking the form of mixed street corridors, retail shops are embedded within the surrounding neighbourhood and within a wider ensemble of uses – shops alongside municipal services, leisure facilities, and places of work and living. As a critical locus of life outside the home, the mixed street corridor is the key to sustainable urban place-shaping in the United Kingdom.

Despite this centrality, the United Kingdom's traditional high streets – these mixed use corridors – are taken for granted and often left to their own devices. In many cases this has led to a spiral of decline where formerly vibrant local commercial streets – once the social and economic centres of their areas – turn into degraded spaces, overrun by traffic, run down with derelict property and empty sites, and populated by ubiquitous signage, fast food establishments, betting shops, and other formulaic or marginal retail. This is a trend seen in towns and cities across the United Kingdom, but also internationally (Figure 12.1).

Despite some redirection of public investment to these streets – in particular to support their retail functions – the mixed street corridor has come under increasing stress. The threats of spiralling traffic growth, concentration of amenities and investment in large, car-dependent development in out-of-town locations, and the seemingly unstoppable growth of e-commerce (Lone, 2018) contribute to these trends (arguably driven on in a neoliberal era by the relative freehand given to the market). They are also a by-product of neglect amongst local policymakers, businesses, and development and management professionals who don't understand such seemingly "everyday" places or know how to manage them.

"Mix" is a characteristic of urbanity and therefore of historic cities across the world – mixing uses, activities, social groups, and built forms

Figure 12.1. Mixed street corridors, a universal typology and a typology often in decline (clockwise from top left: Eilat Street, Tel Aviv; Chestnut Street, Philadelphia; Huang Pi Nan, Shanghai; Trafalgar Road, Greenwich). (Credit: Photos courtesy the author.)

in a fine-grained melange of urban interactions. Mixed street corridors are likely to be vibrant (Schmitz & Scully, 2006), rich in experience (Carr et al., 1992; Montgomery, 1998), efficient in economic exchange (Jacobs, 1994; Grant, 2002; Schwanke, 2003), socially democratic and equitable (Sennett, 1990; Duany et al., 2000), and environmentally sustainable (Llewelyn-Davies, 2000; Clarke, 2009) (see Table 12.1). As a consequence, the pursuit of mix has increasingly driven urban policy agendas from the mid-1990s onwards. Yet, despite its being so good for us (Carmona, 2019), many have observed how regeneration, planning, development, design, and urban management relentlessly undermine mix in all its forms, leading to a more socially, functionally, and physically stratified city. Even new developments conceived and built as "mixed"

Table 12.1. Relating nine precepts of sustainability to urban mix

Sustainability precept	Relation to "mix"
Resource efficiency	Mixing uses vertically in buildings and horizontally along streets can reduce land take and create energy-efficient building forms
Diversity and choice	Mixing uses and activities increases diversity and choice across the range of urban facilities and amenities
Quality of life	Mixing uses and activities enhances vitality and therefore passive enjoyment of the street and contact between diverse social groups
Resilience	Mixing uses can increase building adaptability and enhance the viability of commercial uses by increasing user visits and visit duration
Pollution reduction	Mixing uses and facilities locally can stimulate walking and cycling and reduce the need to travel
Concentration	Mix enables and serves the densification of cities and in the process helps to make public transport more viable
Distinctiveness	Mix increases local street animation and encourages the creation of active building edges and locally distinctive communities of users
Biotic support	Mixing green with built infrastructure enhances biodiversity and aesthetic enjoyment in urban areas and encourages more healthy lifestyles
Self-sufficiency	Mixing uses enables those without personal transport, such as the elderly, to live more self-sufficient lives for longer

Source: Precepts from Carmona (2009).

continually fail the "diversity" test when compared to parts of the historic city (Cooper et al., 2009, p. 239).

A key characteristic of mixed street corridors is their complexity. Retail is but one part of an intricate urban structure that consists of four inter-related dimensions (Figure 12.2):

1. Physical fabric – often historic in origin, degraded but robust in the face of significant change
2. Places – of social activity and economic exchange
3. Movement corridors – channels of communication through the city
4. Real estate – typically in multiple uses and highly fragmented ownerships

This multidimensional nature of such streets is often misunderstood. For example, in the United Kingdom, town centres have been the subject of

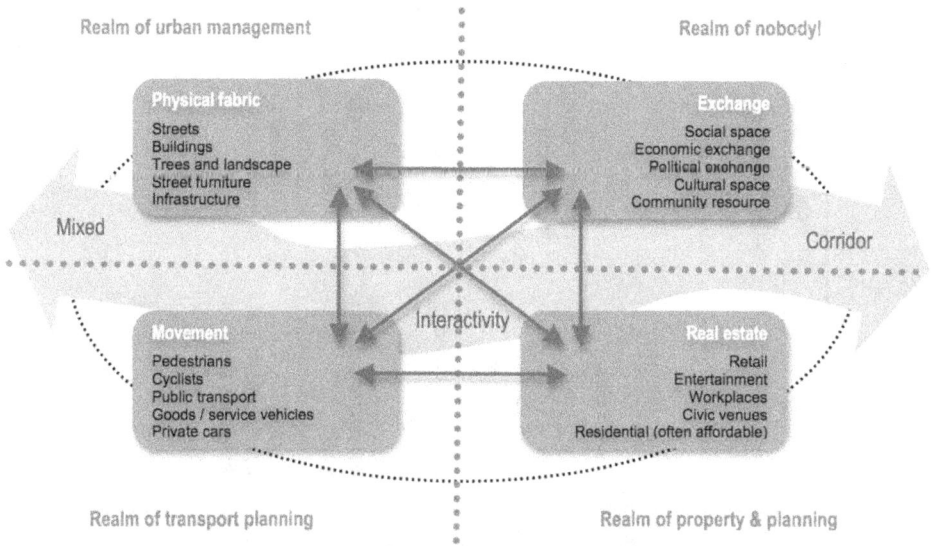

Figure 12.2. Multiple dimensions of mixed street corridors.

a high-profile review commissioned by the government from retail guru Mary Portas (2011). In her analysis, Portas advocated a range of measures to revitalize these spaces – the most high-profile of mixed streets – but her focus was on only a small fraction of the mixed street stock and just one dimension of these complex spaces – retail. Other recent reviews have tended to take a similar very limited perspective. Retail is a vitally important part of the story of these streets and of their future, but as retail comes under increasing strain, it may need to find a different role. Rather than being the driver of footfall, life, and local economies, it may take on a more supportive and leisure-based role – retail in the mix, but no longer the focus of all the attention.

The Challenges of Mixed Streets

The four interrelated dimensions sit within a larger political/economic context that can frustrate local responses; for example, there has been continued growth of out-of-town development (Geoghegan, 2012), despite the town-centre-first policies of successive national governments. A significant challenge lies in understanding the needs, conflicts, and synergies within and between the four dimensions (Table 12.2). Indeed, it is commonly argued that it is precisely the ability of private organizations

Table 12.2. SWOT of mixed street corridors

Issues	Strengths	Weaknesses	Opportunities	Threats
Management	Diversity of interests Multi-disciplinary perspectives	Complex management environment Fragmented governance Ad hoc decisions No holistic vision No coherent user voice	Better coordination of responsibilities Engaging with street-user groups Active management through TCM or BIDs Differentiation strategies	Failure to recognize their value Failure to learn
Real estate	Huge sunk investments (public and private) Diversity of investors Genuine diverse mixed use	Decline in high street retail Decline in independent retailers Closure of key local services Decline in civic amenities Reduced multiplier effects Increased costs – rent, rates, alterations	Reduced planning gain requirements Invest in the catchment Growth in the convenience market Better small business advice Efficient public sector investment vehicle	Danger of reaching a tipping point Crude planning – retail hierarchies Competition from out-of-town retail Vulnerability of chain stores
Physical fabric	Historic and distinctive Robust Adaptable to change	Poor public realm Decline of heritage assets Street furniture clutter Poor cleanliness and maintenance Cloning and loss of local identity Poor lighting	Public realm investment raising economic value Stated user willingness to pay for improvement Reinforce distinct sense of place Simplified streetscape schemes	Failure to reinvest Continued poor management Continued leaching of diversity

Exchange	Natural social venues Diverse range of user groups Low actual crime Diverse economic activity	Lack of responsibility for exchange functions Poor understanding of user profile Chain stores reducing local wealth recycling Conflict with functional concerns High perception of crime	New markets, events, social activities Active management to reduce fear of crime	Eventual decline of community Entrenched social exclusion
Movement	Well connected Natural movement corridors	Conflict for space Culture of separation and traffic flow efficiency High traffic load Poor integration of public transport High accident potential Lack and cost of parking Servicing restrictions Inadequate cycle facilities	Traffic calming Pedestrian-oriented crossing points Parking as a management tool Bus pull-ins	Future growth in traffic Dominance of buses Failure to address pedestrian needs

to think holistically about their assets – the shopping centre, business park, residential complex, leisure park, etc. – that makes them such devastating competitors to the more traditional mixed spaces (Audit Commission, 2002). By contrast, in a context where responsibility for mixed street corridors is so heavily fragmented (Carmona et al., 2008), the long-term stewardship of these structures creates a particularly "wicked" problem that is also low on the agenda of many key stakeholders.

Two problems that result illustrate the wider place-shaping challenges and prospects for sustaining the mixed street corridor.[1] First, too often the different agencies and stakeholders with a role to play will see these spaces from a narrow sectorial viewpoint: planners in terms of concentrations of land uses, transport planners as flows of traffic, property owners in terms of income streams from property assets, developers and designers as individual sites and projects, and so forth. Second, often no one will have a view of the whole, of the "place." Thus although ad hoc decisions will be taken day-in day-out on issues that affect the overall experience of place – on new uses for buildings, maintenance regimes, policing, and so forth – no one will be actively considering: whether the impact is positive or negative on the whole, the externalities decisions give rise to, or the long-term strategy for corridors.

Such problems may be exacerbated by the linear nature of mixed street corridors. This linearity means that, first, mixed strips often contrast dramatically (in character and quality) with the homogenous neighbourhoods through which they pass. Second, they typically have no coherent voice arguing their case, in part because they "inconveniently" cross local administrative and political boundaries. Third, in contrast to town centres, they are perceived as too insubstantial to merit their own dedicated management team or attention.

Moreover, as pieces of "old" and decayed fabric, to policy-makers they may seem to be more part of the "problem of the city" – congested, polluted, deprived, dilapidated, inefficient, etc. – than part of the solution to urban problems. Dines and Cattell (2006), for example, have noted starkly contrasting narratives of old and new in cities, where old is often associated with dirty, ugly, old-fashioned, anti-social space by key development interests, whilst community-based interests will tend to associate the older fabric with diversity, vitality, affordability and social mixing. As such there is a continuing tension between opposing agendas for many high street sites that set total redevelopment against regeneration of the existing urban fabric (Dines, 2009). There are also significant pressures against choosing to invest either public or private resources in such environments when set against the relative ease of investing in large brownfield or even greenfield sites where ownership, construction,

Figure 12.3. Vital social, economic, and cultural exchange function. (Credit: Photo courtesy of the author.)

conservation, regulatory, design, and development challenges are not nearly so complex (challenges that are universal; see Mostafavi, 2010).

Nevertheless, one of the few significant studies of local mixed streets in the United Kingdom suggests that, as urban structures, they command considerable local support from surrounding communities, encouraging sustainable and inclusive patterns of living in the process (Jones et al, 2007). Moreover, recent research in London demonstrates that collectively as complex pieces of physical, social, and economic fabric, and as large-scale sunk investments in urban infrastructure, mixed street corridors continue to play a vital strategic and local role (Figure 12.3).

Understanding Mixed Street Corridors through the London Case

Research on mixed street corridors in London[2] aimed to develop a better understanding and insight into the potential of mixed streets to support the sustainable development of the city – particularly in terms of economic growth (jobs) and housing. The research focused on understanding the 500 kilometres of London's mixed streets outside the Central Activities Zone (CAZ)[3] – mixed streets reflective of those found elsewhere across the United Kingdom. While London's size, scale, real estate profile, and system of governance does make it a special case, at the

Figure 12.4. Mapping mixed street corridors across London shows a dense pattern of accessible business centres along key travel routes. (Source: Created by author.)

local scale it is remarkably similar to any other traditional British city. It features some of the richest and poorest communities in the country, and across its area of 30 miles square, its streets and spaces mirror the diversity and challenges that characterize those in any other town or city across the country.

London's main mixed street corridors grew variously from a combination of development along key historic routes out of the city and from pre-existing historic village centres. Today they represent just 3.6 per cent of London's road network (Figure 12.4), although the research described here revealed that they have a potential and significance that belies their limited physical extent, and far beyond that generally recognized in policy and practice (Table 12.3). Of particular note is the enormous diversity of land uses and physical forms that urban blocks on

Table 12.3. Manifold potential of London's mixed street corridors

Potential	Description
Development opportunities	Three-quarters of London's developable brownfield land is on or within 500 metres of its local high streets, half within a two and a half-minute walk
Boosting quality of life	Thriving local high streets deliver huge quality-of-life benefits to living and working populations, and crime rates are far lower than are often perceived to be
Delivering employment opportunities	On or within 200 metres of its local high streets London has a higher number of employees (1.45 million) in almost double the number of businesses than in the CAZ
Not just retail	Almost half of non-domestic uses on or within 200 metres of mixed streets are in a diverse range of industrial and office uses
Recycling wealth	These small local businesses (averaging just 8.5 employees) have knock-on competitiveness, innovation, local economic development and sustainability benefits, recycling profits into the local economy
Benefiting all Londoners	Two-thirds of Londoners (5 million) live within a five-minute walk and 10% actually on or immediately next to a local high street, which disproportionately benefit many vulnerable, economically disadvantaged, and less mobile groups
Sustainable movement framework	Users are largely local and choose to travel by public transport or on foot (80%); the very essence of a sustainable movement framework
Transport capacity	In about half of London's local high streets potential exists to improve public transport accessibility further, taking advantage of the sizable concentration of activities there and the potential to stimulate development potential
Linear regeneration	The city is therefore not simply a collection of nodal town centres and surrounding residential areas, but a continuous urban fabric, joined by linear mixed use corridors that traverse areas of different administrative and social make-up. Investment in these streets brings significant regeneration benefits to all sections of society.

mixed street corridors have. Hidden behind their facades are a bewildering array of activities that feed off each other and the street itself, and that in turn help to fill it with life; indeed the research revealed that retail uses typically account for only around half of non-domestic uses in and around such local high streets. This diverse crust of activity is usually one block deep along mixed street corridors (Figure 12.5), a source of employment and great vitality that goes some way to explain the impressive employment figures associated with London's local high

Figure 12.5. Mixed street and its "unseen" mixed urban hinterland. (Source: Drawing, Fiona Scott.)

streets (see Table 12.3). This characteristic structure allows for continuity (on to the street itself) and change (within the block), and therefore for great adaptability over time.

However, many of London's mixed street corridors (reflecting similar local high streets across the United Kingdom) have been in long-term decline since the 1970s, most recently affected by the general decline in local retail. An essential point, then, is the need to look beyond retail services to establish a viable future for such streets. Moreover, this network of streets has become saturated by traffic with consequential high levels of pollution, often far in excess of European Union objective levels, representing a key threat both to the health of London's local high streets and, in a very real sense, to their users and residents. The first represents a classic case of market failure (a failure of the high street to adapt to new patterns of retail and technology), and the second a classic "tragedy of the commons" (that, without road pricing,[4] high streets fill with traffic until congestion allows no more). Both will require positive public sector intervention to address the situation.

Mixed local high streets represent substantial sunk investments in fixed public and private assets that have historically been highly sensitive to the different local communities they serve. If invested in, they are likely to lead to sustainable knock-on economic impacts (i.e., multiplier effects) that can be captured over the long term in council tax and business rates increases. This implies that public investment in such streets will be an effective use of resources because new jobs and housing will benefit from and strengthen existing infrastructure, communities, public services, and private businesses, rather than having to create them from scratch. But such interventions will need to transcend physical fabric, real estate, exchange, movement, and management threats (see Table 12.2), and above all address the problems of traffic overload in many of these vitally important local spaces.

The Possible Futures of Mixed Street Corridors

The hypothesis behind the research on London's mixed street corridors was that mixed street corridors continue to play a vital strategic and local role across the city (as they do elsewhere), representing an important element in the urban fabric. As such, they have great potential to accommodate (and generate) future growth, new jobs, and housing in a sustainable manner. Yet, as complex pieces of physical, social, and economic fabric, the public sector has seen such places simply as locations on an abstract retail hierarchy, in transport terms as traffic corridors, and in urban management terms as a low priority. Two contrasting futures might be envisaged: the "canal future" and the "sustainable future."

The Canal Future

The canal future offers an analogy from history – and a warning about what the future may hold. The situation of many mixed street corridors today is not dissimilar to that of the British canal network in the mid-nineteenth century, when canals rapidly lost their core purpose as industrial arteries, first in competition with the railways, and then with roads. The consequence was the decline and demise of the rich life, culture, and infrastructure that had grown up on and around canals. Although canals eventually found a new purpose, for leisure (for which they are now heavily subsidized by the state), today users largely move along and engage with these spaces as detached and transient outsiders, or alternatively overlook them as part of canal-side property developments, enjoying their aesthetic value but failing to generate large-scale positive externalities.

Unlike the canals that declined, mixed street corridors are typically not short of traffic – quite the opposite. However, what *is* similar to today's canals is that traffic largely passes them by with little engagement with the

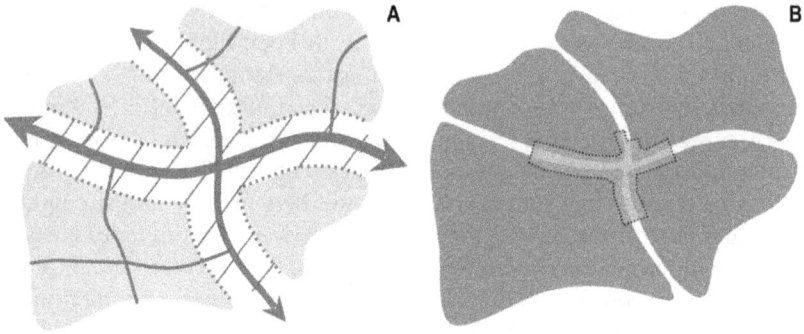

Figure 12.6. Contrasting visions of mixed street corridors. (Source: Created by author.)

other functions of these streets, contributing to their long-term demise because of the negative environmental, health, and social consequences of vehicular traffic. What is less understood are the consequences of the decline of the street environment (and the mixed activity fronting it) and the complex interactions that take place with the surrounding hinterland. At what point does the decline become fatal? Like canals, many mixed street corridors may have to reinvent themselves. If they don't, then for many such spaces the future will portend only further disinvestment, decline, and disengagement.

The Sustainable Future

A sustainable future for mixed street corridors in London, as elsewhere, represents one of the city's great unrealized opportunities for accommodating growth and enhancing sustainability. Rather than placing these (like much of the urban sustainability agenda) in the "too difficult to handle" category, mixed street corridors could be made a strategic priority for public sector policy, investment, and action over a generation.

In the case of mixed streets, this implies a view of the city that sees the street itself embedded in a hinterland that feeds and shapes the space, whilst the resulting corridor is continuous and connective ("A" in Figure 12.6). This contrasts with the spatially defined and localized notions of retail and neighbourhood designations that characterize land use plans today ("B" in Figure 12.6) and falsely see retail as somehow isolated from the hinterland that feeds it. A wide variety of largely bottom-up collaborative initiatives are possible to safeguard the future of these quintessentially urban and sustainable spaces – as summarized in Table 12.4.

Table 12.4. Approaches to rejuvenating mixed street corridors

Physical fabric	Real estate	Exchange	Movement	Management
Distinctiveness initiatives	**Vacant properties initiatives**	**Green and civic spaces**	**Traffic calming**	**Community engagement**
• Reveal heritage where possible	• Encourage temporary uses in vacant buildings	• Upgrade quality of neighbouring green spaces and remove barriers to integration with high streets	• If possible, divert through traffic to bypass roads	• Facilitate community consultation and research to properly understand different communities and users and their long-term needs
• Provide loans to upgrade shopfronts on the basis of adopted design guidance	• Compulsory purchase derelict buildings and land and support site assembly	• Consider opportunities for new incidental/civic spaces and pocket parks, e.g., reclaiming road space at junctions/side streets	• Where possible, make high streets 20 mph (30 km/h) zones	• Consider community-based art and other engagement
• Enhance character through public art, landscaping, and opportunities for new landmark buildings	• Introduce living-over-the-shop grant and advice regime		• Where appropriate, adopt naked streets principles, to encourage changed perception of road/pedestrian balance	**Day-to-day management**
• Encourage unblocking of shopfronts to create display space and active frontages	**Intensification and redevelopment**	• Encourage shops, cafes, and restaurants to spill out onto street space	• Lane reduction where possible to allow space for service bays and short-term parking	• Introduce town centre management to better coordinate management roles and responsibilities
• Consider the parts and the whole, and whether the parts have their own distinctive characters and/or role	• Grow high street catchment by prioritizing sensitive new development along and around high streets	**Crime initiatives**	• Introduce super-crossings, allowing diagonal crossing at junctions	• Encourage BID schemes to raise additional resources for management
	• Compile sites to facilitate redevelopment	• Dedicated street wardens to reduce anxiety	**Improved pedestrian experience**	• Invest in long-term maintenance
Public realm	• Encourage reuse of large sites for temporary purposes, e.g., markets, events, exhibitions, etc.	• Encourage trader watch schemes	• Adopt shared space principles where possible off the main high road run	• Better control and coordinate waste disposal and removal
• Distinguish high street through road surface colour and texture		• Encourage family-based evening economy uses	• Widen pavements where congested	
• Adopt consistent, simple, and high-quality public realm treatments: paving and street furniture			• Remove street clutter and barriers to allow pedestrians to move more freely	
			• Improve way-finding, e.g., adopting legible signage	
			• Where poor, enhance connectivity between the high street and its hinterland	

(Continued)

Table 12.4. (Continued)

Physical fabric	Real estate	Exchange	Movement	Management
Trees and soft landscape • Protect street trees and introduce or replace where required to soften the landscape and filter dust • Introduce seasonal colour through planting displays **Lighting strategy** • Replace roads-based lighting with pedestrian-focused lighting schemes • Floodlighting of landmark buildings and creative lighting to enhance evening economy	**Retail diversity** • Protect diversity through local "well-being" powers (e.g., purchasing threatened local businesses) and planning policy • Encourage street markets and mini-markets • Introduce advice service for small businesses **Big-box initiatives** • Redress relationship to the street, through major redevelopment or wrapping schemes • Allow new big-box developments only if sensitively integrated behind a high street facade	**Civic uses** • Resist pressures to consolidate and relocate civic-type functions to off–high street locations • Consider opportunities for new high-street based civic uses, e.g., libraries, idea stores, citizen advice, housing/ payment office, leisure facilities, etc. **Public toilets** • Better manage existing facilities • Open new high-quality, accessible public toilets	**Public transport improvements** • Upgrade bus shelters • Relocate stops to avoid pedestrian/bus congestion • Allow space for bus pull-ins • Enhance interchange spaces and routes between high streets and stations **Cycle network improvements** • Link up cycle network to stations • Introduce continuous cycle routes along high streets • Upgrade cycle parking **Pollution** • Treat road surfaces to reduce particulates • Carefully control new higher building proposals to avoid canyon-type effects • Reduce traffic loads and speeds	• Prioritize everyday cleaning, cleansing, and maintenance • Consider better marketing, e.g., through a dedicated website, events, and activities • Encourage shop owners or residents to adopt benches, flower beds, etc.

For policymakers, the challenge is to understand the drivers that have transformed many urban areas over the past 20 years and that, with a suitably supportive policy context, continue to make them desirable, viable, and potentially more sustainable places for investment. They need to understand the needs, conflicts, and synergies within and between the four interrelated dimensions of complex urban elements, between their evolving physical fabric, role in social and economic exchange, facilitators of movement, and function as locations of real estate use and investment.

Despite market failure and a tragedy of the commons, mixed street corridors retain unrivalled public support and great potential to suitably accommodate (and generate) future growth, new jobs, and housing. Realizing this will require a very different, more intelligent approach to the management of such spaces, to stimulate their growth potential over the long term, and to manage their day-to-day stewardship. It will require intervention that sees such places not as a set of fragmented responsibilities, but instead in a holistic manner, where issues of place and movement are reconciled within their physical fabric, in a manner that maintains a viable real estate market, the efficient functioning of space, and vital and safe social spaces.

For the public sector, this might begin with investment in the public realm in mix street corridors, which related research in the United Kingdom has shown can deliver very significant dividends across each of the four dimensions discussed above (Carmona et al., 2018). How the COVID-19 pandemic will affect this will only be fully known in time, but just as lockdowns in the United Kingdom and internationally continued weeding out weaker retail from the high street, so an opportunity opens up for mixed street corridors, many of which are very local in their catchment and orientation. COVID has clearly reinforced trends already underway whereby those who can, increasingly choose to work from home using technology to communicate. A greater population at home with more time on its hands may offer the lifeline that many of our local high streets need, although to survive they will certainly need to continue their journey from providers of everyday essential services to leisure services. Making these streets places that people really want to spend time (and money) and giving users more high quality walkable space to do it in is likely to be key.

Conclusion

In the United Kingdom, as elsewhere, retail is embedded in a complex mix of uses and activities that coalesce within mixed street corridors. This "mix" is a characteristic of urbanity and therefore of historic cities across the world – mixing uses, activities, social groups, and built forms in a fine-grained melange of urban interactions. In the United Kingdom, despite the redirection of significant public and private investment over the last 20 years into urban areas, mixed street corridors have struggled to remain viable. Contributing to their decline is the reality that there are no stakeholders responsible for taking a holistic view of them as forms of "place." It is a testament to their intrinsic importance in urban life that, despite their neglect, mixed streets in the United Kingdom continue to command considerable local support from surrounding communities, encouraging sustainable and inclusive patterns of living in the process.

As complex pieces of physical, social, and economic fabric, and as large-scale sunk investments in urban infrastructure, mixed street corridors continue to play a vital strategic and local role. London's 500 kilometres of mixed streets outside the Central Activities Zone, for example, represent just 3.6 per cent of the road network but support employment for 1.5 million Londoners (in retail, office, and light industrial uses). They also host a sustainable public transport network, accommodate half of London's brownfield sites, and serve, within a five-minute walk, a population of 5 million. Despite this and the continuing availability, indeed increase in availability, of capacity for development within urban areas, the public sector has continued to place such spaces in the "too difficult to handle" category.

Rather than a dead-end future for these spaces analogous to that of canals in the first half of the twentieth century (a "canal future"), a "sustainable future" would require that they become a strategic priority for public sector policy, investment, and action over a generation, all the more important in view of changing live/work patterns in the aftermath of COVID-19. This would benefit existing communities as well as countless generations to come.

NOTES

1 Recent research has distinguished between the creation of high-quality places – "place-making" – and their long-term management or stewardship – defined as "place-keeping" (see Dempsey & Burton, 2012, for a comprehensive discussion). The concept of "place-shaping" seeks to encompass both in an integrated manner, and focuses on how places are shaped over time through a range of actions, from new investments in space to their on-going management (Carmona 2014a).

2 Mixed research methods were used, combining detailed GIS analysis of London-wide data sources covering development potential, employment,

transport accessibility, resident population, access to healthcare, and pollution, with detailed morphological and typological analysis of mixed streets across the city, and detailed case studies of six such streets. Data was drawn from the Cities Revealed Land Use dataset, LDA Strategic Housing Land Availability Assessment (SHLAA), GLA brownfield database, Annual Business Inquiry (ABI), TfL Public Transport Accessibility Index, 2001 Census of Population, London Atmospheric Emissions Inventory (LAEI), TfL Access To Opportunities and Services database (ATOS). See Carmona (2014b) for a full write-up of the research.

3 The CAZ is a special case, being an almost continuous network of mixed-use streets. For the purposes of the research, mixed street corridors were understood to mean stretches of street with continuous ground floor retail of at least 250 metres of continuous length, the presence of retail being one of the defining characteristics of such spaces.

4 Only streets in Central London are subject to road pricing through the London Congestion Charge.

REFERENCES

Audit Commission. (2002). *Street scene.* Audit Commission.

Barking and Dagenham Council. (2017). Barking Market: The heart of the local community. YouTube. https://www.youtube.com/watch?v=QyVhHM9RMUI

Carmona, M. (2009). Sustainable urban design: Principles to practice. *International Journal for Sustainable Development, 12*(1), 48–77. https://doi.org/10.1504/ijsd.2009.027528

Carmona, M. (2014a). London's local high streets: The problems, potential and complexities of mixed street corridors. *Progress in Planning, 100*, 1–84. http://www.sciencedirect.com/science/article/pii/S0305900614000439

Carmona, M. (2014b). The place-shaping continuum: A theory of urban design process. *Journal of Urban Design, 19*(1): 2–36. https://doi.org/10.1080/13574809.2013.854695

Carmona, M. (2019). Place value: Place quality and its impact on health, social, economic and environmental outcomes. *Journal of Urban Design, 24*(1): 1–48. https://doi.org/10.1080/13574809.2018.1472523

Carmona, M., de Magalhaes, C., & Hammond, L. (2008). *Public space: The management dimension.* Routledge.

Carmona, M., Gabrieli, T., Hickman, R., Laopoulou, T., & Livingstone, N. (2018). Street appeal: The value of street improvements. *Progress in Planning, 126*, 1–51. https://doi.org/10.1016/j.progress.2017.09.001

Carr, S., Francis, M., Rivlin, L., & Stone, A. (1992). *Public space.* Cambridge University Press.

Clarke, P. (2009). Urban planning and design. In A. Ritchie & R. Thomas (Eds.), *Sustainable urban design: An environmental approach* (pp. 12–20) (2nd ed.). Taylor & Francis.

Cooper, R., Evans, G., & Boyko, C. (2009). *Designing sustainable cities.* Wiley-Blackwell.

Coupland, A. (1996). *Reclaiming the city: Mixed use development.* E & FN Spon.

Dines, N. (2009). The disputed place of ethnic diversity. In R. Imrie, L. Lees, & M. Raco (Eds.), *Regenerating London: Governance, sustainability and community in a global city* (pp. 254–272). Routledge.

Dines, N., & Cattell, V. (2006). *Public spaces, social relations and well-being in East London.* The Policy Press.

Duany, A., & Plater-Zyberk, E., with Speck, J. (2000). *Suburban nation: The rise of sprawl and the decline of the American Dream.* North Point Press.

Geoghegan, J. (2012). Growth in out of town shopping centre floorspace, says report. https://www.planningresource.co.uk/article/1127173/growth-town-shopping-centre-floorspace-says-report

Grant, J. (2002). Mixed use in theory and practice: Canadian experience with implementing a planning principle. *Journal of the American Planning Association, 68*(1): 71–84. https://doi.org/10.1080/01944360208977192

Jacobs, J. (1994). *The death and life of great American cities.* Penguin Books.

Jones, P., Roberts, M., & Morris, L. (2007). *Rediscovering mixed-use streets: The contribution of local high streets to sustainable communities.* Policy Press.

Llewelyn-Davies. (2000). *Urban design compendium 1.* English Partnerships & The Housing Corporation.

Lone, S. (2018). *The United Kingdom 2018 ecommerce report.* Ecommerce Foundation.

Mostafavi, M. (2010). *Why ecological urbanisn, why now?* In M. Mostafavi with G. Doherty (Eds.), *Ecological urbanism* (pp. 12–55). Lars Muller Publishers.

Portas, M. (2011). *The Portas review: An independent review into the future of our high streets.* Department for Business, Innovation & Skills.

Schmitz, A., & Scully, J. (2006). *Creating walkable places.* Urban Land Institute.

Schwanke, D. (2003). *Mixed-use development handbook* (2nd ed.). Urban Land Institute.

Sennett, R. (1990). *The conscience of the eye: The design and social life of cities.* Faber & Faber.

Business Profile: Barking Market

London, United Kingdom

Barking in east London has an ancient right to hold a market dating back to 1175. Located in the town centre, it is operated by the London Borough of Barking and Dagenham and opens four days a week. The market users as well as the traders are hugely diverse, with the market providing discount goods and an essential social milieu to an often under-privileged population. During the COVID pandemic the market was closed for 12 weeks but returned in the summer of 2020 with a revised layout in order to ensure better social distancing. All traders are licenced by the council, with the monies used to manage the facility. The market and its traders give a flavour to the entire district.

Figure B12.1. Barking Market. (Credit: Image courtesy of Matthew Carmona.)

PART FIVE

Toward Solutions

13 Curating Main Streets: The Factors of Success

MICHAEL W. MEHAFFY AND TIGRAN HAAS

Introduction

It is a common theme of this book that the success of retail main streets and their businesses is an increasing challenge for cities and their neighbourhoods – faced with growing online retail, the increasing prevalence of big-box retailers, and other forms of competition (Talen & Jeong, 2019). This challenge has been exacerbated by the COVID-19 pandemic and its aftermath (Bartik et al., 2020). Yet as discussed in other chapters, we have come to understand (in part because of the pandemic) the unique and important role of main streets in providing neighbourhood-scale goods and services to nearby residents, small-scale entrepreneurial opportunities for shop owners, economic development, and employment for surrounding neighbourhoods, sense of community and social capital for residents, and many other benefits (Southworth, 2005; Drabenstott et al., 2003; Pendola & Gen, 2008). Reflecting these benefits, all 193 member states of the United Nations unanimously adopted an urban policy framework in 2016 that stresses the value of main streets in "fostering both formal and informal local markets and commerce, as well as not-for-profit community initiatives, bringing people into public spaces and promoting walkability and cycling with the goal of improving health and well-being" (United Nations, 2017). Nevertheless, the failure rate for many main streets and their businesses can be high, particularly for new or regenerated streets that include new businesses. It is therefore critical to find and share effective tools and strategies to develop, maintain, and improve main streets.

This chapter will focus on the most common factors of success for main streets, including not only their successful location and design, but also the detailed ingredients of business recruitment, development, support, and management – what we will refer to as the "curation" of main streets. Since in most cases we cannot simply hope that main streets will spontaneously arise and become successful, we must establish a process of planning and governance, built on sound understanding and management of main street dynamics.

This curation of main streets requires one or more "curators," who may take different forms. Sometimes they will be large property owners who will act as landlords for the retail businesses, or smaller landlords who will come together to create some kind of unified governance structure (such as a business association) to help to perform the curation and support of the businesses. In other cases, a non-profit or local government entity might establish a governance structure, such as a Business Improvement District (BID), Local Improvement District (LID), or some other entity such as a public-private partnership. The considerations below will apply to all of these different kinds of curators.

Location, Location

The first consideration is the location of the street itself, and whether its proximity, connectivity, and visibility to potential visitors and customers is adequate. It is crucial that a retail street has a sufficient number of people passing by who might see and use the businesses. This can be ascertained empirically with traffic counts of cars, other vehicles, and pedestrians, although they may only represent a snapshot in time. These traffic numbers can change significantly through temporary events or over longer spans of time, particularly as an area develops or redevelops. Over a longer period, traffic tends to be higher, depending on whether the street itself is well situated within the larger network of surrounding streets, as described in the chapters of Rosa Danenberg and Conrad Kickert.

Closely related is the question whether there is sufficient consumer demand for the services offered along a retail street. Retailers will attempt to discover if their services are already offered by competing businesses within their trade area, or whether this area is even saturated, meaning that little or no additional demand is available for new businesses to serve. This is of course a dynamic issue, since surrounding populations can increase or decline, demographics can change, and even street patterns can be altered, compromising the visibility and accessibility of stores (for example, when two-way streets change to one-way). What appears to be a saturated market or a poorly connected street today can be reinvigorated tomorrow as new conditions develop.

Retail competition should be seen not only as a detriment to new business. In fact, the synergy of several businesses often draws more customers, responding to the benefits of clustering into a critical mass of retailers (Kickert & Hofe, 2017). This effect is particularly helpful if the offerings are slightly different but compatible – for example, several different kinds of restaurants, or several varieties of shops. One can observe this phenomenon in many thriving urban retail districts that are dominated by only one kind of business, for example New York's "Garment District" or "Diamond District." At the same time, most neighbourhood main streets are intended to serve residents' routine daily needs and activities, so they might contain a broader mix of shops and services.

Competition from volume discount retailers is far more difficult for urban retailers, as big-box stores might undercut prices and make it difficult for typical neighbourhood retailers to compete. Often these volume discount retailers are farther away in larger stores, requiring more driving or other transportation by residents. It is vital to differentiate the local offerings from the volume discount offerings – for example, fresh vegetables and unique local products can supplement more standardized staples bought at the volume discounter. In this way, the main street business can complement rather than unsuccessfully outcompete the volume discounter (Miller et al., 1999).

Who Is Being Recruited?

Successful main streets hinge on successful businesses, which depend on their offerings, their operations, and how well they align with market demand. This is where it is most critical to curate businesses along a street – a process of selection, preparation, planning, design, and execution as well as experimentation, refinement, and improvement. A key debate in curation is between attracting or resisting chain retailers.

Chain retailers have an advantage over independent retailers in that they have well-established methods to select, develop, and operate successful locations. Chains may signal confidence to other businesses that the market is attractive, or they may provide uniquely desirable goods because of their unique access to certain products.

However, chains are also typically less able to adapt to unique neighbourhood characteristics and market segments, or serve unique neighbourhood needs (Halebsky, 2009; Mehta, 2011). They also tend to offer fewer entrepreneurial opportunities to local residents, and there is some evidence that their "economic multiplier" – their contribution to the local economy – is not as great, particularly for chains that have out-of-town headquarters (Blakely & Bradshaw, 2002; Halebsky, 2010). Finally, there is some evidence that chain stores may contribute less to

the development of business involvement in civic affairs, or in reciprocal supportive social relationships (or "social capital") in the neighbourhood (Tolbert, 2005; Green & Haines, 2002).

For all these reasons, it is generally desirable to have at least a strong
representation of non-chain businesses in main streets. The recruitment
of independent retailers increases the burden on landlords and other
curators to select businesses that can learn and adapt to a new market.
Due diligence must be exercised to find business owners who have the
requisite personal characteristics to make up for less experience. On the
other hand, independent business owners are more agile and motivated
to respond to the local market and its customers, and less constrained by
established corporate practices.

Are the Businesses Visible and Attractive?

The first prerequisite for any retail business is visibility to passers-by, by
foot, bicycle, or car. Visible, attractive, and stylish signage is important
on and near stores, as are adequate display windows without excessive
tinting or shading from overhangs or trees. Stores should present themselves through the street in human-scaled, interesting, colourful, and
enticing ways, for example through attractive architecture, artwork, and
vegetation. Good lighting for night-time operations is also important, as
is adequate shelter from inclement weather.

Once entered, retail spaces should be comfortable and attractive.
The field of "atmospherics" research studies the multisensory cues that
improve or deteriorate consumer experiences, including sights, sound,
and smell (Turley & Milliman, 2000). Furthermore, the offerings and
service of a business need to align with consumer preferences. Chain
businesses often have their own established methods of customer attraction and retention. (Indeed, the challenge may be instead to encourage
them to adapt more specifically to the character and needs of a local
main street.) However, for small entrepreneurial businesses – especially
those who are relatively new to the market – there is often work to be
done in these areas. Thus it is critical that curators apply a support system
to develop and nurture successful small businesses.

Who Is Managing the Street, and How?

This brings us to the next big challenge for main street curation: the governance structure. The formal structure (or structures) can vary enor-
mously, depending in particular on the property ownership structure –
one large institutional owner, multiple owners, public and/or private

ownership, etc. In all cases, however, the governance should typically address several criteria:

1. A clear definition of a "public" realm – both physically, in the form of the streetscape and other public spaces, and institutionally, in the form of public sector participation in governance. This participation can take the form of a public-private partnership, public oversight of privately operated public spaces, or other public-sector governance structures.
2. Clear definition of roles and responsibilities for financing, developing, and maintaining distinct parts of the public realm by distinct entities, for example, business frontages by the business owners, common areas by a shared institution, activities within public rights of way, and so on.
3. A mechanism to encourage the recruitment of businesses that best serve neighbourhood needs and long-term public interests.
4. A mechanism that supports the development and promotion of the businesses and their interests, such as a business association, or a landlord-provided education, marketing, and development office.

As mentioned, it is unlikely that a single organization will address all of these criteria. Typically, a given main street will contain a series of overlapping governance structures, which may include, for example, the local city government, independent associations, public-private partnerships, citizen or neighbourhood groups, or other hybrid entities.

One hybrid institutional governance structure, particularly common in Canada and the United States, is the Business Improvement District (BID). Known under various names in different localities, BIDs are entities created by local government with the legal power to tax owners to pay for public space maintenance and improvements, marketing, lobbying, and at times even new business and real estate development (Slone, 2016; Houstoun, 2003; Briffault, 2010). A related variant is known as a Local Improvement District, although LIDs are typically more focused on larger areas including residential streets, and on improvement of infrastructure such as streets and utilities.

Although there is evidence that BIDs can encourage the success of small main-street businesses and the quality of urban life, they are not without controversy (Mitchell, 2001; Briffault, 1999). The danger that they pose is in the commingling of public and private interests, with the potential that private goals will override public needs, or at best erode the transparency and accountability of governance – a perennial concern for contemporary governance (Unger, 2016; Banerjee, 2001). Hogg et al. (2003) point to four ingredients of BID success: strong leadership, a genuine desire for change, clarity of objectives, and support from the local authority. A key addition would be the clarity of roles and interests, ensuring that the public interest is not compromised in favour of private ones.

A simpler governance structure for main streets is the single private owner. Obviously, single ownership allows for more direct and effective decision-making, clearer lines of internal accountability, and greater ability to adapt quickly to varying market demand and opportunities. On the other hand, a single owner may likely prioritize private interests over the public good, will lack public accountability, and may prevent access to public resources such as taxation. These advantages and drawbacks may be balanced by including single owners in a larger hybrid model such as a public-private partnership (discussed further below). These partnerships have also sometimes proven controversial, particularly when accountability and transparency were not sufficient, or when private interests proved to have strong precedence over public benefit. Similar to BIDs, clear governance structures can ensure a balance between public and private interests.

Some private owners are inherently more able than others to aim for long-term growth of property value. Owners with "patient capital" often align better with the public interest, as they have the time to ensure a more vibrant neighbourhood, offering more jobs and prosperity, with strong growth in economic value. Patient capital is more commonly found in large institutional owners, such as insurance companies or pension-fund partnerships, whose assets are invested typically for many years or decades. By contrast, other private developers typically must provide rapid returns on investment, often well before the real value of a development has had time to mature. The result is often a negative impact on the long-term quality and sustainability of a development (Leinberger, 2005, 2008). However, even institutional entities with more patient capital do not act strictly in the public interest. Their actions that affect the public realm and urban commons therefore still require public oversight. Usually the local government provides this oversight through its function to regulate land use and property development – making it an important partner in the governance and curation of a main street.

Larger Challenges of Governance – and Government

In all these cases, the challenge is to balance public, private, and personal interests. In almost all cases, there is a mix of different public and private entities at work, including formal governments and other, often less formal, kinds of governance. This institutional web-network of governance actors is memorably described by the political economist Elinor Ostrom (2010) as "polycentric governance," which, she argued, is to some extent a feature of all governance and indeed all economic systems.

Polycentricity in governance describes the web of actors that shape places, creating tensions not only between competing public and private interests, but between long-term interests and short-term ones, which are often favoured by institutional dynamics and the limited knowledge of individuals (including consumers). Therefore, one important task is to clearly identify and agree upon longer-term goals, particularly for a larger public (such as a successful and thriving main street, or successful public space). Through strategic financial mechanisms, regulatory powers, and public-private collaborative structures, governments and private stakeholders can balance their mutual interests. In such partnerships it is important to have clear roles and lines of accountability, and at the same time to develop a larger network of participation from all the entities that can contribute to success. These should include not only immediate property owners, but also individual business owners, other institutions, civic groups, residents, and other members of the public. Here Ostrom's "polycentric governance" model applies: neither allowing the dominance of private entities with their own (often unaccountable) interests, nor public-sector bureaucracies with their own internal logic and momentum, but a network of agents working together to produce forms of value at many scales of place and time.

Summing Up the Takeaways: The "Four Cs" of Successful Main Streets

The previous section demonstrates that successful retail main streets all aim for and achieve four characteristics, which can be referred to as "the Four Cs":

- *Connectivity:* The main street is located in a strong position within the larger street network, well connected to other parts of the city and to its own neighbourhood, and with good visibility to customers.
- *Catchment:* The main street is within close proximity to an adequate number of potential visitors and customers, including not only those who pass by, but those who live and/or work nearby; and the potential customers are likely to find services and activities that match their needs and interests.
- *Competition:* The main street does not have excessive competition for the same customers from other businesses, and where such

competition exists, it is complementary and/or helps to draw sufficient new customers.
* *Curation:* The main street businesses are carefully selected, organized,
  and managed in a way that enhances their combined attractiveness,
  neighbourhood suitability, and success.

These characteristics must be secured through a customized governance
structure, which involves actors from the private sector, public sector,
civil society, and adjacent residents.

Case Study: Orenco Station, Oregon

Orenco Station near Portland, Oregon, illustrates the four characteristics of successful main streets in the development of a new main street in
the United States. This grey-field redevelopment of a failed housing subdivision into a new 150-acre mixed-use community included a new main
street extending up from a light rail station to the major employer at the
opposite end of the neighbourhood. This main street was a newcomer
to a sprawling, low-density, auto-dominated suburban area between the
urban centres of Portland and Hillsboro with no other walkable, compact, transit-oriented neighbourhoods nearby.[1] This section discusses
the development of Orenco Station's main street and discusses its successes and challenges through post-occupancy research evidence and
findings.

This particular main street was created by PacTrust, a single institutional developer funded by a partnership of two state employee pension
funds and a Wall Street investment firm. This institutional structure provided patient capital to invest for the long run, as well the ability to find
value at different scales of time and place. A "polycentric governance"
structure was set up as a larger collaboration between PacTrust and its
public and private partners.

Background

The Hillsboro region saw strong growth of high-tech businesses in the
1980s and early 1990s, producing a net increase of over 15,000 jobs, but
also a lack of housing that prompted long commutes and traffic congestion. Around the same time, however, the Portland region had adopted
the 2040 Growth Concept, a framework plan that included a proposed
network of light rail lines and other transportation corridors, connecting a series of targeted development centres. These sites were within the
region's "urban growth boundary" – part of Oregon's innovative land

use system aimed at conserving farmland and providing greater urban compactness and connectivity.

One proposed new centre was at the former site of the Oregon Nursery Company in Hillsboro, an early 1900s exporter of ornamental trees and shrubs. The company had built its own small incorporated town, named Orenco – a contraction of the company name. After the nursery closed down in the late 1920s, it was used for industrial development, including what became one of Intel Corporation's largest campuses. The area between this campus and the original Orenco town was also zoned for industrial development. However, the regional transit agency and its partners had secured funding for a new light rail line and stop on the site, contingent on a mixed-use overlay zone that would require many more housing units to rebalance the area's jobs and residents.

After deliberating on whether to sell to another more experienced residential developer, the owner PacTrust ultimately decided to embark on development of the site into a new kind of "complete" and transit-oriented community following the development model of New Urbanism. The neighbourhood would include many different kinds of housing, walkable streets, parks, and open spaces, built either by the owner directly or with its partners and sub-developers. As its centrepiece, the community would feature a new main street. The initial development included 16 businesses offering unique local neighbourhood services, including a grocery, household goods store, laundry, dentist, accountant, gift shop, wine shop, coffee shop, and four restaurants. Only one business was a national chain – the coffee shop, a Starbucks, which gave an early signal to the market that this was an attractive location.

As predicted, the market expanded over the next two decades, and the later phase to the south added 22 new businesses. Some ten years after development, in the height of the 2008–2010 recession, the occupancy rate for then-existing businesses was over 95 per cent, marking this main street as a clear success. The company was careful to follow the "Four Cs" of successful main streets, as follows.

Step One: Choose the Right Location

Initially, the local and regional planning and transit agencies argued strongly that Orenco Station's main retail street should located around the new station area. PacTrust argued instead that a busy existing automobile arterial to the north was better suited to support the retail businesses, especially in their early stages. The main street should align with this street connection as it existed at the time – including three bus lines that already ran along the northern arterial. They hence built the new main street perpendicular to this arterial, with its initial focus at the inter-section. As the area developed and more residents settled in, additional retail was added farther to the south of this intersection to extend to the station area.

This planning occurred in conjunction with the development of a new zoning code and development agreement, a collaboration between the City and PacTrust. The new zoning code created a single mixed-use zone over the entire site, allowing the developer to respond to market conditions and build out a truly urban main street, with tight setbacks, mix of uses, higher density, live/work units, accessory dwellings, and many other innovations. All of these elements would have been illegal under the original zoning code, as indeed they still are in many Ameri-can suburbs.

The development agreement allowed the private sector to do what it does best – identify the evolving market and respond to it successfully. Meanwhile, the public sector could do what it does best – identify the long-term public interest and assure public oversight, involvement, and transparency. Although there was only one major institutional owner in control of the project, this was nonetheless a "polycentric governance" model.

Step Two: Identify the Catchment Area

PacTrust recognized that in the early years, there would not be enough residents in the development to support the retail, and it would be nec-essary to draw from a wider catchment area. A key reason to focus initial retail development at the northern arterial was to draw nearby employ-ees looking for daily services like drop-off laundry, an appetizing meal, an evening drink with friends, or simply an enjoyable walk through the neighbourhood. Furthermore, the retail development was targeted to attract early residents looking to live within walking distance of daily amenities. This was a successful strategy: in post-occupancy research, the proximity of the town centre and its main street was consistently one of the top reasons given for deciding to live in the neighbourhood. Simul-taneously, PacTrust remained careful to accommodate non-resident visi-tors arriving by car to assure success of the businesses.

As the neighbourhood residential population increased and alternate modes like light rail gained popularity, car travel could be de-emphasized. Indeed, recent research has shown a gradual but significant shift of modes away from car travel in Orenco Station. While in 2008, almost two-thirds of the neighbourhood's residents used their car to commute, this proportion decreased to just under a third a decade later (Pobodnik, 2011; Ewing et al. 2018). The choice to "start where the people are" – in

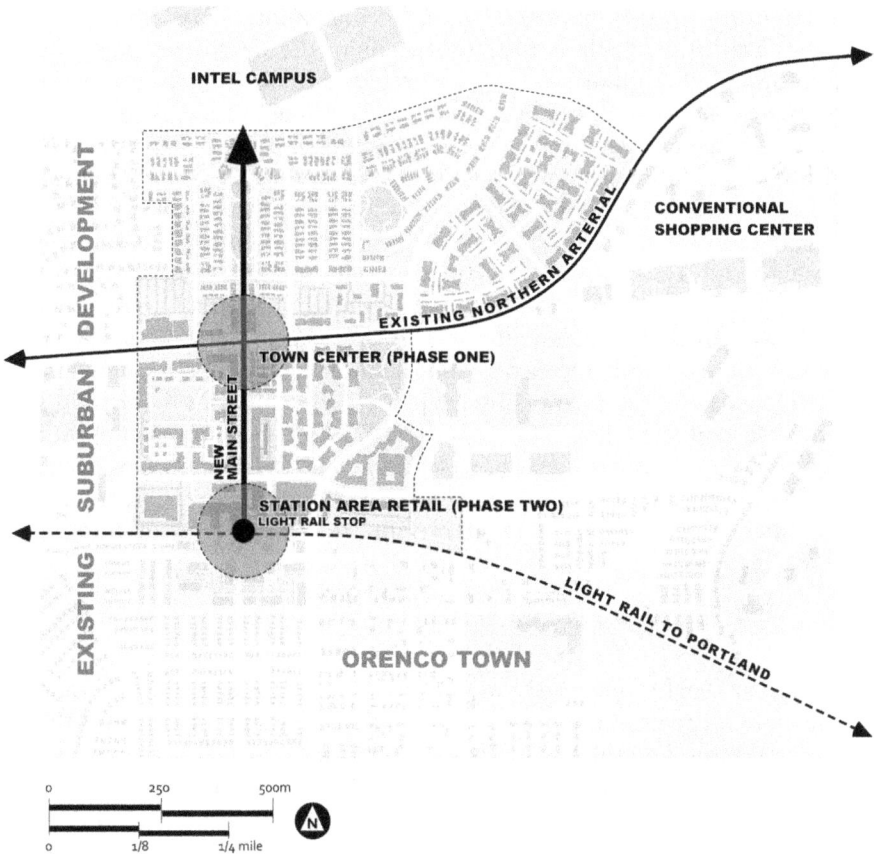

Figure 13.1. The new main street would span the new light rail line at bottom to the major employer at top, crossing a major arterial, where the initial businesses would locate in the "town centre." This strategy of focusing on the most connected "movement economy" proved successful. (Source: Created by authors.)

connectivity and catchment – and then expand and transition from there proved to be the right decision.

Step Three: Plan for the Competition

One of the developer's early decisions was to position the town centre relative to area competitors, who were generally larger and lower-cost "big-box" and "strip-centre" retailers dominated by chains. The new town centre would establish a distinctive niche of unique businesses on a traditional walkable main street, anchoring a more urban neighbour-hood. Market research showed that many of the nearby employees at Intel and other high tech businesses preferred to live in more urban set-tings and buy more distinctive products from time to time. At the same time, they did not want to face the long commutes from urban Portland, and they also still needed to buy generic staple products. The answer was to provide the best of both worlds.

PacTrust owned a 50-acre site at the perimeter of Orenco Station and developed this site into a more conventional shopping centre featuring volume retailers, albeit with walkable tree-lined streets and good access to transit. Instead of competing with such a centre, they made it fully complementary with Orenco Station's town centre in character and merchandise on offer. This strategy proved effective, and there is little evidence that the competition from PacTrust's conventional shopping centre has diminished the success of Orenco Station's main street busi-nesses. Moreover, the more conventional shopping centre generated revenues to operate the main street with greater flexibility in curating its businesses.

Step Four: Provide Curation

A number of master-planned communities have struggled to develop their retail main streets – sometimes in spite of offering subsidies, deeply dis-counted rents, and other concessions. On Orenco Station's main street, rather than providing financial incentives, PacTrust focused on mentor-ing and preparing independent businesses so that they could find their own way to success, which was regarded as an ultimately more sustainable strategy. Some 18 years after their opening, three of the four original restaurants are still in operation, far longer than the 4.5-year median lifespan for American restaurants (Luo & Stark, 2014). The developer's retail leasing specialist developed a desired business mixture, and then met with dozens of small entrepreneurs, tech professionals, homeowners with funding and vision, and existing successful small business owners. There was extensive handholding – consultation about product mix, dis-play, lighting, signage, and myriad other issues.

For example, one of Orenco Station's first restaurants was started by the wife of a senior Intel engineer, who had taught South Asian cuisine cooking classes and even hosted a cooking television program, but had never run her own restaurant. The retail leasing specialist worked closely with her to develop a unique space and suggested operational improve-ments during her operation. Eighteen years later, the restaurant is still in operation, albeit with a new owner. Other businesses worked through

Figure 13.2. Orenco Station's main street in operation. (Credit: Images by author.)

Figure 13.3. New Seasons grocery (*top*) and weekly farmers' market (*bottom*) –
complementarity in competition. (Credit: Images by author.)

challenges with visibility of their wares, variety of offerings, signage, hours of operation, and myriad other issues. The unique nature of the main street and its businesses meant there were no simple formulas to understand what would work, and instead, success was secured by careful assessment, experiment, and refinement, coupled with the leasing specialist's expertise in retail dynamics. One of the most important recruitments was the grocery store, New Seasons – a small local chain with highly individual offerings, well suited to the unique nature of Orenco Station's main street and with distinctive offerings that complement the nearby big-box grocer in PacTrust's conventional shopping centre. New Seasons' understanding of this complementarity is illustrated by their support in recruiting a weekly farmers' market to Orenco Station, which in turn boosts their sales in market days. Furthermore, curation of retail was shared with local residents in frequent discussion and participation meetings.

Conclusions

In a larger sense, the entire region of Portland was responsible for curating the development of Orenco Station – creating the urban growth boundary system, establishing the regional planning authority, building the light rail line, advancing the 2040 Growth Concept with its town centres, and finally producing the overlay zone placed on the Orenco Station site. PacTrust assertively responded to this context through a remarkable set of innovative development practices. Although the project was led by its single large institutional owner, there were dozens of other active partners and participants, including professional consultants, sub-developers, businesses, homeowners, neighbours, the City of Hillsboro, and many other agencies. This complex network of partnerships, along with the subsequent institutional governance structures (HOAs, business associations, etc.) constitute an example in action of what Eleanor Ostrom referred to as "polycentric governance" (Ostrom, 2010). Although the process was certainly not without friction, overall it provided strong added value, with the result that the community and its main street are still thriving.

While the resulting initial phase of the main street was quite successful, more recent events are cautionary. The Starbucks has recently closed, potentially suffering from its image as a chain amidst so many local and personalized businesses. Another recently closed restaurant has been replaced by a real estate office, which may offset the short-term loss of rental income, but may harm the long-term retail curation of the main street. The balance between sustaining long-term value in the face of short-term costs is always ongoing, never secured and never finished.

One can critique Orenco Station's lack of separate, fine-grained owner-ship or the lack of larger regional connectivity – as a result of the persistence of urban sprawl in the region. Much more is required to understand and respond to the dynamics of new main street development – more reforms, new tools, new strategies. A report for the Metro regional planning author-ity on this theme suggests "new tools and approaches for developing centers and corridors" (Center for Portland Metropolitan Studies, 2009).

These lessons apply to other main streets besides new locations in the sprawling suburbs of the United States like Orenco Station.

Of course conditions for existing main streets, inner-city locations, and other kinds of locations will vary greatly from the ones described here, as will conditions in other countries – including legal mechanisms, financial tools, business practices, and other elements of what we might think of as the operating system for growth. However, the deeper lessons discussed here of connectivity, catchment, competition, and curation are universal. Many of the specific tools and strategies are also shareable, although they require local adaptation to fit the context. As we look to implementation of the New Urban Agenda and its goals – and in particular, its goals for healthy and thriving main streets – it is precisely these evidence-based tools and strategies that we must develop and share.

NOTE

1 One of the chapter authors (Michael Mehaffy) worked as the project manager for the master developer of Orenco Station, and therefore has firsthand knowledge of the neighbourhood and its main street.

REFERENCES

Banerjee, T. (2001). The future of public space: Beyond invented streets and reinvented places. *Journal of the American Planning Association, 67*(1), 9–24. https://doi.org/10.1080/01944360108976352

Bartik, A. W., Bertrand, M., Cullen, Z. B., Glaeser, E. L., Luca, M., & Stanton, C. T. (2020). *How are small businesses adjusting to COVID-19? Early evidence from a survey* (Publication No. w26989). National Bureau of Economic Research.

Blakely, E., & Bradshaw, T. K. (2002). *Planning local economic development: Theory and practice* (3rd. ed.). Sage.

Briffault, R. (1999). A government for our time? Business Improvement Districts and urban governance. *Columbia Law Review, 99*(2). 365–477. https://doi.org/10.2307/1123583

Briffault, R. (2010). The business improvement district comes of age. *Drexel Law Review, 3*(19).

Center for Portland Metropolitan Studies. (2009). *Achieving sustainable, compact development in the Portland Metropolitan Area: New tools and approaches for developing centers and corridors.* Oregon Metro. https://www.oregonmetro.gov /centers and corridors report

Drabenstott, M., Novack, N., & Abraham, B. (2003). Main streets of tomorrow: Growing and financing rural entrepreneurs – A conference summary. *Economic Review: Federal Reserve Bank of Kansas City, 88*(3), 73–84.

Ewing, R., Tian, G., Park, K., Stinger, P., & Southgate, J. (2018). *Trip and parking generation study of Orenco Station TOD, Portland region.* National Academy of Sciences, Transportation Research Board. Publication No. 18-05650. https:// trid.trb.org/view/1496885

Green, G., & Haines, A. (2002). *Asset building and community development.* Sage.

Halebsky, S. (2009). *Small towns and big business: Challenging Wal-Mart superstores.* Lexington Books.

Halebsky, S. (2010). Chain stores and local economies: A case study of a rural county in New York. *Community Development, 41*(4), 431–452. https://doi.org /10.1080/15575330903503351

Hogg, S., Medway, D., & Warnaby, G. (2003). Business improvement districts: An opportunity for SME retailing. *International Journal of Retail & Distribution Management, 31*(9), 466–469. https://doi.org/10.1108/09590550310491432

Houstoun, L. O. (2003). Business Improvement Districts (2nd ed.). Urban Land Institute.

Kickert, C., & Hofe, R. v. (2017). Critical mass matters: The long-term benefits of retail agglomeration for establishment survival in downtown Detroit and The Hague. *Urban Studies, 55*(5), 1033–1055. https://doi.org/10.1177 /00420980`17694131

Leinberger, C. B. (2005). The need for alternatives to the nineteen standard real estate product types. *Places, 17*(2), 24–29.

Leinberger, C. B. (2008). *The option of urbanism: Investing in a new American dream.* Island Press.

Luo, T., & Stark, P. B. (2014). *Only the bad die young: Restaurant mortality in the Western US.* US Bureau of Labor Statistics. arXiv preprint arXiv:1410.8603. https://arxiv.org/pdf/1410.8603.pdf.

Mehta, V. (2011). Small businesses and the vitality of main street. *Journal of Architectural and Planning Research 28*(4), 271–291.

Miller, C. E., Reardon, J., & McCorkle, D. E. (1999). The effects of competition on retail structure: An examination of intratype, intertype, and intercategory competition. *Journal of Marketing, 63*(4), 107–120. https://doi.org/10.1177 /002224299906300409

Mitchell, J. (2001). Business improvement districts and the "new" revitalization of downtown. *Economic Development Quarterly, 15*(2), 115–123. https://doi.org/10.1177/089124240101500201

Ostrom, E. (2010). Beyond markets and states: Polycentric governance of complex economic systems. *American Economic Review, 100*(3), 641–72. https://doi.org/10.1257/aer.100.3.641

Pendola, R., & Gen, S. (2008). Does "Main Street" promote sense of community? A comparison of San Francisco neighborhoods. *Environment and Behavior, 40*(4), 545–574. https://doi.org/10.1177/0013916507301399

Podobnik, B. (2011). Assessing the social and environmental achievements of New Urbanism: Evidence from Portland, Oregon. *Journal of Urbanism: International Research on Placemaking and Urban Sustainability, 4*(2), 105–126. https://doi.org/10.1080/17549175.2011.596271

Slone, S. (2016). *Place management in downtown & transit-oriented developments.* Council of State Governments.

Southworth, M. (2005). Reinventing main street: From mall to townscape mall. *Journal of Urban Design, 10*(2), 151–170. https://doi.org/10.1080/13574800500087319

Talen, E., & Jeong, H. (2019). Does the classic American main street still exist? An exploratory look. *Journal of Urban Design, 24*(1), 78–98. https://doi.org/10.1080/13574809.2018.1436962

Tolbert, C. M. (2005). Minding our own business: Local retail establishments and the future of Southern civic community. *Social Forces, 83*(4): 1309–1328. https://doi.org/10.1353/sof.2005.0084

Turley, L. W., & Milliman, R. E. (2000). Atmospheric effects on shopping behavior: A review of the experimental evidence. *Journal of Business Research, 49*(2), 193–211. https://doi.org/10.1016/s0148-2963(99)00010-7

Unger, A. (2016). *Business Improvement Districts in the United States: Private government and public consequences.*

United Nations. (2017, 25 January). *Resolution adopted by the General Assembly on 23 December 2016: 71/256, "New Urban Agenda."* United Nations.

Business Profile: Orenco Station (a Photo Tour)

Hillsboro, Oregon

Figure B13.1. Orenco Station Grill exterior. (Credit: Photo courtesy of the authors.)

Figure B13.2. Orenco Station Grill interior. (Credit: Photo courtesy of the authors.)

Figure B13.3. Orenco Station Grill's Gabe Chavez, Nena Chavez, and Moh Woods. (Credit: Photo courtesy of the authors.)

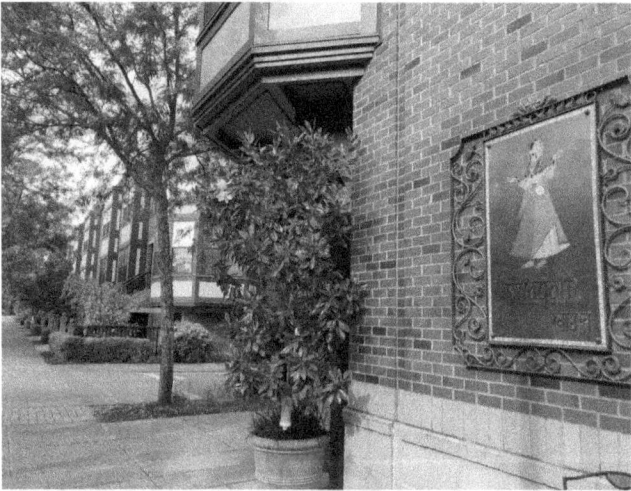

Figure B13.4. Swagath Indian Restaurant. (Credit: Photo courtesy of the authors.)

Figure B13.5. Tous Les Jours Bakery. (Credit: Photo courtesy of the authors.)

14 The Spatial Logic of Urban Retail

CONRAD KICKERT

Retail is undergoing profound changes on both sides of the Atlantic, creating challenges and opportunities. On one hand, retail consolidation and deregulation continue to bolster large suburban formats (see, for example, the chapter by Colin Jones in this book), and health concerns accelerate the virtualization of retail; on the other, the rise of the experience economy and the maker economy – along with urban population growth – are bringing new retail potential to cities (Kickert, 2019). And yet, risk-averse retail organizations remain wary of returning to cities. One reason is an imbalance in retail research: the retail dynamics of suburban locations are far more well known than urban locations. For decades, marketing-based retail research – necessary to substantiate new suburban locations – has been largely a-spatial, especially in North America (Clarke, 1998; Kickert, 2019).

This chapter argues that retail dynamics specific to walkable urban (as opposed to suburban or virtual) locations are indeed knowable. I show this by surveying supply- and demand-side theories and models that apply to urban retail, and by illustrating the validity of this research in three cities in Europe and North America. These theories, models, and illustrations demonstrate the spatial logic of urban retail, from the macro-scale of the region to the micro-scale of the individual retailer. The chapter concludes with a discussion of how more research is needed to further uncover the logic of retail dynamics as it applies to urban locations, thereby supporting urban retail decision making by developers, planners, and retailers.

Supply-Side Models of Retailer Behaviour

The oldest and most persistent models used to describe the retail landscape focus on the behaviour of retailers as they decide to locate in space. Many of these theories have become classics in the field of retail geography, tested and adopted by planning agencies and developers without questioning their validity or applicability. At the root of most regional and citywide theories is the assumption that larger agglomerations of retailers will draw a larger audience, an upward cycle that allows large retail clusters to attract more business and grow further. At the root of this assumption is Central Place Theory, originally formulated by German economist Walter Christaller (1933/1966), and later followed by Lösch (1940). The theory assumes that rational consumers seek to minimize travel as they obtain goods or services, and that travel is stratified by thresholds: someone will travel farther for durable goods like shoes, but less far for convenience goods like groceries. As a result, vendors of different types of goods disperse evenly to serve their customers, stratifying into a system of higher- and lower-order retail centres (Figure 14.1). Subsequent critiques showed that information about available goods and services was asymmetrical, that trips could be linked, that consumer preferences differed, and that centres varied in ways that had little to do with the merchandise they offered (e.g., historical value, social connotation; Shepherd & Thomas, 1980).

Nevertheless, American economists Berry and Garrison (1958a, 1958b) evolved this intra-urban concept to distinguish between different scales of retail clusters within cities, from corner stores and neighbourhood shopping centres to regional malls and downtowns. In a similar vein, Spatial Interaction Theory was used to study the distribution of retail from a relational perspective. Based on Newtonian physics, the theory assumes that consumers gravitate to retail centres on the basis of the size of their merchandise offering, their distance to the centre, and a calibration variable derived from shopping-behaviour studies. The assumed outcome, though more rooted in empirical research, is a similar network of tiered retail clusters not unlike those derived by Christaller (Brown, 1993).

Another perspective on the benefits of retailer clustering comes from merchants. Harold Hotelling's (1929) Principle of Minimum Differentiation suggests that retailers will seek to compete by overlapping as much as possible, including their physical location. By locating next to a competitor, a retailer will be able to serve the competing retailer's trade area. The resulting truism of jostling ice-cream vendors on a beach is a strategy for both retailers and consumers: neighbouring retailers with similar merchandise allows consumers to compare goods, minimizing the risk of failure to find what they need at the best price – an insight that is over a century old (Hurd, 1903, p. 82). Retailers also benefit from clustering by capturing the overflow sales of complementary neighbours. As most retail purchases are part of a multipurpose shopping trip, it is key to locate close to shops that draw a clientele that may visit you afterward.

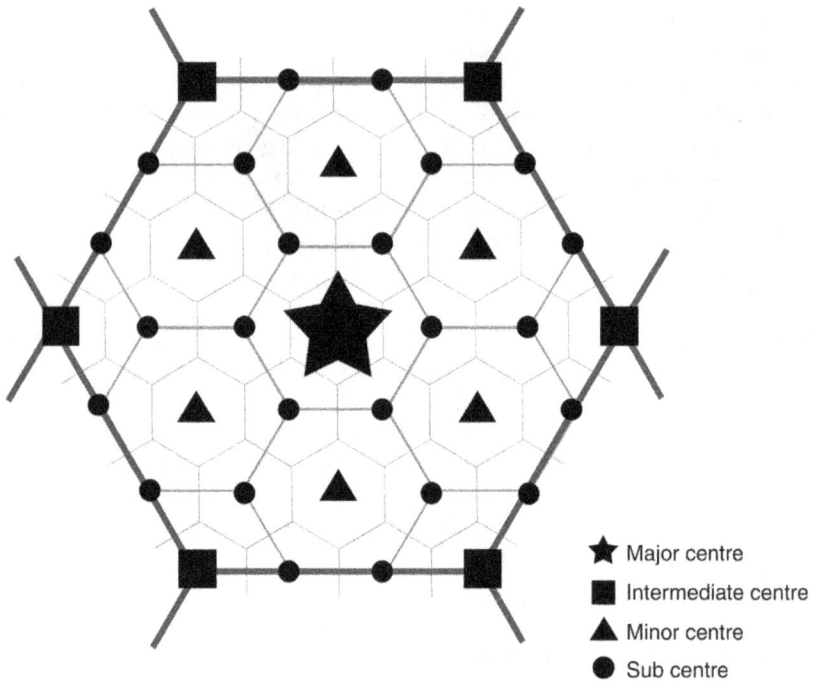

Figure 14.1. Christaller's retail hierarchy, indicating centres and their overlapping service areas.

Examples include fashion and accessory stores, bars and restaurants, and specialty food vendors (Anikeeff, 1996; Sevtsuk, 2020). Economist Richard Nelson empirically corroborated clustering benefits to retailers in his Theory of Cumulative Attraction in 1958, demonstrating that stores that sell similar or complementary merchandise will indeed see more sales as they cluster (Nelson, 1958). A longitudinal study of a century of retail closure in The Hague and Detroit has proven that the size of a retail cluster has a significant influence on individual retailer potential for long-term survival (Kickert & Hofe, 2017).

While retailers benefit from clustering to attract a larger audience, they also stratify in other ways. For example, a small grocer may be content to serve local residents at a small corner location, or a hardware store may be located off the beaten track if customers know where to find it. A shoe store or jeweller, on the other hand, may need to catch the eyes and wallets of the largest possible audience – on the busiest corner in town. Retailers' differing need for central locations determine the rent they are able or willing to pay for this accessibility – with the highest bidder occupying the most central location in the Bid-Rent curve of a city (Alonso, 1960, 1964; Burgess et al., 1925). This assumption of stratification along lines of accessibility is highly scalable, as it applies to the intra-city, inner-city, and even inner-district level. The theory has evolved into one of the more well-substantiated models of urban retail distribution (Brown, 1993). However, like its predecessors, Bid-Rent theory in the retail context oversimplifies retail distribution, assuming perfect consumer knowledge, homogenous historical, geographical, and social conditions, and a lack of planning intervention and organizational bias.

Several economists have attempted to test the classical models of retail distribution *within* cities (most models originally focused on trade *between* cities). After a mostly normative description of retail distribution in cities in the 1930s (Proudfoot, 1937), one of the first empirical attempts was a landmark study of Chicago's internal retail patterns by Brian Berry (1963). He distinguished different types of retail concentrations, formulating a hierarchy of both planned centres and unplanned retail strips at variously accessible locations within the city. In addition, Berry identified "specialized areas" that focus on a single type of retailing with a wider reach, such as auto rows, specialty food markets, and entertainment districts. British economist Ross Davies applied Berry's insights in a study of the urban core of Coventry, combining the duality of centripetal agglomeration economies and centrifugal Bid-Rent stratification into a "complex model" of retail distribution (1972a, 1972b). Like Berry, Davies found three types of retail distribution patterns: stratified rings of different retail types in relation to the main centre of the city, linear retail ribbons that align with accessible urban arterials, and clusters of specialized retailers in various parts of the city (Figure 14.2). Drawing from both Bid-Rent theory and the Principle of Minimum Differentiation, he connected this taxonomy to a classification of retailers by merchandise type, resultant market area, and merchandise quality – an increasingly difficult taxonomy in today's mixture of merchandise categories and qualities (Brown, 1994). Brands now simultaneously cater to price-conscious and boutique consumers, and as described in the chapter by Heather Arnold, retailers increasingly combine merchandise categories.

Many of these retail distribution theories have become staples in the repertoire of public regulators and private retail developers. Widespread acceptance of these classical theories and models has had the effect of a "double hermeneutic cycle," in which retailers, developers, and planners follow these theories and models, consequently further corroborating their validity – whether they had original empirical merit or not (Brown, 1993, p. 209). As retail location decisions are mission critical and involve significant

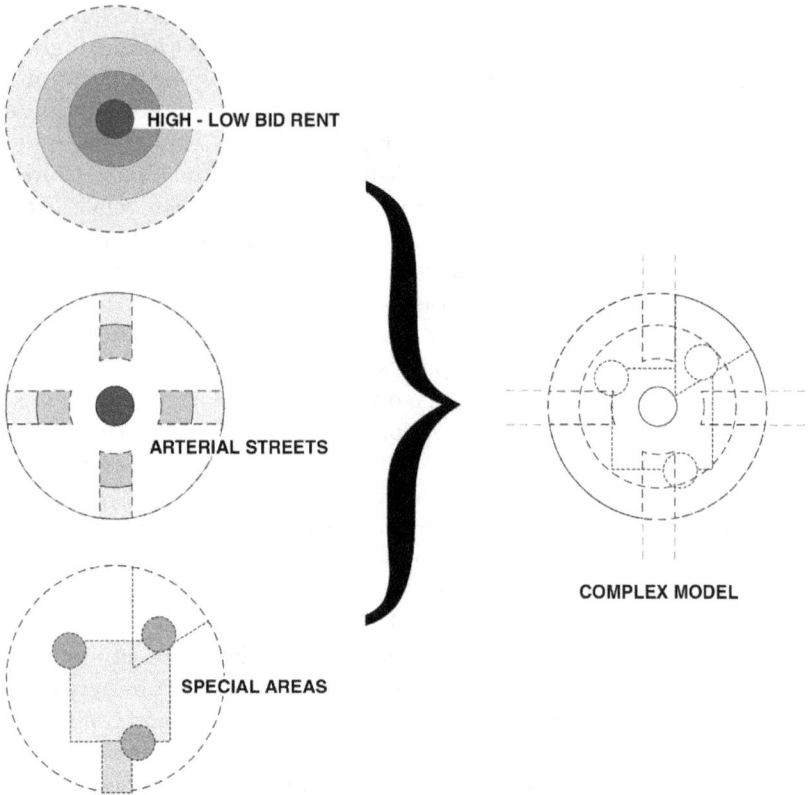

Figure 14.2. Clusters, ribbons, and special areas in the "Complex Model of City Centre Retailing." (Source: Image by the author, after Davies [1972b] and Davies and Bennison [1978].)

investments, the value of accepted theory is so consequential that it is easy to understand the tendency to conform (which is also motivated by the perception that one is reducing risk). In recent decades, computing power and more sophisticated data-gathering methods have been used to further advance the normative theories of the early twentieth century.

But these applications have exposed flaws in the accepted models. While retailing models do offer insights about distribution patterns, there are demand-side oversights that plague current models: the theories assume rational and unbiased decisions of retailers and consumers in a "free market condition" Brown, 1994); they ignore the consumer penchant for experiential or socially compatible (rather than merely central) shopping environments (Bromley & Thomas, 1993; Cohen, 1996; Kooijman, 2000); they lack complete information on retail options (Shepherd & Thomas, 1980); and they fail to address the phenomenon of e-commerce (Kickert, 2019; Smith & Anderson, 2016). Similarly, the supply side of retail organizations and public planning agencies do not have full knowledge of optimal locations, nor do they act as independent agents maximizing their individual benefits (Caplin & Leahy, 1998). Instead, they usually respond to thus far unmodelled external circumstances, including unpredictable political and organizational structures. Locational decisions are increasingly made in council chambers or on Wall Street, not on Main Street.

Demand-Side Models of Consumer Behaviour

Perhaps the greatest shortfall in the classical models of retail distribution is their focus on the supply side: the locational decisions that retailers make to serve a rather vaguely defined and usually oversimplified pattern of consumer demand. While models of vendor clustering and dispersal superficially speak the spatial language of urban planning and design, they fail to delve deeper into the behavioural patterns of customers underlying these models. What is required now – in light of a retail landscape undergoing unprecedented spatial and organizational shifts – is more intelligence about the demand side of the retail equation. And in fact, emerging research is increasingly focussed on the underlying dynamics of retail consumer behaviour, enabled by evolutions in surveying, tracking, and analytical methods. This has been particularly encouraged by retailers who demand greater predictive accuracy in the face of increasingly saturated markets. When opening a new location in a cutthroat market, normative platitudes of clustering and dispersal – even if corroborated by research – no longer suffice. This section will describe consumer behaviour studies that have influenced retail decision making and distribution patterns at scales between and within cities.

Early models predicting consumer behaviour and applying it to retail locational choices focused on the regional and citywide level. Economist David Huff evolved Reilly's gravity-based model to determine likely consumer spending at retail centres (Huff, 1963). William Applebaum soon after developed a popular adaption: his "analogue model" defines trade areas and predicted sales for stores by extrapolating data from a comparable store's customer behaviour, socio-economic milieu and physical setting. The model generates tiered rings of predicted market share for a new store location, cumulatively providing an estimate of its patronage. While originally quite rudimentary, this model has significantly evolved in software packages and the marketing departments of larger developers and retailers, which has in turn redefined the pattern of retail distribution in American and European cities – remaining mostly at the level of entire trade areas (Applebaum, 1966; Hernández & Bennison, 2000; Shepherd & Thomas, 1980). Applebaum's approach to predicting consumer behaviour at the aggregate level has gained traction with evolved data synthesis and computational power, coinciding with a shift from geography to marketing research (Clarke, 1998). It is now the dominant method used in retail locational decision making, which increasingly relies on a combination of network analysis and "geodemographic" consumer profiles to find proximity to the right amount, type, and wealth of potential customers (Birkin et al., 2002; Dunne et al., 2013).

Unfortunately, the transformation of retail locational decision making from spatial patterns to aggregated consumer profiles has dulled retailers' understanding of the micro-scale complexity of urban economics, urban society, and most notably of urban form. This oversight has been detrimental to the rebirth of urban retail, as oversimplified models fail to accurately assess the potential of urban locations, catering solely to risk-aversion rather than urban potential (Kickert, 2019). Fortunately, a recent canon of research models and predicts consumer behaviour at the walkable urban scale – a worthy addition to the vocabulary of urban retail decision making. In general, these models connect the spatial and functional structure of cities with pedestrian activity, based on the century-old understanding by economist Richard Hurd that "increasing [pedestrian] traffic [is certain] ultimately to change any street into a shopping street" (1903, p. 93). Various models approach the relations between space, activity and consumption from a wide variety of perspectives.

Perhaps the earliest attempt to connect space and consumption came from direct pedestrian counts, used in mid-nineteenth-century Berlin (Baumeister, 1876) and in Hurd's 1903 work, gaining traction during the 1920s when they were used by real estate appraisers in the United States and Europe. Increasingly sophisticated pedestrian counts have continued to form the basis of notions of urban vitality (Gehl, 1987, 2010; Whyte, 1988), and retail research companies like Experian and Locatus offer frequently updated pedestrian activity ("footfall") data in many European countries. While many counts are still conducted manually at prescribed points or "gates," GPS, Bluetooth, and WiFi remote-sensing methods have evolved to automate the process and add valuable insights on routes, speed, staying time, and – if privacy regulations allow – consumer identifiers (Ratti & Claudel, 2016; Spek et al., 2013; Van der Spek et al., 2009). Pedestrian counts remain popular, as footfall is strongly linked to retail rent and sales, hence serving as reliable indicators of success of use to developers, investors, and operators (Bolt, 2003; Thornton et al., 1991). As Heather Arnold's chapter describes in this volume, the transition of physical retail stores from transactional to experiential spaces expands the importance of passers-by from fuelling in-store sales to aiding brand recognition in general. As a result, rents are increasingly linked to the number of passing "eyes" on the storefront, whether they make a purchase or not.

On the other hand, counts are not able to predict pedestrian behaviour that might result from future interventions. This makes them less useful for substantiating large interventions like new retail streets, clusters, or districts. To address this, several models have been developed to find the patterns underlying pedestrian behaviour, unlocking predictive capacities under altered conditions. Early analysis of urban pedestrian flows recognized the importance of "anchors" such as major department stores, bus stops, car parks and railway stations (Davies & Bennison, 1978). Subsequent research modelled pedestrian behaviour from the perspective of multipurpose trips (a notorious gap in classical modelling), in which pedestrians are assumed to make continuous spatial decisions to move between functions, whether completely random and exploratory, or purposive with destinations in mind (Borgers & Timmermans, 1986; Timmermans, 2009). This bottom-up or "agent-based" approach to modelling consumer behaviour has gained traction in subsequent decades with the rise of computing power and data availability that corroborates modelled interactions between pedestrians, geometry, and key locations (Batty, 2001).

Another approach to modelling pedestrian behaviour focuses on the geometric configuration of streets and spaces in cities to determine pedestrian behaviour (Hillier & Leaman, 1973). Based on the notion that space and society influence one another in a constant cycle, architectural and urban space becomes a social construct with its own "social logic" (Hillier & Hanson, 1984). Specifically for retail, the configuration of streets in cities creates "movement economies" that facilitate (pedestrian) traffic as destinations and through-routes, which make for fertile ground for retailers (Hillier, 1996). In this "Space Syntax" methodology, cities are analysed as spatial networks, correlating the integration of streets and their connectivity to the surrounding city with the number of people using these streets. These movements generate a subsequent land-use pattern with urban centres that contain retail uses, set in a "background network" of mostly residential streets (Hillier, 1999; Van Nes, 2002). Contrary to all aforementioned models, Space Syntax does not take land-use patterns or other key destinations into account, as these are assumed to derive from the configuration of the city (Penn & Turner, 2002). The street configuration of cities indeed strongly correlates with their retail pattern, as demonstrated for the urban region of Detroit in Figure 14.3. As previously

Figure 14.3. Overlay of most interconnected (choice) streets in Detroit (in a radius of 10 kilometres) and retail land use.

mentioned, contemporary retail locations rely on far more than their physical location in the street network, such as their social, demographic, regulatory, and organizational context. In Detroit, suburban malls are mostly reliant on convenient freeway access to a wealthy hinterland, leaving the relatively well-connected but impoverished inner city devoid of retail (Psarra et al., 2013).

Space Syntax methods are not without their critics. In particular, the lack of land use and density qualifiers to guide pedestrian behaviour has stimulated significant debate (Batty, 2013; Hillier, 1999; Hillier & Penn, 2004; Ratti, 2004). Evolving from Space Syntax, Place Syntax adds an accessibility measure to specific urban destinations, thus enabling the measurement of retailers' access to customers and vice versa. Preliminary correlations between accessibility and ground floor retail uses have been promising (Marcus et al., 2018; Ståhle et al., 2005). In a similar vein, urbanist Andres Sevtsuk expands spatial analysis from the Spacy Syntax presumption of a cyclical relationship between form and function. He combines accessibility and gravity in a series of measurements that have been found to correlate spatial layout with retail activity (Sevtsuk, 2010, 2014, 2020; Sevtsuk & Mekonnen, 2012).

Theories and Models Illustrated

In summary, retail success relies on a combination of supply- and demand-side factors. As some of the classical models describe, retailers thrive in clusters, as long as they are surrounded by compatible peers. These clusters can be defined by mapping the location of retailers and recognizing whether they disperse or combine in urban settings. On the other hand, retailers also stratify according to their need for and ability to afford central locations. Furthermore, retailers depend on accessibility to consumers, which can be quantified through Space Syntax models of street choice or integration values – or more recent evolutions of these models. While social, political, and organizational factors should not be ruled out, retail location in cities therefore follows a multilayered set of spatial patterns that align with consumer demand. The spatial consequence of an alignment between retail supply and demand seems still best described in Davies' Complex Model of City Centre Retailing from the 1970s. However, does this nearly 50-year-old model still hold true today? And does it hold true outside of the United Kingdom, where it was devised? Thus far, the Complex Model and its underlying assumptions have rarely been corroborated with retail location patterns in cities with varying social, economic, and cultural contexts.

To test the validity of modelled assumptions of retail and consumer behaviour, this section presents the retail structures of the urban core of Birmingham (England), The Hague (Netherlands), and Vancouver (Canada) – three cities on two continents that represent different urban conditions. Retail surveys for all three cities were conducted as part of several research projects between 2011 and 2019. All follow the same methodology: a combination of walking observation, government records, and business directories are combined to map the location of retailers that face public streets as well as publicly accessible shopping malls at street level. The resulting pattern of street-level retailers for the three cities are presented in Figure 14.4.

This distribution pattern shows a remarkable similarity to Davies Complex Model shown in Figure 14.2, with higher-tier comparison retail concentrating mostly in a central retail core (A), and destination retailers aligning along main arterial streets (B), and in specialized clusters (C) that may or may not follow centrality or accessibility assumptions. In all three cities, a cluster of department stores, planned urban shopping malls, and mostly chain retailers occupies the most central streets of the city, connecting with a large hinterland of high-capacity public transportation connections.

Arterial retail strips radiate out from this retail core. The makeup of these strips is far more diverse than the central retail core, containing specialty retailers, non-chain stores, and service providers that cannot afford the high retail rents on the main corner of town. They effectively form a hybrid between the highly accessible retail core, and the neighbourhood accessibility of dispersed corner stores, as they remain connected and walkable to both. Vancouver has pronounced retail strips that serve its densely populated northwestern residential district, and The Hague has arterials that connect the urban core to nearby neighbourhoods. However, Birmingham's historical arterial retail streets have been decimated by decades of urban renewal and freeway construction projects, which have effectively cut off its urban core from surrounding neighbourhoods. In all three cities, these arterial streets are the key victims of urban retail decline, due to their less central location and the decline of daily goods stores that serve a dwindling number of nearby residents. Figure 14.5 shows this decline in The Hague between 1911 and 2011, as the number of residents living in the city centre declined by over two-thirds (Kickert, 2014).

Besides the urban retail core and arterial streets, all three cities also have specialty clusters. For example, all three cities have a Chinatown, which is often located on the outskirts of the central retail core, but serves a citywide, regional, or even national audience. In Vancouver,

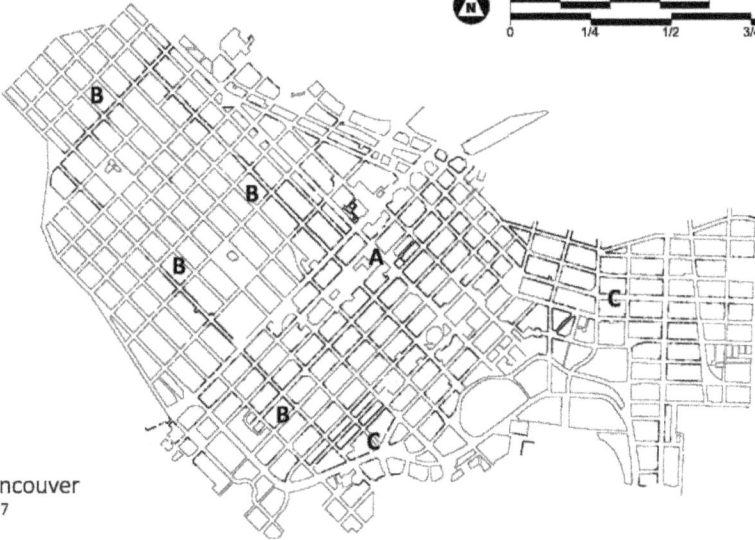

Figure 14.4. Retail structure of Birmingham, The Hague, and Vancouver with taxonomy of retail agglomerations.

The Hague
1911

The Hague
2011

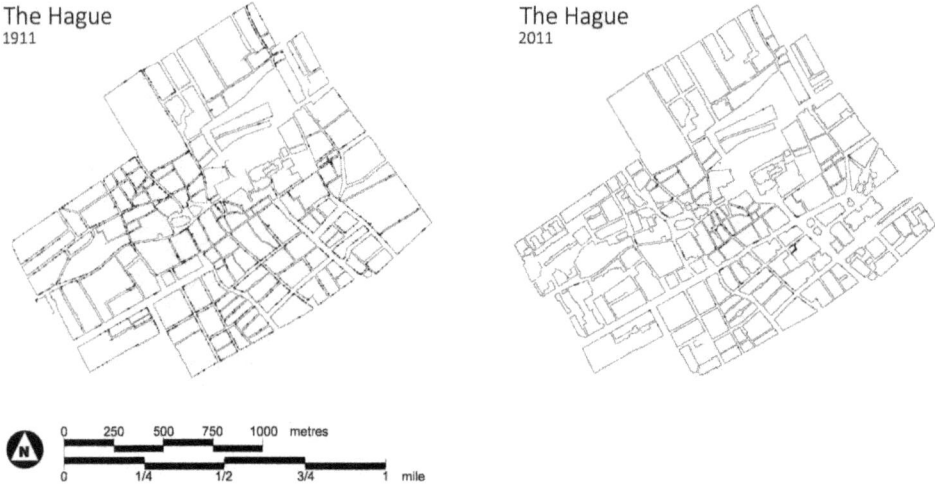

0   250   500   750   1000 metres

0   1/4   1/2   3/4   1 mile

Figure 14.5. A comparison of daily goods stores in The Hague, 1911 and 2011. Number of stores has declined by 88 per cent, especially in peripheral locations.

Chinatown is located in the former retail heart of the city; in The Hague it is located along a southern retail arterial; and in Birmingham, it is located on a former retail arterial now severed by a mid-century urban renewal project. Other specialty clusters focus on experiential retail, like Yaletown in the southern part of downtown Vancouver, a historical warehousing district turned entertainment district, and the Denneweg art gallery, boutique, bar, and restaurant district in the northern part of The Hague's urban core. Often these districts have a consistent internal structure of vendors, mixing dwellings with consumption and creative production (Hutton, 2006). Despite their clear identity, these specialty clusters often blend in with mainstream retailers found in other types of retail areas, like shopping malls.

As described in Davies' Complex Model and as demonstrated in Space Syntax models of retail distribution, the retailers in these three cities choose their location on the basis of their need for centrality and connectivity. In other words, retailers choose the most central location and locate along the busiest street that they can afford. The distribution of retailers in The Hague illustrates these two criteria for retail location. The amount of street-level frontage taken up by retailers[1] decreases with a greater distance from the main, highest-rent corner of the city, and increases with the potential of a street to be used as an urban through route (Figures 14.6 and 14.7).[2] This pattern is not uniform for all stores.

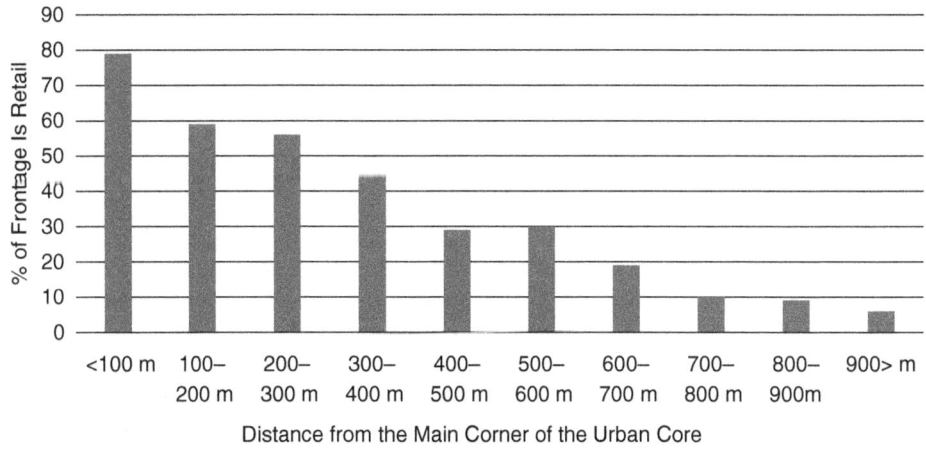

Figure 14.6. Percentage of street-level frontage taken up by retailers in The Hague, 2011, as a function of the distance from the main corner of the urban core – R^2 is 0.95.

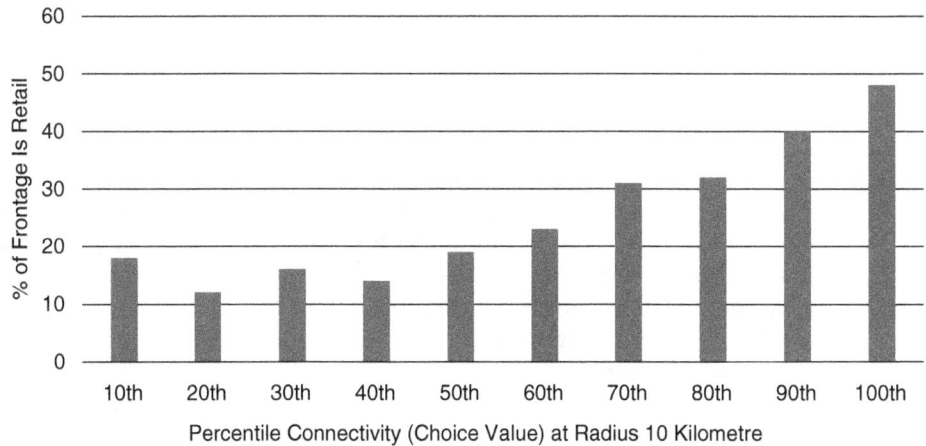

Figure 14.7. Percentage of street-level frontage taken up by retailers in The Hague, 2011, as a function of street connectivity, measured as the Choice value at radius 10 kilometres – R^2 is 0.84.

Different types of retailers have a different ability to pay for these central and well-connected locations. Stores selling comparison goods like fashion, shoes and jewellery have the highest need to induce passers-by to buy their items – and the highest margins to pay for the rent of a location with most passersby. Stores selling niche goods like hardware, books, or even daily groceries depend far less on footfall for serendipitous purchases, and hence locate in quieter downtown locations (Evers et al., 2005; Evers et al., 2011; Kickert, 2014).

Conclusion

This chapter has surveyed and demonstrated the internal spatial logic of urban retail from both the supply- and demand-side perspectives. As the spatial patterns of retail distribution in three cities have illustrated, many of the tenets of the surveyed retail models hold true. This corroboration is important for the future of urban retail for several reasons. North American retailers and developers face increasing risks in investing in future retail outlets, jeopardizing a return to unknown urban territories in lieu of the safety of suburban trade area and site selection data. On the other hand, these suburban markets have become increasingly saturated, and the rise of urban lifestyles and the experience economy signals a resurgence of interest in high-quality urban retail locations. The urban retail research surveyed in this chapter substantiates the viability of new urban retail locations beyond aggregated consumer profiles, taking spatial configuration and interaction into account. Conversely, European cities are seeing an ongoing decline in retail outlets, as the rise of e-commerce, the COVID-19 crisis, and the remnants of the Great Recession challenge the retail hegemony of town and city centres. In such cases, the spatial pattern of retail can inform decision making on retail area consolidation, helping planners, retailers, and developers assess the risk of retail success or failure. On both sides of the Atlantic, the spatial location of retail supply and demand warrants a bigger role in substantiating retail research and decisionmaking.

The retail theories and models included in this chapter highlight only the most salient underpinnings of urban retail research from a spatial perspective, an approach that remains very much alive in Europe and has strong potential for application in North America. While much normative and empirical progress has been made at larger scales, important questions remain about the street-level spatial structure of urban retail. For example, the micro-scale effects of street design on retail viability remain relatively unknown, a curious imbalance with the advanced science of "atmospherics" in shopping centre design (Turley & Milliman, 2000). Similarly, the impact of retail branding, organizational change, and shifts in consumer preferences on urban retail viability and distribution patterns remains largely unknown (Kickert, 2019). As other chapters in this book demonstrate, urban retail has many social, cultural, regulatory, technological, and economic dimensions that affect its viability and spatial distribution – but the extent of this influence needs further investigation.

Nevertheless, the future of research on the spatial patterns of urban retail looks bright. As data availability and computing power increases, researchers are able to generate new retail distribution and consumer behaviour models. New hybrids of retail modelling at macro, meso, and micro scales have strong potential to improve the accuracy of predictions about retail viability and distribution. While a significant amount of contemporary retail research focusses on relatively a-spatial marketing and organizational aspects, many of the key thinkers on urban spatial structures are keenly aware of the profound relevance of retailers to urban vitality. As Bill Hillier and Allan Penn commented, "'where should we put the shops?' is one of the commonest questions asked by designers and planners" (Hillier & Penn, 2004, p. 506). As the retail market continues its turbulent ride, we expect a continued need for scientific navigation.

NOTES

1 This measure is taken for its scalability, as the raw number of retailers depends on the total number of buildings within a centrality or connectivity category, hence skewing any distribution.
2 This potential is defined in Space Syntax as the "choice" value, which measures the propensity of a street to be used as a route to other streets in the urban network. In this case, this choice value is measured within a radius of 10 kilometres.

REFERENCES

Alonso, W. (1960). A theory of the urban land market. *Papers in Regional Science,* 6(1), 149–157. https://doi.org/10.1111/j.1435-5597.1960.tb01710.x

Alonso, W. (1964). Location and land use: Toward a general theory of land rent. *Location and land use. Toward a general theory of land rent.* Harvard University Press.

Anikeeff, M. (1996). Shopping center tenant selection and mix: A review. *Research Issues in Real Estate, 3,* 215–238. https://doi.org/10.1007/978-94-009-1802-3_11

Applebaum, W. (1966). Methods for determining store trade areas, Market penetration, and potential sales. *Journal of Marketing Research, 3*(2), 127–141. https://doi.org/10.1177/002224376600300202

Batty, M. (2001). Agent-based pedestrian modeling. *Environment and Planning B: Planning and Design, 28*(3), 321–326. https://doi.org/10.1068/b2803ed

Batty, M. (2013). *The new science of cities.* MIT Press.

Baumeister, R. (1876). *Stadt-Erweiterungen in technischer, baupolizeilicher und wirthschaftlicher Beziehung.* Ernst & Korn.

Berry, B. J. L. (1963). *Commercial structure and commercial blight: Retail patterns and processes in the City of Chicago.* Department of Geography, University of Chicago.

Berry, B. J. L., & Garrison, W. L. (1958a). The functional bases of the central place hierarchy. *Economic Geography,* 145–154. https://doi.org/10.2307/142299

Berry, B. J. L., & Garrison, W. L. (1958b). Recent developments of central place theory. *Papers in Regional Science, 4*(1), 107–120. https://doi.org/10.1111/j.1435-5597.1958.tb01625.x

Birkin, M., Clarke, G., & Clarke, M. P. (2002). *Retail geography and intelligent network planning.* John Wiley & Sons.

Bolt, E. J. (2003). *Winkelvoorzieningen op waarde geschat.* Merkelbeek: Bakker.

Borgers, A., & Timmermans, H. (1986). A model of pedestrian route choice and demand for retail facilities within inner-city shopping areas. *Geographical Analysis, 18*(2), 115–128. https://doi.org/10.1111/j.1538-4632.1986.tb00086.x

Bromley, R. D. F., & Thomas, C. J. (1993). *Retail change: Contemporary issues.* Routledge.

Brown, S. (1993). Retail location theory: Evolution and evaluation. *International Review of Retail, Distribution and Consumer Research, 3*(2), 185–229. https://doi.org/10.1080/09593969300000014

Brown, S. (1994). Retail location at the micro-scale: Inventory and prospect. *Service Industries Journal, 14*(4), 542–576. https://doi.org/10.1080/02642069400000056

Burgess, E. W., McKenzie, R. D., & Wirth, L. (1925). *The city.* University of Chicago Press.

Caplin, A., & Leahy, J. (1998). Miracle on Sixth Avenue: Information externalities and search. *The Economic Journal, 108*(446), 60–74. https://doi.org/10.1111/1468-0297.00273

Christaller, W. (1966). *Central places in southern Germany.* (C. W. Baskin, Trans.). Prentice Hall. (Original work published 1933)

Clarke, G. (1998). Changing methods of location planning for retail companies. *GeoJournal, 45*(4), 289. https://doi.org/10.1023/a:1006995106736

Cohen, L. (1996). From town center to shopping center: The reconfiguration of community marketplaces in postwar America. *The American Historical Review, 101*(4), 1050–1081. https://doi.org/10.2307/2169634

Davies, R. L. (1972a). The retail pattern of the central area of Coventry. In *The retail structure of cities.* Occasional publication no. 1 (pp. 1–32). Institute of British Geographers.

Davies, R. L. (1972b). Structural models of retail distribution: Analogies with settlement and urban land-use theories. *Transactions of the Institute of British Geographers*, 59–82. https://doi.org/10.2307/621554

Davies, R. L., & Bennison, D. J. (1978). Retailing in the city centre: The characters of shopping streets. *Tijdschrift voor economische en sociale geografie*, 69(5), 270–285. https://doi.org/10.1111/j.1467-9663.1978.tb00864.x

Dunne, P. M., Lusch, R. F., & Carver, J. R. (2013). *Retailing*. Cengage Learning.

Evers, D., Kooijman, D., & Van der Krabben, E. (2011). *Planning van winkellocaties en winkelgebieden in Nederland*. Sdu.

Evers, D., Van Hoorn, A., Van Oort, F., & Noorman, N. (2005). *Winkelen in megaland*. NAi Uitgevers.

Gehl, J. (1987). *Life between buildings: Using public space*. Van Nostrand Reinhold.

Gehl, J. (2010). *Cities for people*. Island Press.

Hernández, T., & Bennison, D. (2000). The art and science of retail location decisions. *International Journal of Retail & Distribution Management*, 28(8), 357–367. https://doi.org/10.1108/09590550010337391

Hillier, B. (1996). *Space is the machine: A configurational theory of architecture*. Cambridge University Press.

Hillier, B. (1999). The hidden geometry of deformed grids: Or, why space syntax works, when it looks as though it shouldn't. *Environment and Planning B: Planning and Design*, 26(2), 169–191. https://doi.org/10.1068/b4125

Hillier, B., & Hanson, J. (1984). *The social logic of space*. Cambridge University Press.

Hillier, B., & Leaman, A. (1973). The man-environment paradigm and its paradoxes. *Architectural Design*, 78(8), 507–511.

Hillier, B., & Penn, A. (2004). Rejoinder to Carlo Ratti. *Environment and Planning B: Planning and Design*, 31(4), 501–511. https://doi.org/10.1068/b3019a

Hotelling, H. (1929). Stability in competition. *The Economic Journal*, 39(153), 41–57. https://doi.org/10.2307/2224214

Huff, D. L. (1963). A probabilistic analysis of shopping center trade areas. *Land Economics*, 39(1), 81–90. https://doi.org/10.2307/3144521

Hurd, R. M. (1903). *Principles of land values*. Record and Guide.

Hutton, T. A. (2006). Spatiality, built form, and creative industry development in the inner city. *Environment and Planning A*, 38(10), 1819. https://doi.org/10.1068/a37285

Kickert, C. (2014). *Active centers: Interactive edges*. [Unpublished doctoral dissertation]. University of Michigan.

Kickert, C. (2019). Retail. In E. Talen (Ed.), *A research agenda for New Urbanism* (pp. 35–48). Edward Elgar.

Kickert, C., & Hofe, R. v. (2017). Critical mass matters: The long-term benefits of retail agglomeration for establishment survival in downtown Detroit and The Hague. *Urban Studies*, 55(5), 1033–1055. https://doi.org/10.1177/0042098017694131

Kooijman, D. (2000). Het recreatieve einde van Christaller. *Rooilijn, 3*, 123–130.

Lösch, A. (1940). The economics of location. Yale University Press.

Marcus, L., Berghauser Pont, M., Stavroulaki, G., & Bobkova, J. (2018). Location-based density and differentiation-adding attraction variables to space syntax. *24th ISUF International Conference. Book of Papers* (pp. 1379–1389). Editorial Universitat Politècnica de València.

Nelson, R. L. (1958). *The selection of retail locations*. FW Dodge Corporation.

Penn, A., & Turner, A. (2002). Space syntax based agent simulation. In M. Schreckenberg & S. D. Sharma (Eds.), *Pedestrian and evacuation dynamics* (pp. 99–114). Springer-Verlag.

Proudfoot, M. J. (1937). City retail structure. *Economic Geography, 13*(4), 425–428. https://doi.org/10.2307/141589

Psarra, S., Kickert, C., & Pluviano, A. (2013). Paradigm lost: Industrial and post-industrial Detroit – An analysis of the street network and its social and economic dimensions from 1796 to the present. *Urban Design International, 18*(4), 257–281. https://doi.org/10.1057/udi.2013.4

Ratti, C. (2004). Space syntax: Some inconsistencies. *Environment and Planning B: Planning and Design, 31*(4), 487–499. https://doi.org/10.1068/b3019

Ratti, C., & Claudel, M. (2016). *The city of tomorrow: Sensors, networks, hackers, and the future of urban life*. Yale University Press.

Sevtsuk, A. (2010). *Path and place: A study of urban geometry and retail activity in Cambridge and Somerville, MA*. MIT Press.

Sevtsuk, A. (2014). Location and agglomeration: The distribution of retail and food businesses in dense urban environments. *Journal of Planning Education and Research, 34*(4), 374–393. https://doi.org/10.1177/0739456x14550401

Sevtsuk, A. (2020). *Street commerce: Creating vibrant urban sidewalks*. University of Pennsylvania Press.

Sevtsuk, A., & Mekonnen, M. (2012). *Urban network analysis: A new toolbox for measuring city form in ArcGIS*. Paper presented at the Proceedings of the 2012 Symposium on Simulation for Architecture and Urban Design, Orlando, FL. https://dl.acm.org/doi/abs/10.5555/2339453.2339471

Shepherd, I. D. H., & Thomas, C. J. (1980). Urban consumer behaviour. In J. A. Dawson (Ed.), *Retail geography* (pp. 7–18). Halstead Press.

Smith, A., & Anderson, M. (2016). *Online shopping and e-commerce*. Pew Research Center. https://www.pewresearch.org/internet/wp-content/uploads/sites/9/2016/12/PI_2016.12.19_Online-Shopping_FINAL.pdf

Spek, S. C. V. d., Langelaar, C. M. V., & Kickert, C. C. (2013). Evidence-based design: Satellite positioning studies of city centre user groups. *Proceedings of the ICE-Urban Design and Planning, 166*(4), 2013. https://doi.org/10.1680/udap.11.00028

Ståhle, A., Marcus, L., & Karlström, A. (2005). *Place syntax: Geographic accessibility with axial lines in GIS*. Paper presented at the Fifth international space syntax symposium. Delft, Netherlands.

Thornton, S., Bradshaw, R., & McCullagh, M. (1991). Pedestrian flows and retail turnover. *British Food Journal, 93*(9), 23–28. https://doi.org/10.1108/00070709110007440

Timmermans, H. (2009). *Pedestrian behavior: Models, data collection and applications.* Emerald Group Publishing Limited.

Turley, L. W., & Milliman, R. E. (2000). Atmospheric effects on shopping behavior: A review of the experimental evidence. *Journal of Business Research, 49*(2), 193–211. https://doi.org/10.1016/s0148-2963(99)00010-7

Van der Spek, S. C., Van Schaick, J., De Bois, P. G., & De Haan, R. (2009). Sensing human activity: GPS tracking. *Sensors, 9*(4), 3033–3055. https://doi.org/10.3390/s90403033

Van Nes, A. (2002). *Road building and urban change: The effect of ring roads on the dispersal of shop and retail in western European towns and cities.* Agricultural University of Norway, Department of Land Use and Landscape Planning.

Whyte, W. H. (1988). *City: Rediscovering the center.* Doubleday.

15 The Future of American Urban Retail Real Estate

HEATHER ARNOLD

The retail industry is changing so rapidly that the future is now. This speedy evolution is confounding – to architects designing for the next century, developers building mixed-use projects, city planners working to designate retail zones in their 10-year comprehensive plans, and commercial lenders anticipating whether retail's demise has been greatly exaggerated. This chapter will provide an industry perspective on the trends in retail, complementing the academic chapters found elsewhere in this volume. Specifically, this chapter will focus on the real estate implications of today's retail transformation.

The retail industry's metamorphosis will result in fundamental changes to four basic elements of retail real estate development and growth: rent rates, lease terms, store sizes, and site selection. Unfortunately for property owners, developers, architects, and tenants, societal shutdowns related to the 2020 outbreak of the COVID-19 pandemic expedited these changes rapidly. Where the industry projected a 10-year transformation, the need to manipulate spaces, leases, and methods of sales became immediate. As we will continue to see, these real estate shifts will not be limited to urban areas or national retailers, but are also encompassing suburban strip centres, small-town main streets, and local mom-and-pop stores.

The Future of Rent Rates

Since merchants sold their goods in stalls in the agora of ancient Greece, retail rents have been determined by a mix of site-defining characteristics such as visibility, building condition, total amount of occupiable space, and, since the past century, parking availability. Taken together, these criteria can be summarized as the ability of this site to attract the largest number of customers and drive high sales volume. Spaces that earn top-of-the-market rents have the potential to drive the most sales. A longstanding industry rule of thumb prescribed that rent should not exceed 8–10 per cent of total sales per square foot. Hence, a furniture store with annual sales of $500 per square foot should pay up to $40 to $50 per square foot in annual rent. That leaves enough money for payroll, taxes, inventory, insurance, and, if needed, security.

The relationship between online sales and physical store locations has complicated this formula. Only a few years ago, a customer might enter a bricks-and-mortar operation to investigate the "look and feel" of an item while discreetly (or indiscreetly) using a smartphone to look at reviews, price-compare, and ultimately purchase the item online – through the same business or elsewhere. As described in the chapter by Daniels and Hernandez in this book, this "showrooming" behaviour is part of an increasingly complex consumer journey from research to purchase and advocacy. For national retailers, this situation has created an accounting problem in tracking sales. Which line of business should receive credit for a sale made through this business? The store, which took the customer's initial interest and converted it into a purchase; or the online portal, which recorded the sale and facilitated its distribution? While this point might seem trivial – revenue is revenue, after all – keep in mind that the physical store is counting on sales to justify its rent, its salespeople, and its continued existence.

Using physical stores and online portals in a coordinated manner to entice, secure, capture, and repeat sales is known as "omnichannel" strategy. Here retail organizations use both offline and online channels to serve consumers in their journey toward a purchase (Verhoef et al., 2015). Loyalty among customers is fostered by a variety of interactions with a company. That explains why, even in this electronic age, retailers are still spending millions of dollars on advertising by mail. They want to appeal to you on your kitchen counter, in your email inbox, in your social media ad space, and on your street corner.

While the demise of the brick-and-mortar retail store has been predicted for almost a decade now, the advance of omnichannel customer outreach has highlighted the vital importance of the physical space and the customer's opportunity to interact with merchandise prior to purchase. However, the store's essential purpose has fundamentally changed, leaving retail real estate without a key metric to determine its value to property owners or tenants. Instead of pricing commercial leases on the basis of the size of a store or a percentage of store sales, experts predict leases to be based on the cost to acquire a customer (CAC), or even a cost per customer seeing or entering stores. This leasing model would resemble the online pay-per-click model, including the opportunity to collect consumer data for later targeting (Harris, 2018).

Essentially, in the omnichannel era, stores are becoming three-dimensional experiential advertisements for brands and retailers, rather than transactional spaces – a phenomenon that is discussed in detail in Daniels and Hernandez's chapter in this book as well. At the same time, stores serve as warehouse space for the increasing number of home deliveries of merchandise. As long-term leases expire and new paradigms emerge for the next round of negotiations, future retail leases could go in several different directions:

- **Value of retail space could be based on formulas related to passing pedestrian and/or vehicular activity, as well as distance from mass transit portals**. This leasing approach is roughly based on a pricing model similar to YouTube advertising, where channels that have videos with more views command higher advertising fees. It elevates two elements of retail real estate valuation above all others: visibility and customer segmentation. Visibility is a simple measurement of eyes on the storefront. Customer segmentation involves identifying a target cohort – morning rush hour cluster of professionals commuting to downtown, for instance – and pricing the retail location on the basis of the value of the exposure to these customers. They may never set foot inside the shop in question but will have its brand (upscale briefcases, say) regularly reinforced in their subconscious as an aspect of their daily travel.
- **Retail space could be valued by the number of people who enter the store, whether they make a purchase onsite or not**. Through beacon technology – the use of wireless transmitters and Bluetooth technology to send ads and coupons to customers right as they shop – rents could be modelled in a manner similar to "pay for clicks" online pricing: Tenants would pay a premium for people who enter the establishment, similar to paying a fee to a search engine for directing a customer to a site. This model prioritizes the importance of physical stores as points of access, not only to products, but to the brand itself.
- **Tenants no longer pay on the basis of the amount of space they occupy, but on the surface area of their storefronts.** As stores morph into places to manage inventory for last-mile deliveries, adopting more of a warehouse function, the value of square footage changes as well. What matters most is the storefront itself – the "billboard" – and rents can be calculated as price per square foot of vertical shop window area rather than selling space.

The most likely strategy for determining rent into the future is a combination of all of the above, with sales-per-square-foot also factored in on some level. As physical retail stores struggle with their changing identity from traditional transactional spaces toward experiential advertisements or warehousing and industrial space, their pricing model will undergo a conversion accordingly.

The Future of Lease Terms

In 1980, the average retail lease length was 10 years. In 2022, it's 3 to 4 years, and deals for periods of less than three years have sharply increased. One driver of this dramatic decrease was the sheer number of retail closures and consolidations that occurred following the 2008 recession, as leases were terminated early. Another reason: the rise of the pop-up store (Morris, 2018).

Following early experiments by independent retailers and artists, the prototypical modern pop-up store became an elaborate construction put up by a national chain, typically during the holiday season. Increasingly, online-native retailers and start-ups have come to use temporary pop-up stores to test the market prior to making a large capital investment in a permanent, physical location, or just to get a burst of customer engagement. Closing sales is almost beside the point. "Similar to any marketing channel, this is just an attempt to make our great experience come to life in a physical format," Bob Sherwin, Wayfair's VP of Marketing, told *Inc.* magazine, describing two pop-up stores the online-native home goods retailer opened during the 2017 holiday season – neither of which actually carried inventory (Albanese, 2018).

From the real estate perspective, not surprisingly, the continued rise of pop-ups hinges on a willingness by landlords to offer short-term leases. Furthermore, shorter leases are in demand beyond pop-up models, as retailers with longer time horizons are also seeking shorter leases than in the past. Customer expectations for delivery times, variety of products, and price tolerance are changing so quickly that businesses are less willing to commit to an investment that might work today but not in a few years. As *Bloomberg Businessweek* notes, landlords are open to this shift:

> Landlords aren't demanding the long-term commitments once considered standard and are instead offering leases as short as a year with extension options. That flexibility extends to temporary locations, too, with spaces dedicated to rotating pop-ups so the tenants can easily test concepts. To further reduce risk, they're offering to help pay for store remodeling and taking a small percentage of sales instead of monthly rent. (Townsend, 2018)

The demand for shorter leases has created a market opportunity for a new kind of broker – online marketplaces that match landlords and temporary tenants. One such company is Appear Here. "Find Space for Your Idea" – a slogan that invites visitors to its website, which lets them search for space by city and duration, ranging from less than 1 month to a year or longer. Developers and property owners are increasingly including these short-term leases into their development budgets and business model from the concept-planning stages. Rather than providing a stop-gap answering to a faltering retail leasing strategy, pop-ups could become the strategy itself – attracting a creative class clientele that is perpetually on the lookout for the next big thing.

The Future of Store Sizes

Many stores are getting smaller, a trend driven partly (though not exclusively) by their new, more focused role in the larger retail ecosystem. Online retailers that choose to establish bricks-and-mortar stores are especially drawn to smaller spaces when their aim is to create experiential "showrooms" to complement their existing order-and-delivery business models. Online-first American menswear retailer Bonobos, for instance, utilizes its inventory-free "guideshops" as places where customers can inspect and try on samples, then order what they like. This format requires little storage space and can nurture repeat customers. Bonobos says new customers who visit its stores typically spend 20 per cent more than they would if starting their experience online. Bonobos CEO Micky Onvural, speaking to *Bloomberg Businessweek*, likened each store to a "ginormous billboard" (Townsend, 2018).

On the other hand, brick-and-mortar retailers understand the value of online fulfilment as they enter the omnichannel retail environment from their angle. Smaller fashion retailers like Gap and Lululemon, but also larger department stores like Target and Walmart and grocery stores like Kroger are aggressively implementing in-store pickup options to leverage the benefits of offline and online channels (Hirsch et al., 2019). In this scenario, retail spaces increasingly serve as fulfilment warehouses, either for Buy-Online-Pickup-In-Store (BOPIS) or for last-mile delivery solutions to consumers at home or at work. This allows retailers to build reduced-footprint stores in areas where larger stores are not possible (such as mixed-use buildings in urban locations), allowing easier connections to their customers.

Increasing demand for smaller stores creates an incongruence with the current standard stock of stores. Since the earliest shops in mixed-use buildings, retail spaces adapted to the 20- to 25-foot width of buildings –

roughly the span that a wooden beam could support (Davis, 2012). Furthermore, stores adapted to a roughly 60-foot depth, aligning with their own operational needs and the dimensions of any floors above them. These consistent store dimensions persisted even when retail spaces began to separate into their own developments, such as Kansas City's Country Club Plaza in the 1920s, although larger stores could take up several retail bays (Coleman, 2006). These dimensions – retail's "golden rectangle" – allowed stores to place their cash wrap (register and sales counter) in a variety of locations that would allow them to monitor comings and goings of browsing customers while also servicing purchasing customers. The depth of these spaces allowed for a primary sales floor up front, an area for discounted merchandise further back, with any offices, storage, and dressing rooms taking up the very back of stores. Except for the rise of larger-format department stores, grocers, and discounters, centuries of urban retail browsing, selection, purchase, and delivery have taken place in these roughly 1,200- to 1,500-square-foot modules (Longstreth, 1987).

Until recently, architects have continued to design mixed-use buildings with ground-floor retail spaces in these dimensions. Not only did these modules make sense to retailers, they were appropriate for the widest variety of tenants. As a consequence, developers had assurance of retail longevity: Tenants might come and go, but the space would remain viable as a vanilla box for the next round of leases. Even before e-commerce showrooming came along, this traditional thinking about retail space had already come under pressure. European chain retailers have struggled for decades to fit their larger operations into these relatively small modules, often forcing them to merge adjacent properties. In the more recent American shift toward convenience and experiential retail, neighbourhood goods and services businesses, along with quick-service restaurants, have come to prefer a shallower 45-foot depth to push the "buzz" of internal shopping and dining activity to the front of the store, visible to passers-by.

Few main street retail spaces have the flexibility to downsize, specifically if they are part of historic properties. One emerging response to the square footage mismatch between retailer need and building size is hybrid stores. These combine two or more retail operations – sometimes with little or no apparent similarity – into one traditional retail space. This isn't entirely new: Barnes & Noble incorporated Starbucks outposts into its stores starting in 1993, which in itself has echoes of soda fountains in emporium stores like Woolworth's. But these days, all sorts of mergers are playing out, among big-name chains and local businesses. Often these hybrids combine the traditional transactional role of stores selling goods or services with more experiential functions like on-site food and drink consumption. The HandleBar Café in Baltimore, Maryland, for example, is half bicycle shop and half restaurant. In Asrai Garden Shop in Chicago, Illinois, a florist and a jewellery store have found common ground as they seek sales from high-income gift shoppers. Raconteur at Third Place Books in Seattle, Washington, allows restaurant patrons to browse through aisles of literature with a cocktail or to read while receiving table service. Sharing expenses is just one reason for sharing space. Looking to appeal to increasingly time-conscious and experience-seeking consumers, AT&T and Capital One Bank are now offering locations with coffee cafes and community meeting spaces as a method to create more than one reason to enter their establishments. AT&T plans to expand its experiential concept, The Lounge, to 1,000 locations across the United States (AT&T Communications, 2018; O'Leary, 2017).

American chain and department stores are similarly looking to shrink their floor-space area, as retail spending is moving online and store visits become more experiential than transactional. Leading chains like Macy's, Kohl's, and JCPenney are transforming their stores, reducing sales floor area toward online pickup and fulfilment space on the one hand, and to experiential business on the other. For example, Kohl's is now allowing Amazon returns in some of their stores and leases space to Planet Fitness in 10 stores in a nationwide "rightsizing" operation (Kohl's Inc., 2019). Furthermore, an increasing number of traditionally suburban retailers are moving into urban markets, prompted by the re-urbanization of the American middle class and the saturation of suburban markets. In its 2018 annual report, Target demonstrated its shift to smaller and more urban store sizes through its new openings. In the year following February 2017, all of Target's more than 20 newly constructed stores were under 50,000 square feet, and the only vacated store was nearly 200,000 square feet. These "mini-Targets" located in cities like Miami, Seattle, Chicago, and Washington, DC, but also in the downtowns of smaller cities including Lexington, Kentucky, and West Lafayette, Indiana – often comprising captive audiences of college students (Target Corporation, 2018). Similarly, suburban stalwart IKEA is moving into smaller urban "planning studios," where their young urban clientele can view products and order them for home delivery (Gualtieri, 2018). The American chain retailer quest for smaller stores yields interesting comparisons with the United Kingdom and Canada, described in other chapters in this volume.

In the future, the ideal store size could easily be smaller yet – especially as automated technology is increasingly able to cater to time-starved consumers. Ordering screens are creating product displays in storefronts or kiosks that focus the customer on a small number of items and provide the opportunity to make a purchase via touch screen. Other companies give a new spin to tested formats to reduce store sizes. For example, in 12 locations across the United States, Sprinkles offers fresh cupcakes via traditional ATM banking systems to delighted customers (Metcalfe, 2012).

The Future of Site Selection

For decades, the most selective retail tenants sought locations in areas with high population densities, high household incomes, high daytime populations of employees, high visibility, and adequate parking. These measurements, and more, were easily and reliably provided by the United States Census Bureau and the Bureau of Labor Statistics. The profound contemporary transformation of retail and the age of Big Data have upended this status quo. These days, a fashion company such as ModCloth can drill down into newly available data gathered online and offline to identify, say, neighbourhoods with concentrations of female shoppers who have a history of purchasing brightly coloured and patterned clothes for a professional wardrobe. That's where they open a storefront. In the case of ModCloth – a clothing company that champions "values of female empowerment and inclusivity" in part by "offering a full range of sizes" – this strategy has meant opening stores in smaller neighbourhood business districts rather than the best-known commercial districts. The company opened its first "Fitshops" in the Shaw neighbourhood of the District of Columbia; the Pacific Heights neighbourhood of San Francisco; Austin's 2nd Street District; and, somewhat more conventionally, SoHo in Manhattan.

With such detailed information, online companies can meet their customers where they live and work. By opening physical stores where their clients are located, these brands seek to stay top-of-mind. This strategy turns the traditional rules of site selection on their head. When most businesses had similar criteria, stores tended to naturally cluster at points of density, wealth, and activity. In the new era of unprecedented data accuracy and nimble locations – both in length of retail leases and store size – an urban neighbourhood of concentrated, similar shoppers could attract a small showroom version of its favourite retailer.

The role of property owners and asset managers has also changed in site selection, as these stakeholders increasingly understand the value of retailers as amenities. In most places, residential and office building development will yield a better return on investment than retail-centric development. However, recent studies have demonstrated that the proximity to stores and cafes improves the ability of the upper floors to lease quickly and above the average market value. Additionally, the brand and culture of the retail businesses on the ground floor broadcast the character and priorities of the people who live and work above. Long-time Seattle developer Maria Barrientos illustrates the importance of retailers for the identity of the building above it: "Retail defines the building at the pedestrian level floor for most people. People are more likely to say 'the building where Skillet Diner is' than to use an apartment name or address" (Reibman, 2016).

A January 2019 report for the Downtown Seattle Association studied the impact of curated and merchandised retail tenant mixes on four office buildings. Performance indicators and property-owner interviews demonstrated that the street-level uses had resulted in upper-floor rent premiums, faster times to achieve stable occupancy, and consistently high occupancy. As stated within the document,

> [B]uilding owner/operators have attracted retail businesses to serve as an amenity that helps secure higher lease rates from the office tenants and residents above. This is a departure from traditional retail leases that exist to generate the highest revenue possible to the building owner. (ECONorthwest, 2019)

The relationship between a building's retail spaces and upper levels has become commonly understood in real estate markets far beyond downtown Seattle.

Smaller chain, regional, and even independent retailers can use developer interest to access premier spaces that they would not have been capable of competing for on the basis of rent rates alone. They can benefit from the value that unusual and locally loved stores add to the rent per square foot of the rest of the building – which is almost always a multitude of the ground floor space. For example, grocery stores can add 5–10 per cent to nearby apartment rents, as noted by David Webb, vice chairman at real estate services firm CBRE (Anderson, 2016). Developers will take a lower rent on the retail to bring in the building's "billboards" that drive the value of everything else, even embracing independent retailers with less credit on the books.

The Future of Independent Urban Retail

Companies like Bonobos and ModCloth (both bought by Walmart Inc. in 2017), Wayfair, Target, Kohl's, and Macy's have the benefit of scale, deep pockets, and some freedom to experiment in their efforts to adjust to the new retail environment. Small retailers have little such leeway.

Where national chains can merge their online and in-store sales revenue into a single omnichannel bottom line, retailers with only a few stores or even just one can fall prey to the shopper who browses in person and then turns to e-commerce to compare prices and, finally, make a purchase. Losing sales this way can be debilitating to an independent retailer who typically operates within small profit margins.

An additional obstacle for these retailers is the cost of catching up to technology. Digitizing a store inventory, creating and maintaining a website while also managing an active presence on social media can present significant time and financial burdens on small businesses. The chapter by Liz Mack in this book describes the struggles that otherwise agile independent business owners face in adopting e-commerce.

However, single-location stores still account for almost 95 per cent all retail businesses, according to the Economic Census. The stores that have survived so far, and will continue to thrive in the future, are those that continue to adhere to timeless principles such as personal connections, service with a smile, convenience, and specialization based on local needs and tastes – fulfilling the desires of a consistently large cohort of consumers in search of authenticity and localism (Andres Coca-Stefaniak et al., 2010; McGee & Peterson, 2000; Megicks & Warnaby, 2008). As the *Harvard Business Review* recognized in 2013, "The infiltration of technology into every part of our lives has made many people seek out personal, low-tech/high-touch experiences and relationships with the companies they patronize" (Yohn, 2013). As Liz Mack recognizes, independent retailers should not rest on their laurels, as chain retailers are deploying increasingly advanced technology to emulate the personalization and connection that independent retailers are known for (Grewal et al., 2017).

Conclusion

The retail apocalypse that has been forecast and feared for more than a decade has claimed a host of victims. Demises that were predicted for the longer-term future have been expedited by COVID-19 shopping and dining restrictions. From a customer interaction standpoint, our customer behaviour and shopping instincts have been permanently changed. However, it has also forced a burst of innovative thinking about why people enjoy shopping in person and how stores can accommodate this enjoyment safely and profitably. Storeowners themselves can't make all the needed adjustments on their own: local governments, landlords, main street advocates, builders, architects, and the customers themselves also have important roles to play in transforming our retail real estate systems. They will need to understand and leverage the implications of rapid changes in retail rents, lease terms, store sizes, and store site selection.

Whatever the future of retail and shopping holds, the responsibility of physical retail stores to spark our imagination and to establish our heartfelt connections to the idea of "community" must be a shared one. Perhaps more than any other time in American history, the importance of retail spaces as a core, communal aspect of our neighbourhoods and our lives has been starkly revealed.

Even without the COVID-19 crisis, America has a clear oversupply of retail space, and the projected continued decline of demand for retail space will prompt difficult choices ahead for shopping centres and main streets alike. Striking a balance between modern society's need for safety and convenience and retail's contribution to the character of commercial districts everywhere will be a neighbourhood character, urban design, and place-making challenge far into the future.

REFERENCES

Albanese, J. (2018). This is the reason every brand is launching a pop-up store. *Inc.* https://www.inc.com/jason-albanese/this-is-reason-every-brand-is-launching-a-pop-up-store.html

Anderson, B. (2016, 15 March). Do first floor retail spaces pencil out for multifamily developers? *National Real Estate Investor.* https://www.wealthmanagement.com/multifamily/do-first-floor-retail-spaces-pencil-out-multifamily-developers

Andres Coca-Stefaniak, J., Parker, C., & Rees, P. (2010). Localisation as a marketing strategy for small retailers. *International Journal of Retail & Distribution Management, 38*(9), 677–697. https://doi.org/10.1108/09590551011062439

AT&T Communications. (2018). *The Lounge by AT&T in Capitol Hill combines retail, coffee and technology* [Press release]. https://about.att.com/story/2018/att_concept_store.html

Coleman, P. (2006). *Shopping environments: Evolution, planning and design.* Architectural Press.

Davis, H. (2012). *Living over the store: Architecture and local urban life.* Routledge.

ECONorthwest. (2019). *Activating ground floors in downtown Seattle's buildings.* Prepared for the Downtown Seattle Association. https://downtownseattle.org/files/advocacy/DSA-GroundFloor-Study_Final_2019.01.08.pdf

Grewal, D., Roggeveen, A. L., & Nordfält, J. (2017). The future of retailing. *Journal of Retailing, 93*(1), 1–6. https://doi.org/10.1016/j.jretai.2016.12.008

Gualtieri, T. (2018, 10 October). Ikea to focus on 30 megacities for future small-store openings. *Bloomberg*. https://www.bloomberg.com/news /articles/2018-10-10/ikea-to-focus-on-30-megacities-for-future-small-store -openings

Harris, M. (2018, September/October). A new formula for retail rents. *Retail Leader*. https://retailleader.com/new-formula-retail-rents

Hirsch, L., Thomas, L., & Lucas, A. (2019, 13 June). Retailers fight back: Bricks-and-mortar brands are figuring out online sales, and their stocks are soaring. *CNBC*. https://www.cnbc.com/2019/06/13/retailers-and-restaurants-flex -online-muscle-in-battle-against-amazon.html

Kohl's Inc. (2019). *Updates on Kohl's store optimization initiatives* [Press release].

Longstreth, R. W. (1987). *The buildings of main street: A guide to American commercial architecture.* Preservation Press.

McGee, J. E., & Peterson, M. (2000). Toward the development of measures of distinctive competencies among small independent retailers. *Journal of Small Business Management, 38*(2), 19–33. https://www.proquest.com/openview /1b91548f81191d2402e610941264c122/1?pq-origsite=gscholar&cbl=49244

Megicks, P., & Warnaby, G. (2008). Market orientation and performance in small independent retailers in the UK. *The International Review of Retail, Distribution and Consumer Research, 18*(1), 105–119. https://doi .org/10.1080/09593960701778192

Metcalfe, J. (2012, 1 March). Sprinkles cupcakes explains its 24-hour "cupcake ATM." *Bloomberg*. https://www.citylab.com/life/2012/03/sprinkles-cupcakes -explains-its-24-hour-cupcake-atm/1382/

Morris, K. (2018, 1 September). Retailers looking to test run stores see opportunity in short-term leases. *The Wall Street Journal.*

O'Leary, K. (2017). More than just a drink. *ABA Banking Journal, 109*(6), 34–35.

Reibman, J. (2016, 23 June). Is dead retail space bringing your building down? *Daily Journal of Commerce.*

Target Corporation. (2018). *2018 annual report.*

Townsend, M. (2018, 22 October). Warby Parker is coming to a suburban mall near you. *Bloomberg Businessweek*. https://www.bloomberg.com/news /articles/2018-10-22/warby-parker-and-casper-are-coming-to-a-strip-mall -near-you#xj4y7vzkg

Verhoef, P. C., Kannan, P. K., & Inman, J. J. (2015). From multi-channel retailing to omni-channel retailing: Introduction to the special issue on multi-channel retailing. *Journal of Retailing, 91*(2), 174–181. https://doi .org/10.1016/j.jretai.2015.02.005

Yohn, D. L. (2013). Megastores want to be like mom-and-pop shops ... sort of. *Harvard Business Review.*

Conclusion: Urban Retail Redefined

CONRAD KICKERT AND EMILY TALEN

This book has made a solid case for the need to support the continued survival of urban retail. As many of our authors have explored, often the case rests on arguments beyond business economics: street-level storefronts house more than businesses – they house the social life of neighbourhoods and cities. Whether they provide coffee, groceries, art, or entertainment, street-level commercial spaces build social capital by functioning as hubs of social activity – the "third places" that sociologist Ray Oldenburg vividly described decades ago (Oldenburg, 1989). An active retail corridor provides entrepreneurial opportunities too, and, where ethnic cuisines, specialty stores, and an elusive "hipness" are featured, a distinct cultural identity. Main streets provide more than just merchandise and services, they provide the precondition for walkable, sociable, and sustainable cities, adding to public life as much as the private bottom line.

Can we expect storefronts to continue this role? After all, merchandise seems to be on a relentless march from the store shelf to the warehouse, and the COVID-19 crisis is accelerating the virtualization of street-level commerce. While the vast majority of retail sales on either side of the Atlantic are still conducted in stores, history has taught us that we may be in the midst of another retail revolution. Just as the department store superseded the independent clothier and milliner, the supermarket killed the corner grocer, and the suburban big box gutted many main streets, online juggernauts are unlikely to take any prisoners. As the "retail apocalypse" chews its way through the retail landscape, what will our urban future look like at eye level?

This is a question that requires us to grapple with the fact that, as much as we value the social capital and cultural identity that storefronts provide, the future of urban retail hinges on the rapidly changing forces that are shaping retail and services markets on either side of the Atlantic.

These closing thoughts connect the on-the-ground findings in this volume with these changing market trends, summarizing the external challenges and opportunities that our storefronts are facing.

As many chapters in this book conclude, e-commerce has a complex effect on urban storefronts. Likely, many items still on our store shelves will shift online. Especially easily shippable items are likely to continue their transition from physical to digital stores, as online fulfilment offers unique advantages of information, selection, pricing, and convenience (Pauwels & Neslin, 2015). Products like music and books – unless housed in a select number of independent retailers – left store shelves decades ago, with items like electronics, home goods, and fashion rapidly following in their wake (Lipsman et al., 2019). While these items have been sold mostly in suburban big-box superstores in North America for decades, the virtualization of durable goods sales will strongly impact traditionally protected European High Streets. The current health crisis has drastically accelerated the virtualization of durable goods retail, and we are unlikely to see a full return to pre-pandemic physical retail presence in this category.

The next e-commerce frontier is the online grocery market, which takes up the largest share of consumer retail spending in Europe and North America. While many consumers may be wary of buying fresh food online, and while order picking and delivery systems still have some distance to go, viable online grocery delivery is considered the next frontier for many large retailers, and significant innovations will likely disrupt this market over the next few years (KPMG International, 2017; Retail News, 2018). The virtualization of grocery sales will thin out urban supermarket and corner store networks even further, exacerbating "food deserts." On the other hand, e-commerce food delivery has the potential to enhance equitable food access via innovative delivery programs – potentially subsidized by governments (Vogel, 2021; US Food and Nutrition Service, 2019). While in countries like the United Kingdom and China, online groceries have been part of the retail landscape for years, it took the COVID-19 pandemic to force a breakthrough for online groceries in the United States and other smaller markets. Whether it will prove permanent or not, the beleaguered corner grocer will face yet another foe.

However, we see several opportunities arise for urban retail stores. First, they offer last-minute convenience that few online delivery services can match. As we need that carton of eggs or allergy pills *right now*, there will always be a place for nearby stores over even the speediest delivery service. Furthermore, urban retailers counter the growth in e-commerce with a strong push toward experiential retailing. As we can buy things that we need online, storefronts increasingly fuel and fulfil our desire to buy things that we *want*. Urban stores are successfully transitioning from being places of mere utilitarian transaction to places offering positive social and personal experiences and fulfilment. As demonstrated in this volume's chapters on Canada, the United Kingdom, and the United States, stores on both sides of the Atlantic are transforming their former sales floors to spaces for brand attachment and physical interaction. The urban store of the future will fulfil a new role in the "customer journey" toward a purchase, as described by Chris Daniels and Tony Hernandez, focusing less on completing a sale and more on attaching a customer to a brand or product – whether in the store or not. Retailers see physical and online sales as complementary "channels" in multi- or omnichannel strategies that use the advantages of both to increase sales (Neslin et al., 2006; Pauwels & Neslin, 2015). As stores increasingly function as advertisements and places for interaction, established brands and online start-ups may seek to connect their message to the authenticity and visibility of urban environments (Lewis & Dart, 2014; Lipsman et al., 2018). Stores are increasingly adopting short-term leases and pop-up formats through platforms like Appear Here, hoping to remain at the top of mind of easily bored consumers (Mull, 2019).

The desire for experiences stretches far beyond product-based retail – the urban main street corridor is quicky transitioning into a place for food, beverages, and entertainment. As it turns out, we are part of the best experiences ourselves, as the need to meet and see one another has propelled storefronts as social spaces. If the COVID-19 crisis has proven anything, it is the innate human desire to be among other people. For years, this desire has fuelled the rise of food-service establishments, profoundly changing the basic makeup of the urban main street. As traditional retailers dwindle, food-service establishments (which comprise everything from full-service restaurants to bars, lunchrooms, coffee shops, and food trucks) in American cities have grown by almost a third since 2000, and this counter-cyclical trend is mirrored in Europe. For the first time in history, American spending on food outside the home now eclipses home-cooked meals (US Census Bureau, 2019). While the long-term trajectory of food-service establishments is sensitive to health scares, recessions, regulations, and labour dynamics, one can trust that this highly innovative sector is likely to adapt to any new conditions. The Great Recession of 2008 fuelled the rise and resilience of fast-casual formats like food trucks and food halls, and new indoor and outdoor dining formats have redefined urban food culture during the COVID-19 pandemic (Cho & Todd, 2018; Henley, 2020; Marshall, 2020; New York City Department of Transportation, 2022). Authors like

Jeff Parker, Terry Clark, and Hyesun Jeong demonstrate in this book that the growth of these establishments reflects more than merely culinary desires, as they fulfil urbanites' desires for belonging and cultural capital. This may explain why the most prominent consumer behind this growth is younger and upwardly mobile, propelling accessible yet distinct formats like gourmet food trucks, food halls, boutique coffee shops, and craft breweries – all at the risk of gentrifying neighbourhoods (Barajas et al., 2017; Walker & Fox Miller, 2019). Similar to that of other retailers, food-service establishments' quest for uniqueness may actually fuel a self-defeating cycle of homogenization and gentrification (New Economics Foundation, 2004; Kasinitz et al., 2015; Zukin et al., 2009). Furthermore, supermarkets are eager to take back market share, and online food-delivery services are looming to virtualize food services as profoundly as other retail categories (Charles, 2018; Group & Portalatin, 2018). At least in the short term, the 2020 pandemic has also proven severely disruptive for food services, as scared patrons and stringent social distancing guidelines decimated revenues for many bars, restaurants, lunchrooms, and coffee shops. While employment has plummeted in food services, some argue that a newly unemployed cohort of chefs is already preparing their next ventures. Beyond food services, beauty, financial, rental, and repair services also take up a significant number of urban storefronts. Their dynamics vary from a relative resilience to economic cycles and online attrition for beauty services, to strong online pressures for financial services and even countercyclical growth for repair services (Kickert, 2019). As Western nations are aging and becoming more conscious of their health, the growth of wellness services in urban storefronts seems all but guaranteed (Roth, 2018).

Where do these market forces leave the small-scaled, often independent urban retailer? On one hand, independent retailers are agile and innovative, outsmarting online juggernauts and offline chains with their ability to change formats and merchandise offerings at the decree of a store owner rather than the corporate boardroom. Unburdened by real estate holdings and corporate inertia, a new wave of independent businesses is redefining the symbiosis between physical storefronts and online sales, combining the experiential sense of place that urban sites offer with the large audience of e-commerce platforms (Anderson, 2012). Heather Arnold illustrates the flower-cum-jewellery shop and the bicycle-cum-coffee shop as examples of hybrid retail that offer complementary products and experiences to consumers. Vikas Mehta demonstrates the confidence and pride that storeowners take in their business innovations. This confidence corroborates transatlantic research demonstrating that independent business owners believe they have an edge over physical chain retailers in the e-commerce onslaught, as their unique advantages of personalized customer service and unique ambiance cannot be replicated online (Dawson et al., 2008, pp. 342–363; McGuinness & Hutchinson, 2013; Wrigley, 2009). Furthermore, they benefit from a growing consumer demand for "localism" – at least among those who can afford it (Andres Coca-Stefani et al., 2010). Yet financial and organizational barriers remain among small businesses to invest in inevitable e-commerce technology, as Elizabeth Mack's chapter demonstrates. Many independent retailers struggle to keep up with chains in bringing their business online through omnichannel strategies. And there is a discrepancy between independent legacy retailers and new independent retail start-ups. The former have been more resistant to embracing e-commerce, while the latter might even choose to forgo permanent physical stores altogether in lieu of online platforms like Etsy, Shopify, and Amazon Marketplace. Over half of Amazon's sales originate from third-party businesses (Bezos, 2017). But can this already fickle symbiosis be sustained in light of further virtualization and consolidation of retail?

This retail jockeying – the ebbs and flows of e-commerce and experiential retail and services – signal the diversification of the urban retail street. The loss of chain stores and the homogeneity they created has given way to an array of uses and independent businesses. These are not the mom-and-pops of yesteryear – pop-up brand flagships, start-up showcases, innovative food-service establishments, and new kinds of services define a streetscape with greater complexity. Some storefronts are shifting from being places of consumption to places of production, where "maker districts" composed of small-batch manufacturers and artisans sell directly to passers-by – or to the world at large through e-commerce platforms (Wolf-Powers et al., 2017). Other storefronts become home to creative production in a fascinating taxonomy of new desktops, from the coffee-table laptop to the shared co-working desk and the full-time small storefront office. Yet other urban storefronts (re)turn into residential spaces, as worldwide housing pressure continues to grow (Kickert & Talen, 2022).

This new composition invites us to expand our notion of what High Streets and main streets are and how storefronts – many of which might no longer contain retail – contribute to urban life. The shift toward the accommodation of not just new forms of retail and food services, but of creative production, services like day-cares, schools and wellness centres, offices and co-working spaces, and residences in all shapes and sizes – is a shift along the exposed, primary infrastructure of the city: the street. While market forces and disruptions have set up main streets for a roller-coaster ride of reinvention, their relevance to urban life remains as strong as ever.

As the microcosm of an urban society predicated on interaction, the new lives of our storefronts may be less commercial than today, but not necessarily less social. As our urban future becomes more walkable and interactive, the stakes for our storefronts have become higher than ever.

Can governments, citizens, and businesses work together to ensure that our recast main streets continue to thrive as the hearts of our cities?

REFERENCES

Anderson, C. (2012). *Makers: The new industrial revolution.* Crown Business.

Andres Coca-Stefaniak, J., Parker, C., & Rees, P. (2010). Localisation as a marketing strategy for small retailers. *International Journal of Retail & Distribution Management, 38*(9), 677–697. https://doi.org/10.1108/09590551011062439

Barajas, J. M., Boeing, G., & Wartell, J. (2017). Neighborhood change, one pint at a time: The impact of local characteristics on craft breweries. In N. G. Chapman, J. S. Lellock, & C. D. Lippard (Eds.), *Untapped: Exploring the cultural dimensions of craft beer* (pp. 155–176). West Virginia University Press.

Bezos, J. P. (2017). [Shareholder letter].

Charles, A. M. (2018). *Delivery: Dining in is the new dining out.* Cowen Equite Research.

Cho, C., & Todd, J. (2018). Food away from home during the Great Recession. In M. J. Saksena, A. M. Okrent, & K. S. Hamrick (Eds.), *America's eating habits: Food away from home.* United States Department of Agriculture.

Dawson, J. A., Findlay, A., & Sparks, L. (2008). *The retailing reader.* Routledge.

Henley, J. (2020, 28 April). Lithuanian capital to be turned into vast open-air cafe. *The Guardian.*

Kasinitz, P., Zukin, S., & Chen, X. (2015). Local shops, global streets. In S. Zukin, P. Kasinitz, & X. Chen (Eds.), *Global cities, local streets* (pp. 195–215). Routledge.

Kickert, C. (2019). What's in store: Prospects and challenges for American street-level commerce. *Journal of Urban Design, 26*(2), 159–177. https://doi.org/10.1080/13574809.2019.1686352

Kickert, C., & Talen, E. (2022). Beyond retail: Envisioning the future of retail. *Built Environment, 48*(1), 5–10. https://doi.org/10.2148/benv.48.1.5

KPMG International. (2017). *The truth about online consumers: 2017 Global Online Consumer Report.*

Lewis, R., & Dart, M. (2014). *The new rules of retail: Competing in the world's toughest marketplace.* St. Martin's Press.

Lipsman, A., Cakebread, C., Cheung, M.-C., Rotondo, A., & Wurmser, Y. (2018). *The future of retail 2019.* eMarketer.

Lipsman, A., Ceurvels, M., Cheung, M.-C., Peart, M., Rotondo, A., Vahle, P., & Abrams, K. v. (2019). *Global ecommerce 2019.*

Marshall, A. (2020, 5 July). As cities reopen, outdoor dining may provide a lifeline. *Wired Magazine.*

McGuinness, D., & Hutchinson, K. (2013). Utilising product knowledge: Competitive advantage for specialist independent grocery retailers. *International Journal of Retail & Distribution Management, 41*(6), 461–476. https://doi.org/10.1108/09590551311330834

Mull, A. (2019, 13 November). The zombie storefronts of America. *The Atlantic.*

Neslin, S. A., Grewal, D., Leghorn, R., Shankar, V., Teerling, M. L., Thomas, J. S., & Verhoef, P. C. (2006). Challenges and opportunities in multichannel customer management. *Journal of Service Research, 9*(2), 95–112. https://doi .org/10.1177/1094670506293559

New Economics Foundation(2004). *Clone town Britain: The loss of local identity on the nation's high streets.* New Economics Foundation.

New York City Department of Transportation. (2022). *Open restaurants policy.* https://www1.nyc.gov/html/dot/html/pedestrians/openrestaurants.shtml

NPD Group & Portalatin, D. (2018). *The future of dinner report.*

Oldenburg, R. (1989). *The great good place: Cafes, coffee shops, bookstores, bars, hair salons, and other hangouts at the heart of a community.* Da Capo Press.

Pauwels, K., & Neslin, S. A. (2015). Building with bricks and mortar: The revenue impact of opening physical stores in a multichannel environment. *Journal of Retailing, 91*(2), 182–197. https://doi.org/10.1016/j.jretai.2015 .02.001

Retail News. (2018). *Retail 2025 shopper study.*

Roth, R. (2018). *Gym, health & fitness clubs in the US.* IBISWorld industry report 71394.

US Census Bureau. (2019). *Advance monthly sales for retail and food services.*

US Food and Nutrition Service (2019). USDA launches SNAP online purchasing pilot [Press release].

Vogels, E. (2021, 22 June). *Digital divide persists even as lower-income Americans make gains in tech adoption.* Pew Research. https://www.pewresearch.org /fact-tank/2021/06/22/digital-divide-persists-even-as-americans-with -lower-incomes-make-gains-in-tech-adoption

Walker, S., & Fox Miller, C. (2019). Have craft breweries followed or led gentrification in Portland, Oregon? An investigation of retail and neighbourhood change. *Geografiska Annaler: Series B, Human Geography, 101*(2), 102–117. https://doi.org/10.1080/04353684.2018.1504223

Wolf-Powers, L., Doussard, M., Schrock, G., Heying, C., Eisenburger, M., & Marotta, S. (2017). The maker movement and urban economic development. *Journal of the American Planning Association, 83*(4), 365–376. https://doi.org /10.1080/01944363.2017.1360787

Wrigley, N., Branson, J., Murdock, A., & Clarke, G. (2009). Extending the Competition Commission's findings on entry and exit of small stores in British high streets: Implications for competition and planning policy. *Environment and Planning A, 41*(9), 2063–2085. https://doi.org/10.1068/a41326

Zukin, S., Trujillo, V., Frase, P., Jackson, D., Recuber, T., & Walker, A. (2009). New retail capital and neighborhood change: Boutiques and gentrification in New York City. *City & Community, 8*(1), 47–64. https://doi.org/10.1111/j.1540-6040.2009.01269.x

Contributors

Luc Anselin is the Stein-Freiler Distinguished Service Professor of Sociology and the College at the University of Chicago. He is also the founding director of the Center for Spatial Data Science at the University of Chicago, a centre established to advance computational and statistical methods for dealing with spatial data. Anselin was trained in economics and econometrics at the Free University of Brussels (Belgium) and earned an MA and PhD in regional science from Cornell University. His research focuses on the development of statistical methods to analyse spatial data in which location, distance, and interaction are explicitly taken into account. He is also the developer of the original SpaceStat and the GeoDa software packages for spatial data analysis (more than 450,000 users worldwide). He was elected to the US National Academy of Sciences in 2008 and the American Academy of Arts and Sciences in 2011.

Heather Arnold is the senior manager of Amazon's Retail Economic Development Division. Prior to working at Amazon, Ms Arnold was the director of research and analysis at Streetsense, a global retail, creative, and design consultancy. She specializes in market analysis and incentive planning for real estate developers and downtown environments. Ms Arnold holds a master's of city and regional planning from Cornell University as well as an undergraduate architectural history degree from the University of Virginia.

Matthew Carmona is professor of planning and urban design at The Bartlett, UCL. He is an architect/planner with research interests in urban design governance, the design and management of public space, and the value of urban design. He chairs the Place Alliance and edits www.place-value-wiki.net. His research can be found at https://matthew -carmona.com.

Terry Nichols Clark is professor of sociology at the University of Chicago. Terry Clark is interested in using decision-making theory to approach urban politics and other social phenomena. Dr. Clark is international coordinator of the Fiscal Austerity and Urban Innovation Project, which is surveying city officials across the United States and in 35 other countries. His project and book *Scenescapes*, co-authored with Daniel Silver, examines the patterns and consequences of the amenities that define our streets and strips in the context of spatial and political environment. They articulate the core dimensions of the theatricality, authenticity, and legitimacy of local scenes – cafes, churches, restaurants, parks, galleries, bowling alleys, and more. *Scenescapes* not only reimagines cities in cultural terms, it details how scenes shape economic development, residential patterns, and political attitudes and actions.

Kevin Credit is an assistant professor at the National Centre for Geocomputation at Maynooth University. Broadly, his research is focused on better understanding how urban spatial structure and transportation systems influence economic, environmental, and social outcomes using quantitative approaches and large open-source datasets. Kevin's recent work looks at topics such as the impact of transit construction on carbon emissions, underlying racial and ethnic disparities in COVID-19 outcomes, recent retail trends, and the development of spatially explicit random forest models.

Rosa Danenberg is a PhD student at the Department of Urban Planning and Environment at KTH Royal Institute of Technology in Stockholm, Sweden. Her research focuses on understanding main streets as public space.

Christopher Daniel is a post-doctoral researcher at the Centre for the Study of Commercial Activity at Toronto Metropolitan University in Toronto, Canada. He has published articles on corporate concentration in the Canadian retailing industry in addition to publications on forecasting commercial space demand. His PhD research studied the effects of the GGHA Growth Plan and e-commerce growth on the future demand for commercial real estate in the GGHA. This research continues under the added lens of the COVID-19 pandemic and the forecasted effects it will have on the demand for commercial space.

Irene Farah is a PhD candidate in city and regional planning at UC Berkeley. Her research interests lie at the intersection of informal employment, urban governance, health, and spatial analytics. In particular, she studies street vendors in Mexico City and how the spatiality and politics of their employment conditions affect their health.

Tigran Haas is the former director of the Center for the Future of Places at KTH – Royal Institute of Technology and current associate professor (tenured) at KTH – Royal Institute of Technology, ABE School. Dr. Haas is also the faculty member of KTH Digital Futures and a research scholar at MIT – Centre for Advanced Urbanism (LCAU), SA+P School. He has written more than 100 scholarly articles and co-authored and edited eight books.

Tony Hernandez is a professor in the School of Retail Management and Department of Real Estate Management at the Ted Rogers School of Management, Toronto Metropolitan University. Dr. Hernandez also serves as the director and Eaton Chair in Retailing at the Centre for the Study of Commercial Activity. Founded in 1992, the CSCA is a not-for-profit academic research unit at Toronto Metropolitan University that primarily studies retail and service sector activities. The centre works with private and public sector organizations, including major retailers, developers, brokerage and leasing firms, consultants, government departments, and industry associations. His research and teaching focuses on retail innovation and strategy, location analysis, and retail planning.

Hyesun Jeong is an assistant professor of urban design at the University of Cincinnati (UC) and a (co-)director of Orville Simpson Center for Urban Futures. Dr. Jeong's interdisciplinary research studies the built and cultural environment of global cities. Topics of her research and design studio include arts and cultural placemaking, Main Street, sustainable transit, and ecological urbanism. Prior to joining UC, she worked at the University of Texas at Arlington and the University of Chicago. She obtained a PhD in architecture from the Illinois Institute of Technology. She has practised at global design firms, including Goettsch Partners (Chicago), Dominique Perrault Architecture (Paris), and POSCO Architects and Consultants (Seoul). Her research has been published in peer-reviewed journals and book chapters. Recently, her research-design project on green infrastructure received the national award from the American Institute of Architecture's Upjohn Grant Initiative program.

Colin Jones has been professor of estate management at Heriot-Watt University since 1998. He worked at the Universities of Manchester, Glasgow, and the West of Scotland. His research interests span commercial, industrial, and housing market economics and policy. Colin has edited or written books. He has published more than 70 papers in academic journals. Colin was winner of the UK Royal Town Planning Institute's 2013 award for excellence in academic spatial planning research.

Conrad Kickert is an assistant professor at the University at Buffalo's School of Architecture and Planning. He has a background in urbanism and architecture from the TU Delft (Netherlands) and holds a PhD in architecture from the University of Michigan. He has worked as an urban researcher and designer for design offices, property developers, and non-profit organizations in The Netherlands, the United Kingdom, and the United States. His research studies the evolving relationships between urban form, urban life, and the urban economy.

Elizabeth Mack is an associate professor in the Department of Geography, the Environment, and Spatial Sciences at Michigan State University, where she teaches courses in economic geography. Dr. Mack's research utilizes mixed methods to understand the evolution of the economy in the face of rapid technological change and climate change. Research on technological change evaluates the impact of information and communications technologies on the development trajectory of regional economies and everyday work. Her work on the environment and climate change evaluates household responses to changing environmental contexts, as well as uses and the ability to pay for water services. Dr. Mack's research has been funded by the National Science Foundation, the United States Department of Agriculture, the National Aeronautics and Space Administration, and the Kauffman Foundation for entrepreneurship research.

Michael W. Mehaffy is a researcher, educator, consultant, and author who has held teaching and/or research appointments at seven graduate institutions in six countries, and he is on the editorial boards of four international journals of urban design. He has consulted internationally on notable projects in which main streets played a key role. Among them, he was the project manager for the master developer of Orenco Station, a widely studied Oregon transit-oriented development. The project featured a new main street offering goods and services meeting the daily needs of local residents.

Vikas Mehta is the Fruth/Gemini Chair, Ohio Eminent Scholar of Urban/Environmental Design, and professor of urban design at the School of Planning, University of Cincinnati. Dr. Mehta's work focuses on the exploration of place as a social and ecological setting and as a sensorial art in creating a more responsive, equitable, stimulating, and communicative environment. Dr. Mehta has written/co-written and edited/co-edited six books and numerous book chapters and journal articles on urban design pedagogy, public space, urban streets, neighbourhoods, retail, signage and visual identity, public space in the Global South and more. His most recent book is *Public Space: Notes on Why It Matters, What We Should Know, and How to Realize its potential* (Routledge, 2022).

Rachel Meltzer is the Plimpton Associate Professor of Planning and Urban Economics at Harvard University's Graduate School of Design. Her research is broadly concerned with urban economies and how market and policy forces can shape disparate outcomes across neighbourhoods. She focuses on economic development, housing, land use, and local public finance.

Jeffrey Nathaniel Parker is an assistant professor of sociology at the University of New Orleans. Dr. Parker is an urban sociologist broadly interested in place reputation at the level of neighbourhood, city, and region. More specifically, his work focuses on how perceptions of place structure action and inequality. Drawing on interviews, ethnography, and archival research, he has written and co-written about the political consequences of racialized perceptions of disorder, administrative strategies of racialized voter suppression, the consequences of neighbourhood stigma on gentrification, the social networks and patterns of trust among poor mothers of young children, place reputation among lesbian, bisexual, and queer women, merchants' subjective relationships to gentrification, the creation of the urban sociological canon, the complicated connections between chain stores and notions of hipness, the economics of indie rock, and the social meaning of potholes.

Emily Talen is professor of urbanism at the University of Chicago, where she teaches urban design and directs the Urbanism Lab. She holds a PhD in urban geography from the University of California, Santa Barbara. She is a fellow of the American Institute of Certified Planners, and the recipient of a Guggenheim Fellowship. Talen has written extensively on the topics of urban design, New Urbanism, and social equity. Her books include *New Urbanism and American Planning, Design for Diversity, Urban Design Reclaimed*, and *City Rules* and *Neighborhood*. She is co-editor of the *Journal of Urbanism*.

Index

www.ingramcontent.com/pod-product-compliance
Lightning Source LLC
Chambersburg PA
CBHW030234030426
42336CB00009B/101